Progress in Molecular and Subcellular Biology

22

Springer

Berlin
Heidelberg
New York
Barcelona
Hong Kong
London
Milan
Paris
Singapore
Tokyo

Philippe Jeanteur (Ed.)

Cytoskeleton and Small G Proteins

With 30 Figures

Springer

Professor Dr. PHILIPPE JEANTEUR
Institute of Molecular Genetics
of Montpellier
CNRS
University of Montpellier II
1919 Route de Mende
F-34293 Montpellier Cedex 5
France

ISSN 0079-6484
ISBN 3-540-64808-9 Springer-Verlag Berlin Heidelberg New York

Library of Congress Cataloging-in-Publication Data.

Cytoskeleton and small G proteins / Philippe Jeanteur (ed.).
p. cm. – (Progress in molecular and subcellular biology; 22)
Includes bibliographical references and index.
ISBN 3-540-64808-9 (hbk.: alk. paper)
1. G proteins. 2. Ras proteins. 3. Cytoskeleton. 4. Cytoskeletal proteins. 5. Cellular signal transduction. I. Jeanteur, Ph. (Philippe). II. Series.
QP552.G16C97 1999
572′.6 – dc21

Cover design: Meta Design, Berlin
Typesetting: Best-set Typesetter Ltd., Hong Kong
SPIN 10631625 39/3137 5 4 3 2 1 0 – Printed on acid-free paper

Preface

Animal cells present an extreme variability in their shapes in relation to their physiological properties. For instance, fibroblastic cells are tightly attached to the extra-cellular matrix and display a flattened, spindle-shaped morphology. Neuronal cells self-organize as a network through a complex branching of dendrites and a long axonal extension. Resting peripheral blood lymphocytes are poorly adhesive and maintain a spherical, smooth shape, while macrophages produce many pseudopodal extensions involved in the recognition of foreign molecules. In addition to the variability of the morphology of the cells that constitute different organs, many cell types also modify dynamically their morphology in response to environmental changes, leading to differential cell motility, migration, adhesion, polarity or intercellular contacts. This wide plasticity of cell morphology is promoted and maintained by the cytoskeleton, which is composed of the three interconnected actin microfilaments, tubulin microtubules and intermediate filaments networks, all capable of assembly and disassembly.

Over the past few years, the Rho family of Ras-like GTPases emerged as key proteins that mediate extracellular signalling pathways leading to the formation of polymerized actin-containing structures such as ruffles, lamellipodia and filopodia. Since the discovery of the first member RhoA in 1985, 13 members have so far been characterized in human cells. Most of Rho proteins are highly conserved between species as distant as yeast, slime mold, insects and mammals, which points to their fundamental role in cellular physiology.

This volume describes several aspects of the implication of Rho proteins in different cellular and physiological processes through the control of the cytoskeleton organization. Ann Ridley reviews the isolation and the biochemical characterization of Rho proteins, and addresses the role of RhoA, Rac and Cdc42 in cell adhesion, cytokinesis and proliferation. Pierre Chardin summarizes the biochemical properties of Rnd proteins, a newly discovered Rho family subgroup.

Like other Ras-related proteins, most of the Rho GTPases adopt either active GTP-bound or inactive GDP-bound conformational states. The transition from the GDP to the GTP state is positively controlled by guanine exchange factors (GEF). A large number of Rho GEFs have been isolated mainly as truncated constitutively active versions that exhibited oncogenic properties. John Collard describes the structural features of the members of the Rho GEF family, and their functions in signal transduction, development, cell proliferation and invasion. Inactivation of Rho GTPases is controlled by two different types or regulators: the GTPase activating proteins (GAP), which bind the GTP-bound Rho protein and stimulate its intrinsic GTPase activity, and the guanine dissociation inhibitors (GDI), which sequester Rho proteins from membranes thereby allowing their solubilization and their redistribution to other subcellular compartments. Gerard Zalcman and colleagues review the current state of knowledge about Rho GAP and Rho GDI families and discuss their biological functions.

In their GTP-bound form, GTPases adopt an active conformation that allows them to interact with multiple downstream effectors, leading to a diversification of the input signal. Characterization of the effectors turned out to be very fruitful for understanding the molecular mechanisms through which Rho GTPases control the organization of the cytoskeleton. By means of yeast two-hybrid selection and affinity chromatography, the hunt for Rho targets led to the identification of several classes of proteins, in particular kinases and proteins with coiled-coil domains. Kozo Kaibuchi and Linda Van Aelst describe the isolation and the biological function of Rho, Rac and Cdc42 targets, while Edward Manser focuses on the role of the p21 associated kinases (PAK).

Biological properties of the Rho GTPases have also been investigated at the physiological level, by analyzing their requirement for the cell cycle and development. Philippe Fort looks at the cross-talks between Ras and Rho proteins, as well as between Rho proteins, that lead to cell transformation and apoptosis. Jeffrey Settleman addresses some functional aspects of the Rho proteins in the developmental biology of the worm, fruitfly and mouse.

A particular aspect of the cytoskeleton biology of Rho GTPases is their implication in microbial pathogenesis. Indeed, Rho proteins have been characterized as major cell targets of numerous bacterial toxins over a wide range of pathogens, as described by Patrice Boquet and Philippe Sansonetti.

Although the precise cellular function of most of the Rho proteins remains to be elucidated, analysis of the Rho family has generated a considerable amount of data leading to a fragmental but yet meaningful picture of the changes in the cytoskeleton involved in fundamental processes such as morphogenesis, cell migration, axonal guidance and cell proliferation.

Montpellier, France
October 1998

Philippe Fort
Philippe Jeanteur

Contents

Rho Family Proteins and Regulation of the Actin Cytoskeleton

Anne J. Ridley

1 Introduction

The actin cytoskeleton plays an essential role in many cellular functions, including cell motility and migration, cell division, intracellular transport processes, and phagocytosis. In order to mediate these diverse processes, the actin cytoskeleton is highly dynamic and can be rapidly reorganized in response to a variety of different signals. Actin reorganization is an early response of cells to many growth factors, cytokines and lipid mediators (Ridley 1994). In addition, the composition of the extracellular matrix, the extent of cell spreading and interactions with other cells play an important role in regulating actin organization (Ruoslahti 1997; Schwartz 1997). These different extrinisic signals combine to regulate cellular responses involving the cytoskeleton, and may also induce long-term changes by altering the transcription of cytoskeletal proteins or regulatory proteins. Finally, intrinsic signals during the cell cycle lead to reorganization of actin filaments during mitosis followed by formation and contraction of the actin-based contractile ring at cytokinesis (Ridley 1995).

The signalling pathways leading to changes in actin organization are complex and involve many different components. It has become clear, however, that members of the Rho family of small GTP-binding proteins act as key coordinators of these pathways. Rho and the related proteins, Rac and Cdc42, have been implicated in regulating actin organization in a variety of different model systems, including *Saccharomyces cerevisiae*, *Drosophila melanogaster*, *Dictyostelium discoideum*, transgenic mice and cultured mammalian cells. In addition, they have been ascribed many other functions, including synthesis of cell wall components in *S. cerevisiae* (Bussey 1996), and regulation of transcription, secretion, endocytosis and mitogenesis in mammalian cells (Ridley 1996; Symons 1996). This chapter introduces the Rho family of proteins, describes their effects on the actin cytoskeleton and associated adhesion sites to the extracellular matrix or to other cells, and briefly reviews other functions

Ludwig Institute for Cancer Research, 91 Riding House Street, London W1P 8BT, UK, and Department of Biochemistry and Molecular Biology, University College of London, Gower Street, London, UK

Progress in Molecular and Subcellular Biology, Vol. 22
Ph. Jeanteur (Ed.)

that have been ascribed to this family of proteins which may or may not be linked to their roles in regulating actin organization.

2
Structure and Regulation of Rho Family Proteins

2.1
Primary Structure of Rho Family Proteins

Rho family proteins are part of the Ras superfamily of GTP-binding proteins, and are highly conserved in eukaryotes, from yeast to plants and mammals. In mammals, they include Rho(A,B,C), Rac(1,2), Cdc42 (two alternatively spliced variants), TC10, RhoD, RhoE, RhoG and TTF (Murphy et al. 1996; Ridley 1996; see also Chardin, this Vol.). Members of this family show at least 40% identity at the amino acid level. The most extensively characterized members are Rho, Rac and Cdc42. *Rho* was the first Rho family gene to be cloned, initially from the seaslug, *Aplysia*, and subsequently from humans (reviewed in Hall 1994). RhoA, RhoB and RhoC are three highly homologous isoforms of Rho, which show 92% identity (RhoA and RhoC) or 85% identity (RhoA/RhoC and RhoB) at the amino acid level. The majority of differences lie within the last 15 amino acids of the C-terminus. Cdc42 was first characterized in *S. cerevisiae*, where it was isolated as a cell cycle mutant defective in budding (Johnson and Pringle 1990). The two mammalian homologues of yeast Cdc42 differ by nine amino acids (Shinjo et al. 1990), and are actually two alternatively spliced variants with different C-terminal sequences. The two mammalian *Rac* genes, *Rac*1 and *Rac*2, were cloned through their homology to *Cdc42* (G25K; Didsbury et al. 1989). Rac1 and Rac2 are 92% identical, and again differ primarily within the C-terminal 15 amino acids, where Rac1 but not Rac2 contains a polybasic sequence.

2.2
Expression of Rho Family Proteins

Where it has been investigated, most *Rho* family genes are widely expressed in different tissues (Olofsson et al. 1988; Moll et al. 1991; Vincent et al. 1992; Foster et al. 1996), although *Rac*2 is only expressed in haematopoietic cells (Shirsat et al. 1990), and in chick embryos the chick *Rac*1B gene appears to be selectively expressed in the developing nervous system (Malosio et al. 1997). RhoB and RhoG are growth factor-inducible genes (Jähner and Hunter 1991; Vincent et al. 1992), but whether this is important for cell cycle- or proliferation-linked changes in actin organization is not known.

2.3 Post-Translational Modifications

As with other Ras superfamily members, Rho family proteins are modified by prenylation and carboxymethylation of a conserved cysteine, four amino acids from the C-terminus, follwed by removal of the last three amino acids (Fig. 1; Adamson et al. 1992a; Marshall 1993). Rho, Rac and Cdc42 proteins are all prenylated by a 20-carbon chain geranylgeranyl group, but RhoB can also be prenylated by a 15-carbon chain farnesyl group (Adamson et al. 1992a). In

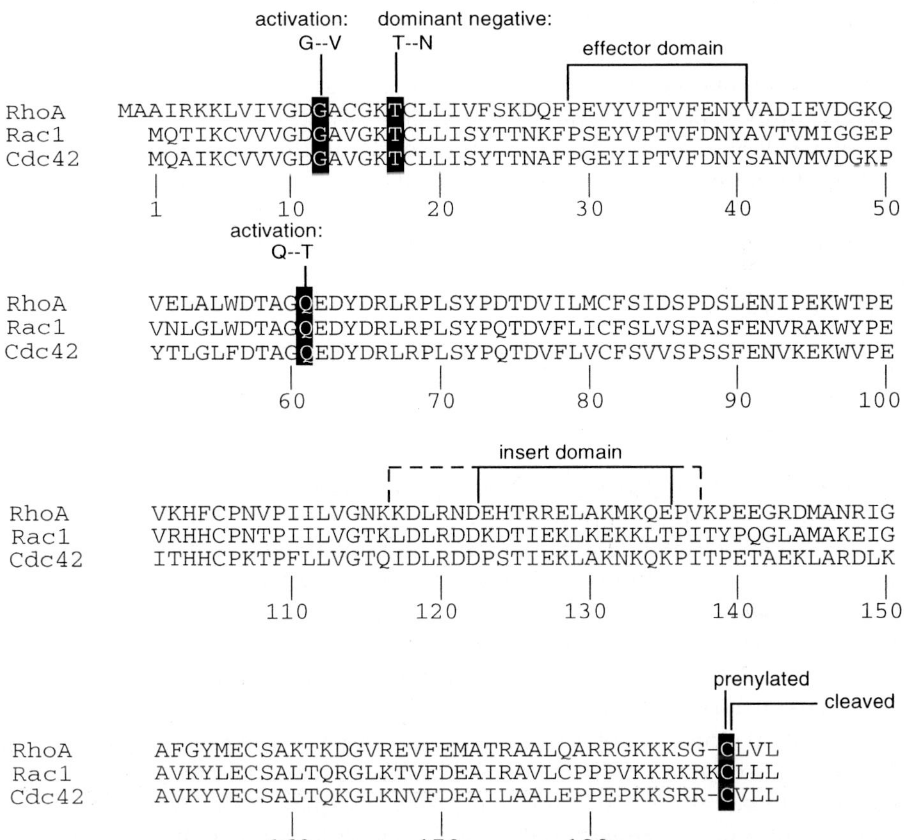

Fig. 1. Sequence alignment of human RhoA, Rac1 and Cdc42 proteins. Amino acids that are commonly mutated to make constitutively active forms of the GTPases or to make a dominant negative form are *highlighted*, together with the C-terminal conserved cysteine that is prenylated in the mature proteins. The effector domain, defined originally by mutations in Ras and subsequently by crystal structures of Rho and Rac (Hirshberg et al. 1997; Wei et al. 1997), is indicated. The *insert domain* indicates amino acids inserted with respect to the Ras sequence, while the *dotted lines* around the insert domain indicate the full extent of the structural domain which includes the inserted amino acids. (Hirshberg et al. 1997)

contrast, RhoE is only farnesylated (Foster et al. 1996; see also Chardin, this Vol.). These modifications have been shown to be important for at least some, although not all of the functions of Rho family proteins (Kreck et al. 1996; Lebowitz et al. 1997), and are required for translocation of RhoA, RhoB and RhoC to membrane compartments (Adamson et al. 1992a). The difference in the modification of RhoB is important for its localization: RhoB localizes predominantly to an endosomal/lysosomal compartment, whereas RhoA/C are principally cytoplasmic, although they also show some plasma membrane localization (Adamson et al. 1992b). RhoB may therefore have a different function in cells to RhoA or RhoC, and appears to be specifically required for Ras transformation of fibroblast cell lines (Lebowitz et al. 1995).

In addition to these modifications at the C-terminus, RhoA has also been reported to be phosphorylated on serine residue 188 (not present in RhoB or RhoC), which may regulate its activity (Lang et al. 1996).

Rho family proteins are directly modified by a number of bacterial toxins and exoenzymes (see Boquet et al., this Vol.), and these have been extremely useful for analysing their function. The most extensively used is C3 transferase, an exoenzyme produced by *Clostridium botulinum* that ADP-ribosylates Rho at amino acid 41 and thereby inactivates it. Whether Rho family proteins are regulated via ADP-ribosylation by endogenous proteins has not been shown, but remains a possibility.

2.4
Structural Analysis of Rho Family Proteins

In general, Rho family proteins bind GDP and GTP, and hydrolyse GTP. They interact with a number of proteins which regulate their GTPase cycle, including exchange factors, GTPase activating proteins (GAPs) and guanine nucleotide dissociation inhibitors (GDIs; see Stam and Collard and Zalcman et al., this Vol.). An exception to this pattern is found in RhoE which does not detectably bind GDP or hydrolyse GTP (Foster et al. 1996), and may be a member of a subgroup of Rho family proteins which are regulated by a different mechanism (see Chardin, this Vol.).

The crystal structures of RhoA complexed with GDP (Wei et al. 1997) and of Rac1 complexed with a GTP analogue (Hirshberg et al. 1997) have been solved, and show that, as expected from their primary sequence, the overall structure is very similar to Ras. However, there are some differences in the coordination of Mg^{2+}, and they have an extra α-helical domain when compared with H-Ras, between amino acids 117 and 137 (Rac numbering), which includes the 13-amino acid insert domain from 123–135 (Fig. 1; see below).

Mutation of a number of conserved amino acids in Ras superfamily proteins can alter their intrinsic and GAP-stimulated GTPase activity, or alter their guanine nucleotide binding affinity. In Ras, several such mutations occur naturally in cancers, and convert Ras to an oncogene (Barbacid 1987; Quillam et al. 1994). Constitutively active proteins either have decreased GTPase activity and are insensitive to GAPs, allowing them to remain predominantly in the GTP-bound, active, form in cells, or they have an enhanced exchange rate (Barbacid 1987; Bourne et al. 1991). Proteins with mutations of amino acid 12 (based on Ras numbering) from Gly to Val, or of amino acid 61 from Thr to Glu, are the most commonly used constitutively activated forms of Rho family proteins (Fig. 1). The crystal structures show that these amino acids are positioned very similarly in Rho, Rac and Ras, and are therefore predicted to function similarly (Hirshberg et al. 1997; Wei et al. 1997). Dominant negative forms of the proteins are normally created by mutating amino acid 17 from Ser/Thr to Asn (Fig. 1). This amino acid coordinates to the essential Mg^{2+} ion required for guanine nucleotide binding in all Ras superfamily GTPases (Bourne et al. 1991; Wei et al. 1997). Mutation of this amino acid is therefore predicted to reduce Mg^{2+} coordination and nucleotide binding, and indeed Rac and Cdc42 proteins with this mutation have a much lower affinity for GDP/GTP than their unmutated counterparts (Self and Hall 1995). These dominant negative proteins are therefore believed to be locked in an inactive state, thus inhibiting the activity of their respective endogenous GTPases by competing for binding to exchange factors (Farnsworth and Feig 1991).

In their active, GTP-bound form, Rho family proteins interact directly with a number of different "target" proteins which are likely to mediate their actions (see Boettner and van Aelst and Kaibuchi, this Vol.). Inactivating mutations in Rac and Cdc42 can reduce their interaction with one or more target proteins, and mutagenesis studies have so far been mostly confined to a region known in Ras as the "effector loop" (amino acids 30–41). Specific mutations in this region can either completely inhibit the activity of the GTPases, or alternatively can partially inhibit, or inhibit some responses induced by one protein but not others. These mutants have been shown to interact with some target proteins but not others (Lamarche et al. 1996; Westwick et al. 1997). Recently, however, the insert domain (amino acids 123–135) in Rho family proteins not found in Ras has been shown to be required for Rac-induced NADPH oxidase activation (Freeman et al. 1996; Joseph and Pick 1996). The sequence for the Rho and Cdc42 insert helices are very different from Rac (Fig. 1), implying that they will interact with a distinct set of proteins and thereby confer specificity to responses mediated by Rho family proteins (Wei et al. 1997).

3 Rho Family Proteins and Growth Factor-Induced Actin Reorganization

3.1 Rho

In a variety of cultured mammalian cell types, microinjection or expression of activated Rho proteins stimulates the accumulation of stress fibres, and Rho is required for the formation of stress fibres induced by growth factors and cytokines (Ridley and Hall 1992; reviewed in Machesky and Hall 1996; Tapon and Hall 1997). These stress fibres consist of bundles of actin filaments associated with myosin filaments and other proteins, and are contractile. The contractile nature of Rho-induced stress fibres in fibroblasts is clearly observed when the cells are plated on deformable substrata (Chrzanowska-Wodnicka and Burridge 1996), and consistent with this, Rho is required for contraction of smooth muscle cells (Hirata et al. 1992). Stress fibres are linked at the plasma membrane to multi-protein complexes known as focal adhesions, where transmembrane integrins are associated directly or indirectly with a large number of structural and signalling proteins (Burridge and Chrzanowska-Wodnicka 1996). The formation of focal adhesions is regulated by Rho and is intimately linked with stress fibre formation, although the two responses are to a certain extent separable (see below, Sect. 4.1).

Many observations on Rho function are consistent with its ability to stimulate the formation of actomyosin-based structures and to regulate their contractility. These structures are not always stress fibres. For example, Rho is required in vivo for the healing of small wounds in chick embryos, which are closed through the contraction of an actin-base purse-string (Brock et al. 1996). This purse-string appears morphologically similar to the Rho-regulated peripheral bundles of actin filaments observed in isolated epithelial cell colonies (Ridley et al. 1995). In polarized epithelial cells, where the apical pole consists of a perijunctional ring of F-actin and myosin II, Rho is required for the integrity of this ring (Nusrat et al. 1995). In macrophages, microinjection of activated Rho proteins rapidly stimulates cell contraction, and although macrophages do not have stress fibres, they possess fine Rho-regulated actin cables in the cytoplasm which are presumably contractile (Allen et al. 1997). In neuronal cell lines, Rho also stimulates cell contraction and mediates neurite retraction in response to a variety of extracellular agents (Jalink et al. 1994; Kozma et al. 1997).

As well as stimulating actomyosin-based contractility, Rho can stimulate an overall increase in F-actin in cells (Norman et al. 1994; Wócjiak-Stothard et al. 1998), presumably by stimulating actin polymerization. Consistent with this, Rho is required for membrane "folding" during *Shigella* entry into HeLa cells, and appears to mediate actin polymerization from new actin nucleation sites, but appears not to be involved in forming the nucleation sites (Adam et al. 1996). The *Shigella*-induced membrane folds may be functionally similar to the HGF-induced membrane folds observed in KB cells, which also require Rho and are where Rho is localized (Nishiyama et al. 1994). However, in Swiss 3T3 fibroblasts Rho did not significantly induce new actin polymerization as an early response, as measured by the incorporation of fluorescently labelled actin monomers into stress fibres (Machesky and Hall 1997). The ability of Rho to stimulate increased actin polymerization may therefore vary between cell types and/or the status of cells.

Our understanding of Rho function has been considerably enhanced by studies in *Drosophila* (see Settleman, this Vol.) and yeast. In *S. cerevisiae*, four

RHO genes have been isolated, *RHO1*, *RHO2*, *RHO3* and *RHO4*. Of these, *RHO1* shows the closest homology to mammalian RhoA, and is essential for cell viability. It is required for bud growth, and is localized to the growing bud tip and to the mother-daughter neck at cytokinesis (Bussey 1996). Rho1p appears to coordinate bud construction by regulating glucan synthase, a major enzyme involved in cell wall synthesis, and by concomitantly organizing the actin cytoskeleton during budding, possibly via its interaction with the profilin-binding protein Bni1p (Imamura et al. 1997; see also Sect. 3.3 below).

A number of target proteins interacting with Rho have been implicated in mediating actin reorganization (see Kaibuchi, this Vol.), and it seems likely that Rho interacts with several different effector proteins to allow the coordination of actin polymerization with the formation of stress fibres. For example, Rho-kinase (also known as ROCK or ROKα) is involved in regulating actomyosin-based contractility via increased phosphorylation of myosin II light chain (reviewed in Ridley 1996; Burridge and Chrzanowska-Wodnicka 1996). A target for Rho that may be involved in stimulating actin polymerization is p140mDia, which can bind to the actin-binding protein profilin and induces actin filament formation when overexpressed in COS cells (Watanabe et al. 1997). Rho and other family members could also stimulate actin polymerization by enhancing synthesis of polyphosphoinositides, which bind to a number of actin-binding proteins, including profilin, vinculin and gelsolin (reviewed in Stossel 1993; Tapon and Hall 1997).

3.2 Rac

Microinjection and transfection studies have revealed that, in contrast to Rho, Rac regulates the formation of lamellipodia and membrane ruffles in a variety of cell types (reviewed in Tapon and Hall 1997). Lamellipodia are plasma membrane protrusions containing a meshwork of actin filaments, and extend over the substratum to form new adhesive contacts known as focal complexes (Welch et al. 1997). They are commonly found at the leading edge of migrating cells, driving the forward extension of cells. On adherent cells, membrane ruffles are similar in structure to lamellipodia, but protrude upwards from the dorsal surface of the cells. Both lamellipodium extension and membrane ruffling involve active actin polymerization occurring adjacent to the plasma membrane.

Rac was first shown to stimulate lamellipodium formation in quiescent fibroblasts, and to mediate lamellipodium formation induced by a variety of growth factors (Ridley et al. 1992). In addition, Rac can also stimulate membrane ruffling under certain conditions, for example in macrophages microinjected with activated Rac1 protein (Allen et al. 1997). PI 3-kinase is required for lamellipodium formation and membrane ruffling induced by many but not all growth factors, and has been shown to act upstream of Rac (Nobes et al. 1995;

Ridley 1995; Parker 1995). Rac1 microinjection also stimulates lamellipodium formation on growth cones and along neurites of N1E-115 neuroblastoma cells (Kozma et al. 1997). In PC12 neuronal cells, Rac is required for NGF-induced neurite elongation, and is hypothesized to be required for the adhesive and motile function of growth cones rather than for the actual assembly process (Lamoureux et al. 1997). Interestingly, activated Rac protein is unable to stimulate lamellipodium formation in epithelial cells where intercellular junctions are present: instead, it actually appears to enhance actin filament accumulation at cell-cell junctions (Ridley et al. 1995; Takaishi et al. 1997). Dominant negative Rac does, however, inhibit lamellipodium formation induced by either HGF/SF or activated Ras. This suggests that the signalling pathways regulating lamellipodium formation are more complex in epithelial cells than in fibroblasts or macrophages.

Lamellipodium extension is believed to be driven by new actin polymerisation, which occurs at the plasma membrane (Mitchison and Cramer 1996). Consistent with this, Rac stimulates the incorporation of fluorescently labelled actin monomers into filaments in lamellipodia (Machesky and Hall 1997). Rac can also stimulate actin polymerization in permeabilized platelets, and concomitantly augments $PI(4,5)P_2$ synthesis (Hartwig et al. 1995). Addition of a peptide that binds to polyphosphoinositides inhibits Rac-induced actin polymerization, suggesting that Rac-induced actin polymerization may be mediated at least in part by the production of PIP_2. It is likely, however, that other downstream targets of Rac are also involved in regulating actin reorganization (see Boettner and van Aelst, this Vol.).

No Rac homologue exists in *S. cerevisiae*, but studies of *Rac*-related genes in *Drosophila*, *C. elegans* and *Dictyostelium* have been very informative about Rac function in multicellular organisms, and suggest that it plays a key role during morphogenetic changes by regulating actin organization (see Settlemen, this Vol.).

3.3
Cdc42

Microinjection and transfection studies have demonstrated that Cdc42 induces the formation of filopodia in several mammalian cell types, including fibroblasts and macrophages (Allen et al. 1997; Tapon and Hall 1997). As with lamellipodia, filopodia in fibroblasts and macrophages are associated with focal complexes located approximately at the base of each filopodium (Nobes and Hall 1995; Allen et al. 1997). Filopodia are fine plasma membrane protrusions containing bundles of actin filaments cross-linked by actin-binding proteins, and extend from the leading edge of migrating cells, forming new contacts with the substratum (Lauffenburger and Horwitz 1996). Nerve growth cone extension is also dependent on filopodia, and in a neuroblastoma cell line, Cdc42 has been shown to promote filopodium formation on growth cones

(Kozma et al. 1997). Under different conditions or in other cell types, however, Cdc42 does not detectably induce filopodium formation and instead other actin-containing structures are observed. For example, in confluent fibroblasts some stress fibre formation but not filopodia are observed (Nobes and Hall 1995).

Recombinant Cdc42 can directly stimulate actin polymerization in neutrophil and *D. discoideum* extracts (Zigmond et al. 1997). This action is specific to Cdc42 and is not mimicked by Rac or Rho, although it is possible that an effect of Rac is not detected because the coupling between Rac and actin polymerization is disrupted during cell lysis. Interestingly, the stimulation of actin polymerization does not correlate with $PI(4,5)P_2$ or $PI(3,4,5)P_3$ synthesis, nor is it inhibited by an antibody to polyphosphoinositides, suggesting that neither $PI(4,5)P_2$ nor other polyphosphoinositides act downstream of Cdc42 in this response.

In the yeast *S. cerevisiae*, Cdc42 is essential for coordinating polarized cell surface growth and reorganization of the actin cytoskeleton during both budding and mating (Chant 1996; Leberer et al. 1997). It is also involved in controlling polarized cell growth in the fission yeast *Schizosaccharomyces pombe* (Miller and Johnson 1994). The *S. cerevisiae* actin cytoskeleton consists of cortical actin patches associated with the plasma membrane, and actin cables running through the body of the cell. These actin cables become polarized along the axis of the cell as a new bud forms, and they are believed to be important for the directed delivery of membrane components to the bud site or mating projection (Welch et al. 1994). Cdc42 is involved in the polarization of the actin cytoskeleton and is localized to bud sites, although actual bud site selection is dependent on many other gene products (Chant 1996). Exactly how Cdc42 regulates actin organization in *S. cerevisiae* is unknown, although two potential targets for Cdc42, Ste20 (a PAK homologue) and Bni1p have been implicated in mediating the Cdc42 response (Cvrcková et al. 1995; Evangelista et al. 1997). The concept that Cdc42 may play a more general role in establishing cell polarity via its effects on the actin cytoskeleton is suggested by studies in *Drosophila*. In the wing disc epithelium, Cdc42 is required for epithelial cell elongation, which is turn is essential for the generation of apico-basal polarity (Eaton et al. 1995, see also Settleman, this Vol.).

3.4 Other Rho Family Proteins

So far, the abilities of Rho, Rac and Cdc42 to regulate actin organization have been most thoroughly characterized, but there is increasing evidence that other Rho family members can influence cell morphology and cytoskeletal arrangements. Expression of activated RhoD leads to actin rearrangement in several different mammalian cell types, and concomitantly alters the motility and distribution of early endosomes, suggesting a link between vesicular transport and the actin cytoskeleton (Murphy et al. 1996). In *Drosophila*, a novel member of the Rho family, RhoL, has been identified (Murphy and Montell 1996). Expression of activated RhoL induces breakdown of cortical actin filaments in nurse cells associated with oocytes. Whether these proteins influence actin organization directly or via Cdc42, Rac and Rho remains to be established.

3.5 Links Between Different Rho Family Proteins: A Role in Cell Migration?

Studies in quiescent, serum-starved Swiss 3T3 fibroblasts and in macrophages have shown that Cdc42, Rac and Rho can act in a cascade, where activation of Cdc42 leads to Rac activation, and Rac activation leads to Rho activation (Ridley et al. 1992; Nobes and Hall 1995; Allen et al. 1997). In fact, in several cell types, the predominant morphological change induced by activated Cdc42 is Rac-mediated lamellipodium extension, and in Swiss 3T3 cells, the ability of Cdc42 to induce filopodia is most apparent when Rac activation is inhibited (Nobes and Hall 1995). These links between Cdc42, Rac and Rho may well be functionally relevant to cell migration, where extension of filopodia and lamellipodia at the leading edge of a cell has to be coordinated with active contraction in the cell body to pull the bulk of the cell forward (Lauffenburger and Horwitz 1996). If the cell body contraction is mediated by Rho, it may well be activated periodically in response to a critical level of lamellipodial extension. Consistently high levels of Rho activity and the resulting focal adhesions and stress fibres, however, inhibit cell migration (Ridley et al. 1995), presumably by increasing adhesion to the substratum. In cells already containing many stress fibres, such as growing MDCK epithelial cells, microinjection of Rac actually leads to a decrease in stress fibres (Ridley et al. 1995). This action of Rac may facilitate migration by decreasing integrin-mediated adhesion, although periodic Rho activation may still be required to mediate cell body contraction. In neuronal cells, a decrease in Rho activity may similarly be important for neurite outgrowth: Rho mediates neurite retraction and acts antagonistically to Rac and Cdc42, which promote neurite outgrowth (Kozma et al. 1997). Little is known about the molecular links between Cdc42, Rac and Rho: although it is presumed that specific exchange factors will be involved, the pathways may well be indirect, as it has been shown for example, that Rac-induced leukotriene release can contribute to Rho activation (Peppelenbosch et al. 1995).

4 Rho Family Proteins and Cell Adhesion

The actin cytoskeleton is structurally linked both to integrin-mediated adhesion sites where transmembrane integrins interact with either the extracellular matrix or with receptors on other cells, and to cadherin-based adherens junctions and tight junctions. In altering actin organization, Rho family proteins inevitably affect these linked adhesion sites. In addition, they may act directly to modify adhesion site organization, rather than indirectly via changes in the actin cytoskeleton.

4.1 Adhesion to the Extracellular Matrix

In addition to regulating responses to cytokines and growth factors, Rho is also required for signalling to the actin cytoskeleton induced by activation of integrins. When fibroblasts are plated on the extracellular matrix protein fibronectin, they can form stress fibres and focal adhesions in the absence of growth factors. In addition, tyrosine phosphorylation of a number of focal adhesion proteins is stimulated, and in some cell types, activation of MAPK pathways and gene transcription is induced (reviewed in Burridge and Chrzanowska-Wodnicka 1996; Schwartz 1997). Rho is required for the fibronectin-induced formation of stress fibres and for the tyrosine phosphorylation of focal adhesion components (Barry et al. 1997). Rho stimulates actin stress fibre formation in cells plated on fibronectin but not in cells plated on the non-specific substrate, poly-L-lysine, showing that integrin engagement is important for Rho-induced organization of the actin cytoskeleton into actomyosin bundles (Hotchin and Hall 1995; Flinn and Ridley 1996). As expected, overexpression of Dbl, an exchange factor for Rho and Cdc42, stimulates stress fibre formation in cells plated on fibronectin, but interestingly actually inhibits stress fibre formation in cells plated on gelatin, and does not stimulate stress fibre formation on either collagen I or collagen IV (Defilippi et al. 1997). This suggests that Rho requires engagement of specific fibronectin-binding integrins in order to induce stress fibre formation.

Rac and Cdc42 also mediate the formation of adhesion sites to the extracellular matrix. In fibroblasts, these sites are much smaller than Rho-mediated focal adhesions, and are predominantly localized in lamellipodia or associated with filopodia (Nobes and Hall 1995). In macrophages, which do not have focal adhesions, Cdc42 acts upstream of Rac to regulate the formation of small focal complexes, which contain integrins and other focal adhesion-associated proteins (Allen et al. 1997). It is presumed that these structures are linked to the actin cytoskeleton, although this has not formally been proven.

4.2 Intercellular Adhesions

Recently, evidence has accumulated to suggest that Rho family proteins are involved in controlling the formation of adherens and tight junctions between epithelial cells. Rho and Rac are both required for the formation of adherens junctions between keratinocytes, induced by increasing extracellular [Ca^{2+}] from low to high (Braga et al. 1997). Similar results were obtained in MDCK epithelial cells, where expression of activated Rac enhanced the formation of adherens junctions (Takaishi et al. 1997). Conversely, C3 transferase, which inhibits Rho, induced the disappearance of both adherens junctions and tight junctions (Nusrat et al. 1995; Takaishi et al. 1997), although whether this reflects a direct role for Rho in these junctions or an indirect consequence of its effects on actin cytoskeletal organization and adhesion to the extracellular matrix is not known.

In epithelial cells, Rho function has recently been linked to that of the ERM (ezrin/moesin/radixin) family of actin-binding proteins, which link actin filaments to the plasma membrane via transmembrane binding partners such as CD44 or ICAM-1 (Tsukita et al. 1997). ERM proteins are primarily located in microvilli on the apical surface of epithelial cells, although they have also been observed at intercellular boundaries (Bretscher et al. 1997). Rho can colocalize to the plasma membrane with ERM proteins, and Rho inactivation prevents the localization of ERM proteins and vinculin to the plasma membrane (Kotani et al. 1997). RhoGDI apparently binds in a complex with CD44 and ERM proteins, and could therefore act as an intermediary to target Rho to ERM proteins, or alternatively ERM proteins could increase Rho activity by titrating out RhoGDI (Takahashi et al. 1997). Finally, ERM protein interaction with actin is required for Rho and Rac to induce actin reorganization and focal complex assembly in permeabilized fibroblasts (Mackay et al. 1997). A Cdc42-interacting protein, CIP4, shows sequence homology to a small region of the ERM proteins and may act as a transducer to the actin cytoskeleton as it induces actin reorganization when overexpressed in Swiss 3T3 cells (Aspenström 1997).

Rho is also required for interactions between cells involving integrins on one cell and integrin ligands on the surface of another cell. For example, in B lymphocytes, intercellular aggregation mediated via the integrin LFA-1 and its interaction partner ICAM-1 is dependent on Rho (Tominaga et al. 1993). In addition, C3 transferase inhibits thrombin-induced platelet aggregation, which is dependent on activation of the integrin GPIIb–IIA (Morii et al. 1992). Rho may therefore be involved in mediating “inside-out” signalling involving conversion of integrins to an active conformation capable of binding to ligands.

5
Rho Family Proteins and Cytokinesis

Dramatic changes in actin organization occur in cells during mitosis and cytokinesis (Salmon 1989; Fishkind and Wang 1993). At mitosis, stress fibres disperse in adherent cultured cells and lamellipodia and ruffles are lost. As a consequence of these changes, cells round up and become very loosely attached to the substratum. Subsequently, the actomyosin-based contractile ring assembles and then constricts to form the cleavage furrow, finally separating each cell into its two daughter cells.

Both RhoA and Cdc42 have been shown to play a role during cytokinesis. RhoA is localized to the cleavage furrow, and is required for actin filament assembly and constriction of the contractile ring in *Xenopus* eggs and embryos and in sand dollar eggs (Kishi et al. 1993; Drechsel et al. 1996). *S. cerevisiae* Rho1p localizes to the mother-daughter neck at cytokinesis (Bussey 1996) suggesting that it too is involved in regulating cytokinesis. Rho could induce constriction via the action of its downstream target Rho-kinase on myosin light chain (see Kaibuchi this Vol.). In addition, Rho-kinase can phosphorylate the intermediate filament protein glial fibrillary acidic protein (GFAP), at the same sites that are phosphorylated during cytokinesis (Kosako et al. 1997). As GFAP is localized to the cleavage furrow, these results suggest that Rho could act via Rho-kinase to coordinate changes in the actomyosin cytoskeleton and intermediate filament network during cytokinesis.

Cdc42 plays a role in cleavage furrow formation in *Xenopus* embryos (Drechsel et al. 1996), and mammalian cells expressing activated Cdc42 accumulate multiple nuclei, suggesting that cytokinesis in inhibited (Dutartre et al. 1996). In *D. discoideum*, correct cytokinesis is dependent on a Rac/Cdc42-related protein, RacE, and on two members of the IQGAP family, which in mammalian systems are targets for Rac and Cdc42 (see Boettner and van Aelst, this Vol.; Faix and Dittrich 1996; Larochelle et al. 1996; Adachi et al. 1997; Lee et al. 1997). Interestingly, the effects of disrupting each *IQGAP* gene on cytokinesis are phenotypically distinguishable, suggesting that cytokinesis may involve several steps regulated by different GTPases (Adachi et al. 1997). As IQGAPs can bind actin directly, they may provide a link between Cdc42/Rac and the actin cytoskeleton (Bashour et al. 1997).

6
Other Responses Regulated by Rho Family Proteins

In addition to inducing actin reorganization, several other functions have been attributed to Rho family proteins, some of which may be linked to their actions on the actin cytoskeleton, although others are apparently independent of this.

6.1
Activation of the NADPH Oxidase and Phagocytosis

The requirement for Rac in regulating the NADPH oxidase of phagocytes has been known for several years, as has its involvement in actin reorganization. Rac is one of three cytoplasmic components regulating activation of the membrane bound cytochrome which produces superoxide ions (Ridley 1995).

Superoxide ions are highly toxic and thus activation of the NADPH oxidase should occur subsequent to formation of the phagosome, in order to limit the effects of the ions to invading micro-organisms. Actin polymerization is necessary for phagocytosis (Allen and Aderem 1996), and Rac and Cdc42 are required for Fcγ receptor-mediated phagocytosis (Cox et al. 1997). It is not known precisely which stage of phagocytosis involves Rac and Cdc42, although it is intriguing that PI 3-kinase has been shown to play a role in the final membrane fusion step of phagocytosis (Araki et al. 1996). Given that PI 3-kinase can act upstream of Rac (Parker 1995), it is possible that Rac also plays a role at this stage. In contrast to Rac and Cdc42, Rho is required during phagocytosis for the initial clustering of Fcγ receptors on macrophages (Hackam et al. 1997).

Both Rho and Cdc42 have been implicated in regulating the direct entry of different microorganisms into cells. Entry of *Shigella* requires Rho proteins, which localize directly to the membrane folds surrounding the bacterium (Adam et al. 1996). Interestingly, RhoA shows a slightly different localization to RhoB and RhoC within the folds, suggesting that the isoforms may carry out distinct functions. In contrast, entry of *Salmonella typhimurium* requires Cdc42 but not Rac or Rho (Jones et al. 1993; Chen et al. 1996). Binding of *S. typhimurium* to epithelial cells rapidly stimulates actin reorganization and bacteria are internalized by subsequent macropinocytosis. Whether bacterial uptake involves filopodium extension is not clear.

6.2 Secretion and Endocytosis

It is now clear that trafficking of vesicles within cells involves interaction with the actin cytoskeleton as well as with microtubules, although the molecular links between vesicles and actin filaments are largely unknown. There is evidence, however, that Rho family proteins can affect vesicle trafficking, and this may or may not be linked to their ability to alter actin organisation.

In permeabilized mast cells, recombinant Rac and Rho proteins enhance secretion, whereas C3 transferase and dominant negative Rac inhibit GTPγS-induced secretion (Price et al. 1995). Rac has also been purified from mast cells as a factor that can enhance secretion (O'Sullivan et al. 1996). Secretion is accompanied by actin reorganization and these changes are also mediated by Rac and Rho, but by using inhibitors of actin reorganization it has been shown that the two responses can be independently regulated (Norman et al. 1996).

Rho and Rac have also been implicated in the regulation of both constitutive and receptor-mediated endocytosis (Schmalzing et al. 1995; Lamaze et al. 1996). In *Xenopus* oocytes, Rho can enhance constitutive endocytosis/pinocytosis, whereas C3 transferase inhibits the process (Schmalzing et al. 1995). Conversely, in mammalian cells and in a cell-free endocytosis system,

Rho and Rac inhibit receptor-mediated endocytosis of clathrin-coated vesicles (Lamaze et al. 1996). Interestingly, one of the three Rho isoforms, RhoB, is localized to early endosomes, suggesting a role in endocytosis (Adamson et al. 1992b). RhoD also localizes to early endosomes, and overexpression of GTPase-defective RhoD decreases organelle motility, impeding movement and fusion of endosomes (Murphy et al. 1996). Cycling of RhoD between a GTP- and GDP-bound form is therefore likely to be important for endosomal motility.

The mechanisms underlying the involvement of Rho and Rac in endocytosis and secretion have not been established, but one possibility is that they act by altering membrane phospholipid composition, for example through activation of PIP 5-kinase, PI 3-kinase or phospholipase D (Frohmann and Morris 1996). These enzymes are potential targets for Rho family proteins and are known to play a role in vesicle trafficking, where it is likely that regulated changes in phosphoinositide composition of vesicles is important for vesicle budding and/or fusion (de Camilli et al. 1996; Frohmann and Morris 1996). Alternatively, at least some of the effects observed could be due to reorganization of the actin cytoskeleton such that it is diverted away from sites where it normally interacts with intracellular vesicles.

6.3
Mitogenesis and Transformation

In contrast to Ras, none of the Rho family proteins have been isolated as oncogenes or tumour suppressors from human cancers. There is evidence, however, that they can contribute to the transformation of rodent fibroblast lines: activated Rho, Rac and Cdc42 are weakly oncogenic and can synergise with other oncogenes to induce transformation, and several exchange factors for Rho, Rac and/or Cdc42 have been isolated as oncogenes in NIH-3T3 fibroblast transformation assays (Symons 1996; Qui et al. 1997). In addition, suppression of Ras-induced transformation by the farnesyltransferase inhibitor, L-739,749, appears to be primarily due to its effects on RhoB, rather than on Ras itself (Lebowitz et al. 1995, 1997). In accord with their transforming potentials, microinjection of Rho, Rac and Cdc42 proteins into quiescent Swiss 3T3 fibroblasts has been reported to stimulate DNA synthesis (Olsen et al. 1995). However, Cdc42 can also inhibit DNA synthesis (Molnár et al. 1997), and when expressed in the thymus of mice, Rac enhances apoptosis (Lorès et al. 1997). In these cases, Rac and Cdc42 may act by stimulating the serine/threonine kinase JNK, which has been implicated in inducing apoptosis (Fanger et al. 1997). In fact, it seems likely that the long-term effects of activating these proteins vary considerably depending on the status of the cells, as has also been observed for Ras (Kauffmann-Zeh et al. 1997).

Whether the contributions of Rho, Rac and Cdc42 to cell transformation involve their effects on actin organization and/or cell adhesion has not been fully established. Studies with various Rac effector domain mutants suggest that its ability to induce lamellipodium formation in fibroblasts can be separated from its transforming activity (Westwick et al. 1997). In these experiments, the definition of transformation has been predominantly one of increased growth rate in the absence of adhesion to a substratum. In vivo, however, cancer progression involves invasion and metastasis, which are processes of cell migration and are therefore likely to be linked with changes in actin organization and cell adhesion regulated by Rho family proteins. Thus, for example, Tiam-1, a Rac exchange factor, enhances the invasion of T-lymphoma cells (Habets et al. 1994), a property that may well be due at least in part to changes in the actin cytoskeleton or cell adhesion.

7 Conclusions

Rho, Rac and Cdc42 each regulate a diverse range of cellular functions through their ability to regulate the formation of different actin-based structures and associated adhesion sites. They act specifically by directing new actin polymerization to precise locations, and coordinating this with recruitment of appropriate actin-binding proteins to allow the correct organization of these actin filaments. It is likely that this complex process requires each GTPase to interact with several target proteins (see Kaibuchi and Boettner and van Aelst, this Vol.).

Although it is clear that Rho family proteins can contribute to the transformed phenotype of rodent fibroblasts, whether they play a regulatory role in the development of human cancers is not known. An important question for the future therefore is whether Rho family proteins contribute to the transformed phenotype of human cancers, and whether this involves their ability to regulate actin organization and cell adhesion. Their functions suggest, however, that they will play an active role in processes involving cell migration, and this is borne out by studies on the development of model organisms such as *Drosophila* (see Settleman, this Vol.). They may therefore be potential targets for therapeutic treatment of diseases such as cancer which involve aberrant cell migration.

References

Adachi H, Takahashi Y, Hasebe T, Shirouzu M, Yokoyama S, Sutoh K (1997) *Dictyostelium* IQGAP-related protein specifically involved in the completion of cytokinesis. J Cell Biol 137:891–898

Adam T, Giry M, Boquet P, Sansonetti P (1996) Rho-dependent membrane folding causes *Shigella* entry into epithelial cells. EMBO J 15:3315–3321

Adamson P, Marshall CJ, Hall A, Tilbrook PA (1992a) Post-translational modifications of $p21^{rho}$ proteins. J Biol Chem 267:20033–20038

Adamson P, Paterson HF, Hall A (1992b) Intracellular localization of the p21^{rho} proteins. J Cell Biol 119:617–627
Allen LA, Aderem A (1996) Mechanisms of phagocytosis. Curr Opin Immunol 8:36–40
Allen WE, Jones GE, Pollard JW, Ridley AJ (1997) Rho, Rac and Cdc42 regulate actin organization and cell adhesion in macrophages. J Cell Sci 110:707–720
Araki N, Johnson MT, Swanson JA (1996) A role for phosphoinositide 3-kinase in the completion of macropinocytosis and phagocytosis by macrophages. J Cell Biol 135:1249–1260
Aspenström P (1997) A Cdc42 target protein with homology to the non-kinase domain of FER has a potential role in regulating the actin cytoskeleton. Curr Biol 7:479–487
Barbacid M (1987) *ras* genes. Annu Rev Biochem 56:779–827
Barry ST, Flinn HM, Humphries M, Critchley DR, Ridley AJ (1997) Requirement for Rho in integrin signalling. Cell Adhes Commun 4:387–398
Bashour A-M, Fullerton AT, Hart MJ, Blook GS (1997) IQGAP1, a Rac- and Cdc42-binding protein, directly binds and cross-links microfilaments. J Cell Biol 137:1555–1566
Bourne HA, Sanders DA, McCormick F (1991) The GTPase superfamily: conserved structure and molecular mechanism. Nature 349:117–127
Braga VMM, Machesky LM, Hall A, Hotchin NA (1997) The small GTPases Rho and Rac are required for the establishment of cadherin-dependent cell-cell contacts. J Cell Biol 137:1421–1431
Bretscher A, Reczek D, Berryman K (1997) Ezrin: a protein requiring conformational activation to link microfilaments to the plasma membrane in the assembly of cell surface structures. J Cell Sci 110:3011–3018
Brock J, Midwinter K, Lewis J, Martin P (1996) Healing of incisional wounds in the embryonic chick wing bud: characterization of the actin purse-string and demonstration of a requirement for Rho activation. J Cell Biol 135:1097–1107
Burridge K, Chrzanowska-Wodnicka M (1996) Focal adhesions, contractility, and signaling. Annu Rev Cell Dev Biol 12:463–519
Bussey H (1996) Cell shape determination: a pivotal role for Rho. Science 272:224–226
Chant J (1996) Generation of cell polarity in yeast. Curr Opin Cell Biol 8:557–565
Chen LM, Hobbie S, Galan JE (1996) Requirement of Cdc42 for *Salmonella*-induced cytoskeletal and nuclear responses. Science 274:2115–2118
Chrzanowska-Wodnicka M, Burridge K (1996) Rho-stimulated contractility drives the formation of stress fibres and focal adhesions. J Cell Biol 133:1403–1415
Cox D, Chang P, Zhang Q, Reddy PG, Bokoch GM, Greenberg S (1997) Requirements for both Rac1 and Cdc42 in membrane ruffling and phagocytosis in leukocytes. J Exp Med 186:1487–1494
Cvrcková F, De Virgilio C, Manser E, Pringle JR, Nasmyth K (1995) Ste-20-like protein kinases are required for normal localization of cell growth and for cytokinesis in budding yeast. Genes Dev 9:1817–1830
De Camilli P, Emr SD, McPherson PS, Novick P (1996) Phosphoinositides as regulators in membrane traffic. Science 271:1533–1539
Defilppi P, Olivo C, Tarone G, Mancini P, Rosaria Torrisi M, Eva A (1997) Actin cytoskeleton polymerization in Dbl-transformed NIH3T3 fibroblasts is dependent on cell adhesion to specific extracellular matrix proteins. Oncogene 14:1933–1943
Didsbury J, Weber RF, Bokoch GM, Evans T, Snyderman R (1989) *Rac*, a novel *ras*-related family of proteins that are botulinum toxin substrates. J Biol Chem 264:16378–16382
Drechsel DN, Hyman AA, Hall A, Glotzer M (1996) A requirement for Rho and Cdc42 during cytokinesis in *Xenopus* embryos. Curr Biol 7:12–23
Dutartre H, Davoust J, Gorvel J-P, Chavrier P (1996) Cytokinesis arrest and redistribution of actin-cytoskeleton regulatory components in cells expressing the Rho GTPase CDC42Hs. J Cell Sci 109:367–377
Eaton S, Auvinen P, Luo L, Jan YN, Simons K (1995) Cdc42 and Rac1 control different actin-dependent processes in the *Drosophila* wing disc epithelium. J Cell Biol 131:151–164

Evangelista M, Blundell K, Longtine MS, Chow CJ, Adames J, Pringle JR, Peter M, Boone C (1997) Bni1p, a yeast formin linking Cdc42p and the actin cytoskeleton during polarized morphogenesis. Science 276:118–122

Faix J, Dittrich W (1996) DGAP1, a homologue of RasGTPawe activatin proteins that controls growth, cytokinesis, and development in *Dictyostelium discoideum*. FEBS Lett 394:251–257

Fanger GR, Gerwins P, Widmann C, Jarpe MB, Johnson GL (1997) MEKKs, GCKs, MLKs, PAKs, TAKs, and Tpls: upstream regulators of the c-Jun amino-terminal kinases? Curr Opin Genet Dev 7:67–74

Farnsworth CL, Feig LA (1991) Dominant inhibitory mutations in the Mg^{2+}-binding site of Ras^{H} prevent its activation by GTP. Mol Cell Biol 11:4822–4829

Fishkind DJ, Wang Y-L (1993) Orientation and three-dimensional organization of actin filaments in dividing cultured cells. J Cell Biol 123:837–848

Flinn HM, Ridley AJ (1996) The small GTP-binding protein Rho stimulates tyrosine phosphorylation of focal adhesion kinase, p130 and paxillin. J Cell Sci 109:1133–1141

Foster R, Hu K-Q, Lu Y, Nolan KM, Thissen J, Settleman J (1996) Identification of a novel human Rho protein with unusual properties: GTPase deficiency and in vivo farnesylation. Mol Cell Biol 16:2689–2699

Freeman JL, Abo A, Lambeth JD (1996) Rac "insert region" is a novel effector region that is implicated in the activation of NADPH oxidase, but not PAK65. J Biol Chem 271:19794–19801

Frohman MA, Morris AJ (1996) Rho is only ARF the story. Curr Biol 6:945–947

Habets GGM, Scholtes EHM, Zuydgeest D, van der Kammen RA, Stam JC, Berns A, Collard JG (1994) Identification of an invasion-inducing gene, *Tiam*-1, that encodes a protein with homology to GDP-GTP exchangers for Rho-like proteins. Cell 77:537–549

Hackam DJ, Rotstein OD, Schreiber A, Zhang WJ, Grinstein S (1997) Rho is required for the initiation of calcium signaling and phagocytosis by Fcγ receptors in macrophages. J Exp Med 186:55–66

Hall A (1994) Small GTP-binding proteins and the regulation of the actin cytoskeleton. Annu Rev Cell Biol 10:31–54

Hartwig JH, Bokoch GM, Carpenter CL, Janmey PA, Taylor LA, Toker A, Stossel TP (1995) Thrombin receptor ligation and activated Rac uncap actin filament barbed ends through phosphoinositide synthesis in permeabilized human platelets. Cell 82:643–653

Hirata K, Kikuchi A, Sasaki T, Kuroda S, Kaibuchi K, Matsuura Y, Seki H, Saida K, Takai Y (1992) Involvement of rho p21 in the GTP-enhanced calcium ion sensitivity of smooth muscle contraction. J Biol Chem 267:8719–8722

Hirshberg M, Stockley RW, Dodson G, Webb MR (1997) The crystal structure of human rac1, a member of the rho-family complexed with a GTP analogue. Nat Struct Biol 4:147–153

Hotchin NA, Hall A (1995) The assembly of integrin adhesion complexes requires both extracellular matrix and intracellular rho/rac GTPases. J Cell Biol 131:1857–1865

Imamura H, Tanaka K, Hihara T, Umikawa M, Kamei T, Takahashi K, Sasaki T, Takai Y (1997) Bni1p and Bnr1p: downstream targets of the Rho family small G-proteins which interact with profilin and regulate actin cytoskeleton in *Saccharomyces cerevisiae*. EMBO J 16:2745–2755

Jähner D, Hunter T (1991) The ras-related gene rhoB is an immediate-early gene inducible by v-fps, epidermal growth factor and platelet-derived growth factor in rat fibroblasts. Mol Cell Biol 11:3682–3690

Jalink K, van Corven EJ, Hengeveld T, Morii N, Narumiya S, Moolenaar WH (1994) Inhibition of lysophosphatidate- and thrombin-induced neurite retraction and neuronal cell rounding by ADP ribosylation of the small GTP-binding protein rho. J Cell Biol 126:801–810

Johnson DI, Pringle JR (1990) Molecular characterization of CDC42, a *Saccharomyces cerevisiae* gene involved in the development of cell polarity. J Cell Biol 111:143–152

Jones BD, Paterson HF, Hall A, Falkow S (1993) *Salmonella typhimurium* induces membrane ruffling by a growth factor-receptor-independent mechanism. Proc Natl Acad Sci USA 90:10390–10394

Joseph G, Pick E (1996) "Peptide walking" is a novel method for mapping functional domains in proteins. Its application to the Rac1-dependent activation of NADPH oxidase. J Biol Chem 270:29079–29082
Kauffmann-Zeh A, Rodriguez-Viciana P, Ulrich E, Gilbert C, Coffer P, Downward J, Evan G (1997) Suppression of c-Myc-induced apoptosis by Ras signalling through PI(3)K and PKB. Nature 385:544–547
Kishi K, Sasaki T, Kuroda S, Itoh T, Takai Y (1993) Regulation of cytoplasmic division of *Xenopus* embryo by *rho* p21 and its inhibitory GDP/GTP exchange protein (*rho* GDI). J Cell Biol 120:1187–1195
Kosako H, Amano M, Yanagida M, Tanabe K, Nishi Y, Kaibuchi K, Inagaki M (1997) Phosphorylation of glial fibrillary acidic protein at the same sites by cleavage furrow kinase and Rho-associated kinase. J Biol Chem 272:10333–10336
Kotani H, Takaishi K, Sasaki T, Takai Y (1997) Rho regulates association of both the ERM family and vinculin with the plasma membrane in MDCK cells. Oncogene 14:1705–1713
Kozma R, Sarner S, Ahmed S, Lim L (1997) Rho family GTPases and neuronal growth cone remodelling: relationship between increased complexity induced by Cdc42Hs, Rac1, and acetylcholine and collapse induced by RhoA and lysophosphatidic acid. Mol Cell Biol 17:1201–1211
Kreck ML, Freeman JL, Abo A, Lambeth JD (1996) Membrane association of Rac is required for high activity of the respiratory burst oxidase. Biochemistry 35:15683–15692
Lamarche N, Tapon N, Stowers L, Burbelo PD, Aspenström P, Bridges T, Chant J, Hall A (1996) Rac and Cdc42 induce actin polymerization and G1 cell cycle progression independently of p65PAK and JNK/SAPK MAP kinase cascade. Cell 87:519–529
Lamaze C, Chuang T-H, Terlecky LJ, Bokoch GM, Schmid SL (1996) Regulation of receptor-mediated endocytosis by Rho and Rac. Nature 382:177–179
Lamoureux P, Altun-Gultekin SF, Lin C, Wagner JA, Heidemann SR (1997) Rac is required for growth cone function but not neurite assembly. J Cell Sci 110:635–641
Lang P, Gesbert F, Delespine-Carmagnat M, Stancou R, Pouchelet M, Bertoglio J (1996) Protein kinase A phosphorylation of RhoA mediates the morphological and functional effects of cyclic AMP in cytotoxic lymphocytes. EMBO J 15:510–519
Larochelle DA, Vithalani KK, De Lozanne A (1996) A novel member of the rho family of small GTP-binding proteins is specifically required for cytokinesis. J Cell Biol 133:1321–1329
Lauffenburger DA, Horwitz AF (1996) Cell migration: a physically integrated molecular process. Cell 84:359–369
Leberer E, Thomas DY, Whiteway M (1997) Pheromone signalling and polarized morphogenesis in yeast. Curr Opin Genet Dev 7:59–66
Lebowitz P, Davide J, Prendergast GC (1995) Evidence that farnesyltransferase inhibitors suppress Ras transformation by interfering with Rho activity. Mol Cell Biol 15:6613–6622
Lebowitz PF, Du W, Prendergast GC (1997) Prenylation of RhoB is required for its cell transforming function but not its ability to activate serum response element-dependent transcription. J Biol Chem 272:16093–16095
Lee S, Escalante R, Firtel RA (1997) A RasGAP is essential for cytokinesis and spatial patterning in *Dictyostelium*. Development 124:983–996
Lorès P, Morin L, Luna R, Gacon G (1997) Enhanced apoptosis in the thymus of transgenic mice expressing constitutively activated forms of human Rac2GTPase. Oncogene 15:601–605
Machesky LM, Hall A (1996) Rho: a connection between membrane receptor signalling and the actin cytoskeleton. Trends Cell Biol 6:304–310
Machesky LM, Hall A (1997) Role of actin polymerization and adhesion to extracellular matrix in Rac- and Rho-induced cytoskeletal reorganization. J Cell Biol 138:913–926
Mackay DJG, Esch F, Furthmayr H, Hall A (1997) Rho- and Rac-dependent assembly of focal a dhesion complexes and actin filaments in permeabilized fibroblasts: as essential role for ezrin/radixin/moesin proteins. J Cell Biol 138:927–938

Malosio ML, Gilardelli D, Paris S, Albertinazzi C, de Curtis I (1997) Differential expression of distinct members of Rho family GTP-binding proteins during neuronal development: identification of Rac1B, a new neural-specific member of the family. J Neurosci 17:6717–6728

Marshall CJ (1993) Protein prenylation: a mediator of protein-protein interactions. Science 259:1865–1866

Miller PJ, Johnson DI (1994) Cdc42p GTPase is involved in controlling polarized cell growth in *Schizosaccharomyces pombe*. Mol Cell Biol 14:1075–1083

Mitchison TJ, Cramer LP (1996) Actin-based cell motility and cell locomotion. Cell 84:371–379

Moll J, Sansig G, Fattori E, van der Putten H (1991) The murine *rac*1 gene: cDNA cloning, tissue distribution and regulated expression of *rac*1 mRNA by disassembly of actin microfilaments. Oncogene 6:863–866

Mólnar A, Theodoras AM, Zon LI, Kyriakis JM (1997) Cdc42Hs, but not Rac1, inhibits serum-stimulated cell cycle progression at G1/S through a mechanism requiring p38/RK. J Biol Chem 272:13229–23235

Morii N, Teru-uchi T, Tominaga T, Kumagai N, Kozaki S, Ushikubi F, Narumiya S (1992) A *rho* gene product in human blood platelets. II. Effects of the ADP-ribosylation by botulinum C3 ADP-ribosyltransferase on platelet aggregation. J Biol Chem 267:20921–20926

Murphy AM, Montell DJ (1996) Cell type-specific roles for Cdc42, Rac and RhoL in *Drosophila* oogenesis. J Cell Biol 133:617–630

Murphy C, Saffrich R, Grummt M, Gournier H, Rybin V, Rubino M, Auvinen P, Lütcke A, Parton RG, Zerial M (1996) Endosomal dynamics regulated by a Rho protein. Nature 384:427–432

Nishiyama T, Sasaki T, Takaishi K, Masaki K, Hideaki Y, Araki K, Matsuura Y, Takai Y (1994) rac p21 is involved in insulin-induced membrane ruffling and rho p21 is involved in hepatocyte growth factor- and 12-*O*-tetradecanoylphorbol-13-acetate (TPA)-induced membrane ruffling in KB cells. Mol Cell Biol 14:2447–2456

Nobes CD, Hall A (1995) Rho, Rac, and Cdc42 GTPases regulate the assembly of multimolecular focal adhesion complexes associated with actin stress fibers, lamellipodia, and filopodia. Cell 81:53–62

Nobes CD, Hawkins P, Stephens L, Hall A (1995) Activation of the small GTP-binding proteins rho and rac by growth factor receptors. J Cell Sci 108:225–233

Norman JC, Price LS, Ridley AJ, Hall A, Koffer A (1994) Actin filament organization in activated mast cells is regulated by heterotrimeric and small GTP-binding proteins. J Cell Biol 126:1005–1015

Norman JC, Price LS, Ridley AJ, Koffer A (1996) The small GTP-binding proteins, Rac and Rho, regulate cytoskeletal organisation and exocytosis in mast cells by parallel pathways. Mol Biol Cell 7:1429–1442

Nusrat A, Giry M, Turner JR, Colgan SP, Parkos CA, Carnes D, Lemichez E, Bouquet P, Madara JL (1995) Rho protein regulates tight junctions and perijunctional actin organization in polarized epithelia. Proc Natl Acad Sci USA 92:10629–10633

Olofsson B, Chardin P, Touchot N, Zahraoui A, Tavitian A (1988) Expression of the ras-related ralA, rho12 and rab genes in adult mouse tissues. Oncogene 3:231–234

Olsen MF, Ashworth A, Hall A (1995) An essential role for Rho, Rac and Cdc42 GTPases in cell cycle progression through G_1. Science 269:1270–1272

O'Sullivan AJ, Brown AM, Freeman HNM, Gomperts BD (1996) Purification and identification of FOAD-II, a cytosolic protein that regulates secretion in streptolysin-O-permeabilized mast cells, as a rac/rhoGDI complex. Mol Biol Cell 7:397–408

Parker P (1995) PI 3-kinase puts GTP on the Rac. Curr Biol 5:577–579

Peppelenbosch MP, Qui R-G, de Vries-Smits AMM, Tertoolen LGJ, de Laat SW, McCormick F, Hall A, Symons MH, Bos JL (1995) Rac mediates growth factor-induced arachidonic acid release. Cell 81:849–856

Price LS, Norman JC, Ridley AJ, Koffer A (1995) Rac and rho as regulators of secretion in mast cells. Curr Biol 5:68–73

Qui R-G, Abo A, McCormick F, Symons M (1997) Cdc42 regulates anchorage-independent growth and is necessary for Ras transformation. Mol Cell Biol 17:3449–3463
Quilliam LA, Kato K, Rabun KM, Hisaka MM, Huff SY, Campbell-Burk S, Der CJ (1994) Identification of residues critical for Ras(17N) growth-inhibitory phenotype and for Ras interaction with guanine nucleotide exchange factors. Mol Cell Biol 14:1113–1121
Ridley AJ (1994) Membrane ruffling and signal transduction. BioEssays 16:321–327
Ridley AJ (1995a) Rho-related proteins: actin cytoskeleton and cell cycle. Curr Opin Genet Dev 5:24–30
Ridley AJ (1995b) Rac and Bcr regulate phagocytic phoxes. Curr Biol 5:710–712
Ridley AJ (1996) Rho: theme and variations. Curr Biol 6:1256–1264
Ridley AJ, Hall A (1992) The small GTP-binding protein rho regulates the assembly of focal adhesions and actin stress fibers in response to growth factors. Cell 70:401–410
Ridley AJ, Paterson HF, Johnston CL, Diekmann D, Hall A (1992) The small GTP-binding protein rac regulates growth factor/induced membrane ruffling. Cell 70:401–410
Ridley AJ, Comoglio PM, Hall A (1995) Regulation of scatter factor/hepatocyte growth factor responses by Ras, Rac and Rho proteins in MDCK cells. Mol Cell Biol 15:1110–1122
Ruoslahti E (1997) Stretching is good for a cell. Science 276:1345–1346
Salmon ED (1989) Cytokinesis in animal cells. Curr Opin Cell Biol 1:541–547
Schmalzing G, Richter H-P, Hansen A, Schwartz W, Just I, Aktories K (1995) Involvement of the GTP binding protein Rho in constitutive endocytosis in *Xenopus laevis* oocytes. J Cell Biol 130:1319–1332
Schwartz MA (1997) Integrins, oncogenes and anchorage independence. J Cell Biol 139:575–578
Self AJ, Hall A (1995) Purification of recombination Rho/Rac/G25K from *Escherichia coli.* Methods Enzymol 256:3–10
Shinjo K, Koland JG, Hart MJ, Narasimhan V, Johnson DI, Evans T, Cerione RA (1990) Molecular cloning of the gene for the human placental GTP-binding protein Gp (G25K): identification of this GTP-binding protein as the human homolog of the yeast cell-division-cycle protein Cdc42. Proc Natl Acad Sci USA 87:9853–9857
Shirsat NV, Pignolo RJ, Kreider BL, Rovera G (1990) A member of the *ras* gene superfamily is specifically expressed in T, B and myeloid hemopoietic cells. Oncogene 5:769–772
Stossel TP (1993) On the crawling of mammalian cells. Science 260:1086–1094
Symons M (1996) Rho family GTPases: the cytoskeleton and beyond. Trends Biochem Sci 21:178–181
Takahashi K, Sasaki T, Mammoto A, Takaishi K, Kameyama T, Tsukita S, Tsukita S, Takai Y (1997) Direct interaction of the Rho GDP dissociation inhibitor with ezrin/radixin/moesin initiates the activation of the Rho small G protein. J Biol Chem 272:23371–23375
Takaishi K, Sasaki T, Kotanin H, Nishioka H, Takai Y (1997) Regulation of cell-cell adhesion by Rac and Rho small G proteins in MDCK cells. J Cell Biol 139:1047–1059
Tapon N, Hall A (1997) Rho, Rac and Cdc42 GTPases regulate the organization of the actin cytoskeleton. Curr Opin Cell Biol 9:86–92
Tominaga T, Sugie K, Hirata M, Morii N, Fukata J, Uchida A, Imura H, Narumiya S (1993) Inhibition of PMA-induced, LFA-1-dependent lymphocyte aggregation by ADP-ribosylation of the small molecular weight GTP-binding protein, Rho. J Cell Biol 111:2097–2108
Tsukita S, Yonemura S, Tsukita S (1997) ERM proteins: head-to-tail regulation of actin-plasma membrane interaction. Trends Biochem Sci 22:53–58
Vincent S, Jeanteur P, Fort P (1992) Growth-regulated expression of *rhoG*, a new member of the ras homolog gene family. Mol Cell Biol 12:3138–3148
Watanabe N, Madaule P, Reid T, Ishizaki T, Watanabe GG, Kakizuka A, Saito Y, Nakao K, Jockusch BM, Narumiya S (1997) p140mDia, a mammalian homolog of *Drosophila* diaphanous, is a target protein for Rho smal GTPase and is a ligand for profilin. EMBO J 16:3044–3056
Wei Y, Zhang Y, Derewenda U, Liu X, Minor W, Nakamoto RK, Somlyo AV, Somlyo AP, Derewenda ZS (1997) Crystal structure of RhoA-GDP and its functional implicatins. Nat Struct Biol 4:699–703

Welch MD, Holtzman DA, Drubin DG (1994) The yeast actin cytoskeleton. Curr Opin Cell Biol 6:110–119

Welch MD, Mallavarapu A, Rosenblatt J, Mitchison TJ (1997) Actin dynamics in vivo. Curr Opin Cell Biol 9:54–61

Westwick JK, Lambert QT, Clark GJ, Symons M, van Aelst L, Pestell RG, Der CJ (1997) Rac regualtion of transformation, gene expression, and actin organization by multiple, PAK-independent pathways. Mol Cell Biol 17:1324–1335

Wójciak-Stothard B, Entwistle A, Garg R, Ridley AJ (1998) Regulation of TFN-α-induced reorganization of the actin cytoskeleton and cell-cell junctions by Rho, Rac and Cdc42 in human endothelial cells. J Cell Physiol (in press)

Zigmond SH, Joyce M, Borleis J, Bokoch GM, Devreotes PN (1997) Regulation of actin polymerization in cell-free systems by GTPγS and Cdc42. J Cell Biol 138:363–374

Regulation of Cytoskeleton and Cell Adhesion by Rho Targets

Kozo Kaibuchi[1]

1 Introduction

Accumulating evidence indicates that cells undergo temporal and spatial reorganization of the cytoskeleton and cell adhesions upon stimulation by extracellular signals including growth factors, hormones and bioactive lipids or during cell movement and mitosis (Stossel 1993; Zigmond 1996). Cytoskeletons mainly consist of actin filaments, microtubules and intermediate filaments (Nobes and Hall 1994). These filaments are dynamically reorganized during cell movement and mitosis. Cell adhesions are mediated by cell adhesion molecules such as cadherin and integrin, and their associated cytoplasmic proteins such as catenins and talin (Takeichi 1990; Gumbiner 1996). Cell adhesions are also dynamically reorganized during cell movement and mitosis. However, until recently little was known about the signal transduction pathways which regulate the reorganization of the cytoskeleton and cell adhesions.

Rho is implicated in signaling pathways that regulate actin cytoskeletons, such as stress fibers, and cell-substratum adhesions, such as focal adhesions in fibroblasts (Ridley and Hall 1992). Rho is also involved in the regulation of cell morphology (Paterson et al. 1990), cell aggregation (Tominaga et al. 1993), cadherin-mediated cell-cell adhesion (Braga et al. 1997), cell motility (Takaishi et al. 1994; Ridley et al. 1995), membrane ruffling (Nishiyama et al. 1994), smooth muscle contraction (Hirata et al. 1992; Gong et al. 1996), neurite retraction in neuronal cells (Nishiki et al. 1990; Jalink et al. 1994) and cytokinesis (Kishi et al. 1993; Mabuchi et al. 1993). However, the molecular mechanism by which Rho regulates such functions was not clarified until recently. Recent studies have established molecular pathways connecting Rho to cytoskeletons and cell adhesions.

[1]Division of Signal Transduction, Nara Institute of Science and Technology, 8916-5 Takayama, Ikoma 630-01, Japan

Progress in Molecular and Subcellular Biology, Vol. 22
Ph. Jeanteur (Ed.)

2 Regulation of Rho Activity

Members of the Rho family, including RhoA, B, C, Rac1, 2, and Cdc42, share more than 50% sequence identity (Nobes and Hall 1994). RhoA, B and C (Rho) have the same amino acid sequence in the effector domain, and are thought to be regulated in a similar manner and to have similar functions. The Rho family, like other GTP-binding proteins, exhibits both GDP/GTP-binding and GTPase activities (Nobes and Hall 1994). They have GDP-bound inactive and GTP-bound active forms which are interconvertible by GDP/GTP exchange and GTPase reactions (Nobes and Hall 1994). The GDP/GTP exchange reaction is regulated by stimulatory proteins such as Smg GDS (Mizuno et al. 1991), Dbl (Hart et al. 1991), Ost (Horii et al. 1994), and Tiam-1 (Habets et al. 1994) and by inhibitory proteins such as Rho GDI (Fukumoto et al. 1990). The GTPase reaction is regulated by Rho GTPase-activating proteins (GAP), such as Ras GAP-associated p190 (Settleman et al. 1992), Rho GAP (Lancaster et al. 1994), and Rho GAP p122 (Homma and Emori 1995). When cells are stimulated with certain extracellular signals, such as lysophosphatidic acid (LPA), GDP·Rho is thought to be converted to GTP·Rho, which binds to specific targets and then exerts its biological functions (Fig. 1). However, it is not known how the activity of Rho is regulated by extracellular signals.

3 Functions of Rho

3.1 Formation of Stress Fiber and Focal Adhesion

Botulinum toxin C3 is regarded as a useful tool for dissecting the specific functions of Rho. C3 specifically ADP-ribosylates Rho at Asn^{41} (Nobes and Hall 1994). The ADP-ribosylation of Rho is thought to result in interference between the interaction of Rho and its putative targets, thereby inhibiting the functions of Rho. C3 toxin poorly ADP-ribosylates other Rho family members such as Rac and Cdc42. Rho is believed to regulate the formation of actin stress fibers and focal adhesions based on the following observations. Confluent, serum-starved Swiss 3T3 cells have very few actin stress fibers composed of actin filaments and their associated myosin filaments, which give contractility to the fibers. When the cells are stimulated by LPA, new stress fibers appear, which increase in number and diameter (Ridley and Hall 1992). Prior treatment of the cells with C3 toxin inhibits the LPA-induced stress fiber formation. Microinjection of dominant active RhoA ($RhoA^{V14}$) into the cells also induces stress fiber formation. Rho is also required for the LPA-induced formation of focal adhesions, the site at which stress fibers are linked via integrin to the extracellular matrix (Ridley and Hall 1992). Turnover of stress fibers and their

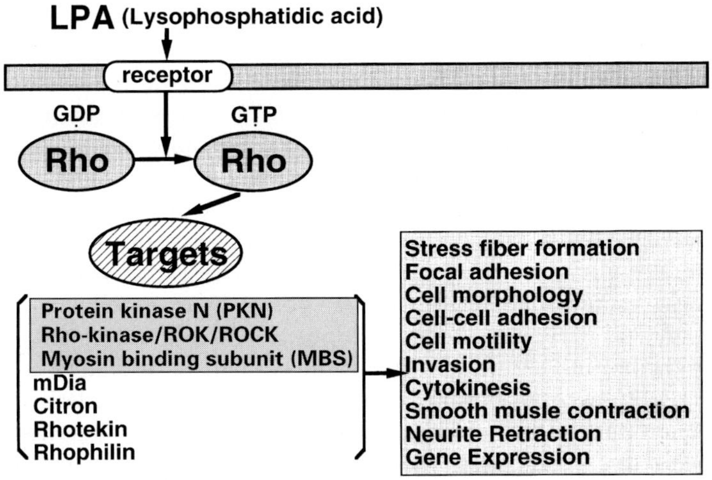

Fig. 1. Physiological functions of Rho

associated focal adhesions appears to be essential for cell motility. Consistently, Rho has been reported in the regulation of cell motility (Ridley and Hall 1992; Takaishi et al. 1994).

3.2 Smooth Muscle Contraction

Stimulation of smooth muscle and non-muscle cells by specific agonists induces Ca^{2+} mobilization and activation of MLC kinase which phosphorylates MLC. This then activates myosin ATPase, and thereby induces contraction of smooth muscles (Hartshorne 1987; Sellers and Adelstein 1987) and of actin-myosin bundles in non-muscle cells (Giuliano and Taylor 1995). However, since the cytosolic Ca^{2+} level does not always correlate with the levels of MLC phosphorylation and contraction, an additional mechanism which regulates the Ca^{2+} sensitivity of levels of MLC phosphorylation and contraction has been proposed (Bradley and Morgan 1987). Since agonists cause the MLC phosphorylation and contraction of permeabilized smooth muscles at fixed submaximal concentrations of Ca^{2+} in a GTP-dependent manner, a GTP-binding protein is thought to regulate the receptor-mediated sensitization of MLC phosphorylation (Kitazawa et al. 1991). Rho is implicated in the GTP-enhanced Ca^{2+} sensitivity of smooth muscle contraction (Hirata et al. 1992; Gong et al. 1996). Guanosine 5′-(3-*O*-thio)-triphosphate (GTPγS; a non-hydrolyzable GTP analogue) increases MLC phosphorylation at submaximal Ca^{2+} concentrations in permeabilized smooth muscle cells, presumably by inhibiting myosin phosphatase through Rho (Noda et al. 1995).

3.3 Neurite Retraction

Molecules such as extracellular matrix components, growth factors, neurotransmitters, bioactive lipids such as lysophosphatidic acid (LPA) and secreted proteases such as thrombin can exert significant effects on neural architecture, ranging from the stimulation of neurite outgrowth to the induction of growth cone collapse and neurite retraction (Tanaka and Sabry 1995; Tessier-Lavigne and Goodman 1996). The rapid withdrawal of developing neurites and their growth cones evoked by extracellular molecules is fundamental to nervous system development and neural plasticity. Rho is implicated in thrombin-, serum-, or LPA-induced neurite retraction and cell rounding in N1E-115 neuroblastoma or PC12 cells (Nishiki et al. 1990; Jalink et al. 1994). When deprived of serum, N1E-115 cells acquire a flattened morphology and begin to extend neurites, in which the activities of Rac and Cdc42 are thought to be necessary (Kozma et al. 1997). Rho-induced neurite retraction is presumed to be driven by a contraction of the cortical actomyosin system (Jalink and Moolenaar 1992), which is known to be regulated by Rho.

3.4 Cytokinesis

Cells undergo cytokinesis in M-phase through the formation of a contractile ring beneath the plasma membrane. The contractile ring is mainly composed of actin filaments and myosin. Since inhibitors of actin polymerization and MLC kinase block the cytoplasmic division of mammalian cells and amphibian eggs, a contractile ring is thought to divide the cells into two by pulling the membrane inward (cytoplasmic division) through its contractility. The sperm-induced cytoplasmic division of *Xenopus* embryos is inhibited by microinjection of Rho GDI or C3 toxin, whereas nuclear division occurs normally (Kishi et al. 1993; Mabuchi et al. 1993). Co-injection of Rho GDI with the activated RhoA prevents the Rho GDI action. Thus, Rho appears to control cytoplasmic division through the regulation of the contractility of the contractile ring.

3.5 Cell-Cell Adhesion

Cell-cell adhesions are categorized into several groups including tight and adherens junctions (Tsukita et al. 1992). Tight junctions, the most apical component of the junctional complex, form a diffusion barrier which regulates the flux of ions and hydrophilic molecules. Tight junctions are mediated by molecules such as occludin and its associated ZO-1 and ZO-2. Adherens junctions are characterized by a well-developed plaque structure in which actin filaments are densely associated. Adherens junctions are mediated by adhesion molecules such as cadherin and its associated catenins. Recent work has shown that Rho, Rac and Cdc42 are involved in the regulation of cadherin-mediated cell-cell adhesion (Braga et al. 1997; Kuroda et al. 1997). For example, microinjection of C3 or Rac^{N17} inhibits cadherin-mediated adhesion in keratinocytes. Rho is also thought to be required to maintain the tight junction through the actin network beneath the plasma membranes of the cell-cell contact sites in epithelial cells (Nusrat et al. 1995).

3.6 Actin Filaments Beneath Plasma Membrane

Actin filaments are linked to membrane integral proteins at the plasma membrane through certain cross-linking proteins. The linkage between actin filaments and plasma membranes undergoes dynamic remodeling during cell movement and mitotic phase, though the molecular mechanism remains unclear (Gumbiner 1996). The ERM family (ezrin, radixin and moesin) is thought to function as one of the cross-linking proteins between the plasma membrane and actin filaments (Tsukita et al. 1992). The ERM family is localized at the specific regions where actin filaments are associated with plasma membranes such as microvilli, cleavage furrows, ruffling membranes, cell-cell adhesion sites and cell-substratum adhesion sites (Tsukita et al. 1992). The NH_2 (N)-terminal and COOH (C)-terminal domains of the ERM family are thought to directly bind to some integral membrane proteins such as CD44 and actin filaments respectively (Tsukita et al. 1994; Turunen et al. 1994). The N- and C-terminal domains of the ERM family oppose each other, presumably through the intramolecular and/or intermolecular head to tail association between the N- and C-terminal domains (Hirao et al. 1996). This intramolecular or intermolecular suppression is released by acidic phospholipids such as 4,5-PIP_2, and Rho enhances the formation of the ERM family-CD44 complex, presumably through 4,5-PIP_2 synthesis (Hirao et al. 1996). Once the ERM family is activated, it appears to be translocated from the cytosol to the plasma membrane, where it serves as a cross-linker (Tsukita et al. 1994). Consistently, with a permeable cell reconstitution assay, it has been shown that the ERM family is essential for Rho-induced cytoskeletal reorganization (Mackay et al. 1997).

3.7 Other Functions

Rho has been reported to regulate various cell functions, including membrane ruffling (Nishiyama et al. 1994), transcriptional activation of the *c-fos* SRE mediated by SRF (Hill et al. 1995), Ras-induced cell transformation (Qiu et al.

1995), cancer cell invasion (Yoshioka et al. 1995), endocytosis of vesicles in *Ocytes* (Schmalzing et al. 1995), synthesis of 4,5-PIP$_2$ via phosphatidylinositol 5-kinase (PI5-kinase; Chong et al. 1994), and planar polarity in epidermal structure in *Drosophila* (Strutt et al. 1997). Since this review aims to focus on regulations of the cytoskeleton and cell adhesions by Rho and its targets, the above functions regulated by Rho will not be described in detail: for such information, the reader is refered to excellent reviews (Nobes and Hall 1994; Van Aelst and D'Souza-Schorey 1997).

4
Rho Targets

We identified three Rho targets: protein kinase N (PKN) (Amano et al. 1996b; Watanabe et al. 1996), Rho-kinase (Matsui et al. 1996), which was also identified as ROKα (Leung et al. 1995), and the myosin-binding subunit (MBS) of myosin phosphatase (Kimura et al. 1996; Fig. 2). PRK2 is an isoform of PKN (Vincent and Settleman 1997). p160 ROCK is an isoform of Rho-kinase (Ishizaki et al. 1996). Other investigators identified different Rho targets: rhophilin (Watanabe et al. 1996), rhotekin (Reid et al. 1996), citron (Madaule et al. 1995) and mDia (Watanabe et al. 1997). PI5-kinase has been shown to be activated by GTP·Rho (Chong et al. 1994).

PKN consists of an N-terminal regulatory domain and a C-terminal catalytic domain (Fig. 2). The catalytic domain is highly homologous to that of protein kinase C (PKC), whereas the N-terminal domain shows no obvious

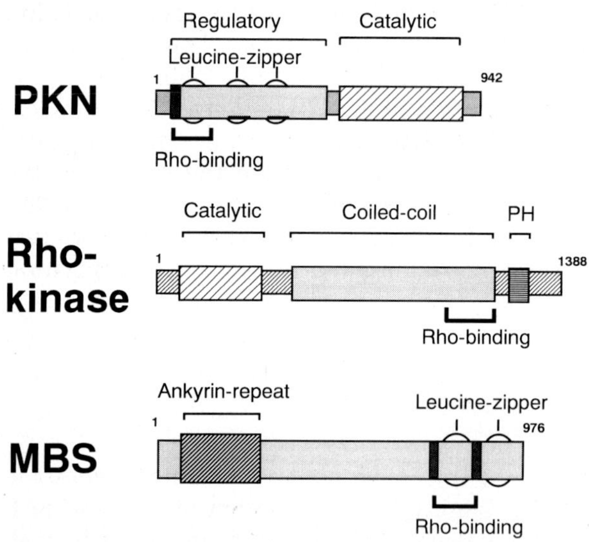

Fig. 2. Structures of Rho targets

homology with other protein kinases, including PKC. The regulatory domain contains three leucine zipper-like motifs. Activated Rho interacts with the N-terminal domain of PKN and activates the kinase activity of PKN (Amano et al. 1996b).

Rho-kinase has a kinase domain in the N-terminal portion, and it shares 72% sequence homology with myotonic dystrophy kinase within the kinase domain (Fig. 2; Matsui et al. 1996). The Rho-kinase has a putative coiled-coil domain in the middle portion and a pleckstrin homology (PH) domain in the C-terminal portion. Activated Rho interacts with the C-terminal portion of the coiled-coil domain and activates the kinase activity of Rho-kinase.

Myosin phosphatase consists of a myosin-binding subunit (MBS), a 37-kDa type 1 phosphatase catalytic subunit, and a 20-kDa regulatory subunit (Fig. 2). MBS enhances the phosphatase activity of the catalytic subunit toward myosin by directly binding to phosphorylated MLC. The C-terminal domain of MBS and the N-terminal domain of rhophilin have a leucine zipper-like motif in the Rho-interacting site similar to that of PKN; this may be one of consensus motifs for Rho-interacting interfaces.

mDia belongs to a family of formin-related proteins including *Drosophila* diaphanous and has two formin homology domains. Diaphanous has been shown to be involved in cytokinesis, cell polarity and cell morphology.

5 Functions of Rho Targets

5.1 Rho-Kinase and MBS

Rho regulates MLC phosphorylation through its targets, Rho-kinase and MBS, depicted in Fig. 3 (Amano et al. 1996a; Kimura et al. 1996). Activated Rho interacts with Rho-kinase and MBS of myosin phosphatase, to activate Rho-kinase and translocate MBS. The activated Rho-kinase subsequently phosphorylates MBS, thereby inactivating myosin phosphatase (Kimura et al. 1996). Rho-kinase by itself phosphorylates MLC at the same site that is phosphorylated by MLC kinase, and activates myosin ATPase (Amano et al. 1996a). Both pathways appear to be necessary for an increase in MLC phosphorylation in permeabilized vascular smooth muscle (Kureishi et al. 1997). MBS is phosphorylated and the myosin phosphatase activity is inactivated during the action of thromboxane A_2 in platelets, and both reactions are reversed by prior treatment of platelets with C3 toxin (Nakai et al. 1997). Phosphorylation of MLC results in the contraction of smooth muscle (Hartshorne 1987; Sellers and Adelstein 1987) or the interaction of actin and myosin in non-muscle cells, leading to stress fiber formation (Giuliano and Taylor 1995). Consistently, the expression of the dominant active form of Rho-kinase induces MLC phosphorylation in COS7 cells (Chihara et al. 1997) and NIH3T3 cells (Amano et al.

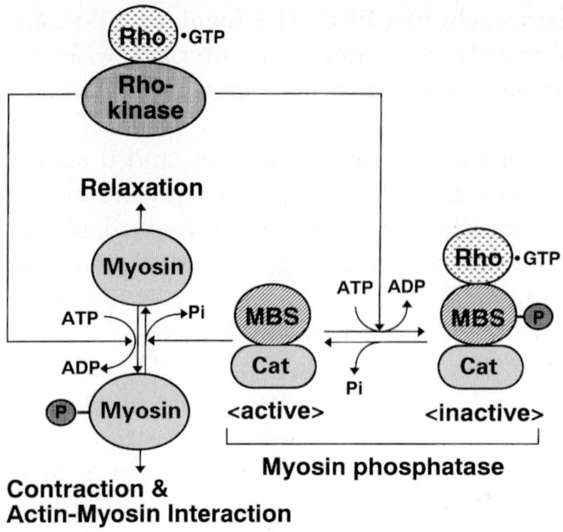

Fig. 3. Model for the regulation of MLC phosphorylation by Rho, Rho-kinase and myosin phosphatase. *Cat* Catalytic subunit of myosin phosphatase

1998), the formation of stress fibers and focal adhesions in Swiss 3T3 and MDCK cells (Leung et al. 1996; Amano et al. 1997; Ishizaki et al. 1997) and neurite retraction (Amano et al. 1998), whereas the expression of the dominant negative form of Rho-kinase inhibits the agonist-induced MLC phosphorylation, the formation of stress fibers and focal adhesions and neurite retraction. The addition of the dominant active form of Rho-kinase to permeabilized vascular smooth muscle induces contraction through MLC phosphorylation (Kureishi et al. 1997). The expression of mutant $MLC^{T18D,S19D}$ (substitution of Thr and Ser by Asp), which is known to lead to the activation of myosin ATPase and a conformational change of myosin II when reconstituted with myosin heavy chain in vitro, also enhances the formation of stress fibers, focal adhesions and neurite retraction (Amano et al. 1998). Thus, the signal pathways depicted in Fig. 2 may account for the mechanism by which Rho regulates smooth muscle contraction, neurite retraction, cytokinesis and cell motility. A specific inhibitor for ROCK/Rho-kinase (Y27632) has recently been developed (Uehata et al. 1997). Y27632 selectively inhibits smooth muscle contraction by inhibiting Ca^{2+} sensitization and corrects hypertension in several hypertensive rat models (Uehata et al. 1997). Thus, Rho-kinase-mediated Ca^{2+} sensitization of smooth muscle contraction may be involved in the pathophysiology of hypertension.

To understand roles of Rho-kinase, we searched for substrates and found that ERM and adducin are good substrates of Rho-kinase both in vitro and in vivo (Table 1). Moesin is phosphorylated at Thr^{558} in thrombin-activated platelets (Nakamura et al. 1995). After further analysis using a phosphorylation site-specific antibody, it is speculated that the phosphorylation at Thr^{558} of

Table 1. Substrates of Rho-kinase

Substrates	Changes of functions	MBS binding
Myosin	Enhancement of binding of myosin to F-actin	+
MBS	Inhibition of myosin phosphatase	–
Adducin	Enhancement of binding of adducin to F-actin	+
ERM (ezrin, radixin, moesin)	Activation of ERM	+
Intermediate filament (GFAP, vimentin)	Disassembly of filaments	N.D.

moesin is required for the stable interaction of moesin with actin at filopodia (Nakamura et al. 1996). Rho-kinase phosphorylates the ERM family (Matsui et al. 1998). The phosphorylation site of moesin by Rho-kinase is identified as Thr^{558}. The phosphorylation of radixin by Rho-kinase inhibits the interdomain association between the N- and C-terminal domains of radixin (Matsui et al. 1998). This result suggests that the phosphorylation of the ERM family by Rho-kinase may release the intramolecular or intermolecular suppression between the N- and C-terminal domains, or may maintain the activity of the ERM family by interfering with the intramolecular or intermolecular suppression. MBS specifically interacts with the ERM family, and myosin phosphatase dephosphorylates moesin, which is phosphorylated by Rho-kinase (Fukata et al. 1998). The phosphatase activity of myosin phosphatase toward moesin is inhibited by the thiophosphorylation of MBS with Rho-kinase. Thus, it is likely that Rho can regulate the activity of the ERM family, at least by regulating the phosphorylation state of the ERM family, through MBS and Rho-kinase in a manner similar to MLC.

Adducin was originally isolated from erythrocytes as calmodulin-binding protein (Bennett 1989). Adducin is thought to stabilize the actin-spectrin network beneath plasma membranes and serves as one of the components of the network. Adducin consists of at least three domains; head, neck and tail. Adducin binds to F-actin and the spectrin-F-actin complex. Adducin subsequently promotes the binding of spectrin to F-actin or the spectrin-F-actin complex to form the spectrin-F-actin network which supports the plasma membrane. In epithelial cells, adducin is concentrated within the spectrin-F-actin network at cell-cell contact sites and the membrane ruffling area. Rho-kinase phosphorylates the neck domain of adducin, and the phosphorylation of adducin by Rho-kinase enhances its F-actin-binding activity (Kimura et al. 1998). MBS specifically interacts with adducin, and myosin phosphatase dephosphorylates adducin which is phosphorylated by Rho-kinase. The phosphatase activity of myosin phosphatase toward adducin is inhibited by the thiophosphorylation of MBS with Rho-kinase. Thus, the phosphorylation of

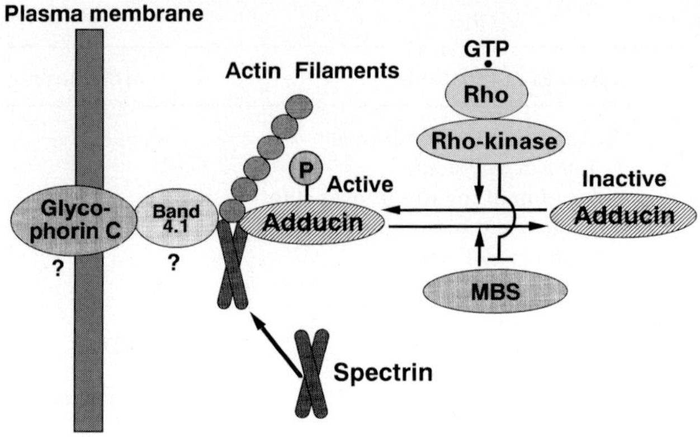

Fig. 4. Model for the regulation of adducin phosphorylation by Rho, Rho-kinase and myosin phosphatase

adducin appears to be dually regulated by Rho-kinase and MBS as described for myosin, and this phosphorylation may lead to the formation of the spectrin-F-actin network (Fig. 4).

Rho-kinase phosphorylates intermediate filaments such as glial fibrillary acidic protein (GFAP) and vimentin (Kosako et al. 1997). This phosphorylation results in disassembly of the filaments. The phosphorylation and disassembly of intermediate filaments appears to occur through the action of Rho-kinase during cytokinesis (Kosako et al. 1997). Thus, it is likely that Rho-kinase is activated during cytokinesis and is responsible for the phosphorylation of MLC as well as intermediate filaments.

5.2
mDia and Bni1

p140mDia, profilin and RhoA are co-localized in the spreading lamellae of cultured fibroblasts and in the membrane ruffling area of phorbol ester-stimulated MDCK2 cells (Watanabe et al. 1997). Profilin has been shown to promote actin polymerization (Nishida 1985). mDia, profilin and RhoA are recruited around phagocytic cups induced by fibronectin-coated beads. Their recruitment is not induced after Rho is inactivated by microinjection of C3 toxin. Overexpression of mDia in COS7 cells induces actin filament formation. Thus, it is likely that Rho regulates actin polymerization by targeting profilin via mDia beneath the specific plasma membranes.

In *S. cerevisiae*, Rho1 is required for bud formation and is localized at a bud tip or a cytokinesis site. Bni1 has been identified as a potential target of Rho1 (Kohno et al. 1996). Bni1 has a domain organization similar to that of mDia and diaphanous, and the formin homology domains. Bnr1 is a homolog of bni1 in *S. cerevisiae* (Imamura et al. 1997). Disruption of bni1 or bnr1 does not show any effect on cell growth, but the bni1/bnr1 double mutant shows a severe temperature-sensitive growth phenotype. Cells of the bni1/bnr1 double mutant arrested at the restrictive temperature are deficient in bud emergence, exhibit a random distribution of cortical actin patches and often become multinucleate. These phenotypes are similar to those of the mutant of PFY1, which encodes profilin. Bni1 and bnr1 interact with profilin at the formin homology domain. Therefore, it is likely that bni1 interacts with profilin and regulates the reorganization of the actin cytoskeleton downstream of Rho1.

5.3 Other Rho Targets

Although PI5-kinase is thought to be activated by Rho, its molecular mechanism is not known at present. Previous work has shown that PIP2, the product of PI5-kinase, can regulate the interaction of a number of actin-binding proteins including profilin, α-actinin, gelsolin and p39CapZ in vitro. Thus, it is possible that increased PIP2 synthesis results in increased actin polymerization and association of actin filaments with focal adhesion. However, the role of PIP2 in controling the cytoskeleton during the action of Rho in intact cells remains controversial. Further studies are necessary to address this isuue. PKN, PRK2, rhophilin, rhotekin and citron have been identified as putative Rho targets. However, the physiological functions of these proteins remain to be clarified.

6 Conclusion

When Rho is activated, it interacts with various targets, including PKN, PRK2, Rho-kinase (ROKα), p160 ROCK (ROKβ), MBS, rhophilin, citron, rhotekin and mDia (Fig. 5). Among them, Rho-kinase has critical roles in divergent pathways downstream of Rho. Rho-kinase is involved in MLC phosphorylation, leading to the formation of stress fiber and focal adhesion, and neurite retraction. Rho-kinase is also involved in ERM phosphorylation, leading to its activation, and in adducin phosphorylation, leading to its binding to actin filaments. The phosphorylation of these substrates may be necessary for reorganization of the cytoskeleton and cell adhesions. Rho-kinase also participates in the transcriptional activation of the *c-fos* SRE (Chihara et al. 1997). MBS forms a complex with the Rho-kinase substrates and the phosphorylation state of these substrates is dually regulated by Rho-kinase and MBS. This means that the phosphorylation state of the substrates are tightly regulated when Rho receives temporal and spatial signals, and is transiently activated. In addition to Rho-kinase, other Rho targets such as mDia and PI5-kinase may be necessary for the formation of an organized actin cytoskeleton through actin poly-

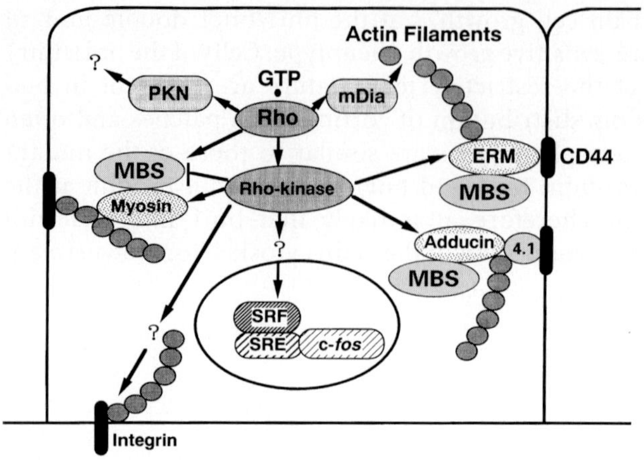

Fig. 5. Signaling pathways downstream of Rho

merization. Further studies are necessary for a better understanding of how temporal and spatial signals are delivered to the cytoskeleton and cell adhesions through Rho and its targets.

References

Amano M, Ito M, Kimura K, Fukata Y, Chihara K, Nakano T, Matsuura Y, Kaibuchi K (1996a) Phosphorylation and activation of myosin by Rho-associated kinase (Rho-kinase). J Biol Chem 271:20246–20249

Amano M, Mukai H, Ono Y, Chihara K, Matsui T, Hamajima Y, Okawa K, Iwamatsu A, Kaibuchi K (1996b) Identification of a putative target for Rho as a serine-threonine kinase, PKN. Science 271:648–650

Amano M, Chihara K, Kimura K, Fukata Y, Nakamura N, Matsuura Y, Kaibucni K (1997) Formation of actin stress fibers and focal adhesions enhanced by Rho-kinase. Science 275:1308–1311

Amano M, Chihara K, Nakamura N, Fukata Y, Yano T, Shibata M, Ikebe M, Kaibuchi K (1998) Myosin II activation promotes neurite retraction during the action of Rho and Rho-kinase. Genes Cell 3:177–188

Bennett V (1989) The spectrin-actin junction of erythrocyte membrane skeletons. Biochim Biophys Acta 988:107–121

Bradley AB, Morgan KG (1987) Alterations in cytoplasmic calcium sensitivity during porcine coronary artery contractions as detected by aequorin. J Physiol 385:437–448

Braga V, Machesky LM, Hall A, Hotchin NA (1997) The small GTPases rho and rac are required for the establishment of cadherin-dependent cell-cell contacts (In Process Citation). J Cell Biol 137:1421–1431

Chihara K, Amano M, Nakamura N, Yano T, Shibata M, Tokui T, Ichikawa H, Ikebe R, Ikebe M, Kaibuchi K (1997) Cytoskeletal rearrangements and transcriptional activation of c-fos serum response element by Rho-kinase. J Biol Chem 272:25121–25127

Chong LD, Traynor-Kaplan A, Bokoch GM, Schwartz MA (1994) The small GTP-binding protein Rho regulates a phosphatidylinositol 4-phosphate 5-kinase in mammalian cells. Cell 79:507–513

Fukata Y, Kimura K, Oshiro N, Saya H, Matsuura Y, Kaibuchi K (1998) Association of the myosin-binding subunit of myosin phosphatase and moesin: dual regulation of moesin phosphorylation by Rho-kinase and myosin phosphatase. J Cell Biol 141:409–418

Fukumoto Y, Kaibuchi K, Hori Y, Fujioka H, Araki S, Ueda T, Kikuchi A, Takai Y (1990) Molecular cloning and characterization of a novel type of regulatory protein (GDI) for the rho proteins, ras p21-like small GTP-binding proteins. Oncogene 5:1321–1328

Giuliano KA, Taylor DL (1995) Measurement and manipulation of cytoskeletal dynamics in living cells. Curr Opin Cell Biol 7:4–12

Gong MC, Iizuka K, Nixon G, Browne JP, Hall A, Eccleston JF, Sugai M, Kobayashi S, Somlyo AV, Somlyo AP (1996) Role of guanine nucleotide-binding proteins – ras-family or trimeric proteins or both – in Ca^{2+} sensitization of smooth muscle. Proc Natl Acad Sci USA 93:1340–1345

Gumbiner BM (1996) Cell adhesion: the molecular basis of tissue architecture and morphogenesis. Cell 84:345–357

Habets GG, Scholtes EH, Zuydgeest D, van der Kammen RA, Stam JC, Berns A, Collard JG (1994) Identification of an invasion-inducing gene, Tiam-1, that encodes a protein with homology to GDP-GTP exchangers for Rho-like proteins. Cell 77:537–549

Hart MJ, Eva A, Evans T, Aaronson SA, Cerione RA (1991) Catalysis of guanine nucleotide exchange on the CDC42Hs protein by the dbl oncogene product. Nature 354:311–314

Hartshorne DJ (1987) In: Physiology of the gastrointestinal tract. Johnson DR (ed) Raven, New York, pp 423–482

Hill CS, Wynne J, Treisman R (1995) The Rho family GTPases RhoA Rac1, CDC42Hs regulate transcriptional activation by SRF. Cell 81:1159–1170

Hirao M, Sato N, Kondo T, Yonemura S, Monden M, Sasaki T, Takai Y, Tsukita S, Tsukita S (1996) Regulation mechanism of ERM (ezrin/radixin/moesin) protein/plasma membrane association: possible involvement of phosphatidylinositol turnover and rho-dependent signaling pathway. J Cell Biol 135:37–51

Hirata K, Kikuchi A, Sasaki T, Kuroda S, Kaibuchi K, Matsuura Y, Seki H, Saida K, Takai Y (1992) Involvement of rho p21 in the GTP-enhanced calcium ion sensitivity of smooth muscle contraction. J Biol Chem 267:8719–8722

Homma Y, Emori Y (1995) A dual functional signal mediator showing RhoGAP and phospholipase C-delta stimulating activities. EMBO J 14:286–291

Horii Y, Beeler JF, Sakaguchi K, Tachibana M, Miki T (1994) A novel oncogene, ost, encodes a guanine nucleotide exchange factor that potentially links Rho and Rac signaling pathways. EMBO J 1354:4776–4786

Imamura H, Tanaka K, Hihara T, Umikawa M, Kamei T, Takahashi K, Sasaki T, Takai Y (1997) Bni1p and bnr1p: downstream targets of the Rho family small G-proteins which interact with profilin and regulate actin cytoskeleton in *Saccharomyces cerevisiae*. EMBO J 16:2745–2755

Ishizaki T, Maekawa M, Fujisawa K, Okawa K, Iwamatsu A, Fujita A, Watanabe N, Saito Y, Kakizuka A, Morii N, Narumiya S (1996) The small GTP-binding protein Rho binds to and activates a 160 kDa Ser/Thr protein kinase homologous to myotonic dystrophy kinase. EMBO J 15:1885–1893

Ishizaki T, Naito M, Fujisawa K, Maekawa M, Watanabe N, Saito Y, Narumiya S (1997) p160ROCK a Rho-associated coiled-coil forming protein kinase works downstream of Rho and induces focal adhesions. FEBS Lett 404:118–124

Jalink K, Moolenaar WH (1992) Thrombin receptor activation causes rapid neural cell rounding and neurite retraction independent of classic second messengers. J Cell Biol 118:411–419

Jalink K, van Corven EJ, Hengeveld T, Morii N, Narumiya S, Moolenaar WH (1994) Inhibition of lysophosphatidate- and thrombin-induced neurite retraction and neuronal cell rounding by ADP ribosylation of the small GTP-binding protein Rho. J Cell Biol 126:801–810

Kimura K, Ito M, Amano M, Chihara K, Fukata Y, Nakafuku M, Yamamori B, Feng J, Nakano T, Okawa K, Iwamatsu A, Kaibuchi K (1996) Regulation of myosin phosphatase by Rho and Rho-associated kinase (Rho-kinase). Science 273:245–248

Kimura K, Fukata Y, Matsuoka Y, Bennett V, Matsuura Y, Okawa K, Iwamatsu A, Kaibuchi K (1998) Regulation of association of adducin with actin filaments by Rho-kinase and myosin phosphatase. J Biol Chem 273:5542–5548

Kishi K, Sasaki T, Kuroda S, Itoh T, Takai Y (1993) Regulation of cytoplasmic division of *Xenopus* embryo by rho p21 and its inhibitory GDP/GTP exchange protein (rho GDI). J Cell Biol 120:1187–1195

Kitazawa T, Masuo M, Somlyo AP (1991) G protein-mediated inhibition of myosin light-chain phosphatase in vascular smooth muscle. Proc Natl Acad Sci USA 88:9307–9310

Kohno H, Tanaka K, Mino A, Umikawa M, Imamura H, Fujiwara T, Fujita Y, Hotta K, Qadota H, Watanabe T, Ohya Y, Takai Y (1996) Bni1p implicated in cytoskeletal control is a putative target of Rho1p small GTP binding protein in *Saccharomyces cerevisiae.* EMBO J 15:6060–6068

Kosako H, Amano M, Yanagida M, Tanabe K, Nishi Y, Kaibuchi K, Inagaki M (1997) Phosphorylation of glial fibrillary acidic protein at the same sites by cleavage furrow kinase and Rho-associated kinase. J Biol Chem 272:10333–10336

Kozma R, Sarner S, Ahmed S, Lim L (1997) Rho family GTPases and neuronal growth cone remodelling: relationship between increased complexity induced by Cdc42Hs Rac1, and acetylcholine and collapse induced by RhoA and lysophosphatidic acid. Mol Cell Biol 17:1201–1211

Kureishi Y, Kobayashi S, Amano M, Kimura K, Kanaide H, Nakano T, Kaibuchi K, Ito M (1997) Rho-associated kinase directly induces smooth muscle contraction through myosin light chain phosphorylation. J Biol Chem 272:12257–12260

Kuroda S, Fukata M, Fujii K, Nakamura T, Izawa I, Kaibuchi K (1997) Regulation of cell-cell adhesion of MDCK cells by Cdc42 and Rac1 small GTPases. Biochem Biophys Res Commun 240:430–435

Lancaster CA, Taylor-Harris PM, Self AJ, Brill S, van Erp HE, Hall A (1994) Characterization of rhoGAP. A GTPase-activating protein for rho-related small GTPases. J Biol Chem 269:1137–1142

Leung T, Manser E, Tan L, Lim L (1995) A novel serine/threonine kinase binding the Ras-related RhoA GTPase which translocates the kinase to peripheral membranes. J Biol Chem 270:29051–29054

Leung T, Chen XQ, Manser E, Lim L (1996) The p160 RhoA-binding kinase ROKα is a member of a kinase family and is involved in the reorganization of the cytoskeleton. Mol Cell Biol 16:5313–5327

Mabuchi I, Hamaguchi Y, Fujimoto H, Morii N, Mishima M, Narumiya S (1993) A rho-like protein is involved in the organisation of the contractile ring in dividing sand dollar eggs. Zygote 1:325–331

Mackay DJ, Esch F, Furthmayr H, Hall A (1997) Rho- and rac-dependent assembly of focal adhesion complexes and actin filaments in permeabilized fibroblasts: an essential role for ezrin/radixin/moesin proteins. J Cell Biol 138:927–938

Madaule P, Furuyashiki T, Reid T, Ishizaki T, Watanabe G, Morii N, Narumiya S (1995) A novel partner for the GTP-bound forms of rho and rac. FEBS Lett 377:243–248

Matsui T, Amano M, Yamamoto T, Chihara K, Nakafuku M, Ito M, Nakano T, Okawa K, Iwamatsu A, Kaibuchi K (1996) Rho-associated kinase, a novel serine/threonine kinase, as a putative target for the small GTP-binding protein Rho. EMBO J 15:2208–2216

Matsui T, Maeda M, Doi Y, Yonemura S, Amano M, Kaibuchi K, Tsukita S, Tsukita S (1998) Rho-Kinase phosphorylates COOH-terminal threonines of ezrin/radixin/moesin (ERM) proteins and regulates their head-to-tail association. J Cell Biol 140:647–657

Mizuno T, Kaibuchi K, Yamamoto T, Kawamura M, Sakoda T, Fujioka H, Matsuura Y, Takai Y (1991) A stimulatory GDP/GTP exchange protein for smg p21 is active on the post-translationally processed form of c-Ki-ras p21 and rhoA p21. Proc Natl Acad Sci USA 88:6442–6446

Nakai K, Suzuki Y, Kihira H, Wada H, Fujioka M, Ito M, Nakano T, Kaibuchi K, Shiku H, Nishikawa M (1997) Regulation of myosin phosphatase through phosphorylation of the myosin-binding subunit in platelet activation. Blood 90:3936–3942

Nakamura F, Amieva MR, Furthmayr H (1995) Phosphorylation of threonine 558 in the carboxyl-terminal actin-binding domain of moesin by thrombin activation of human platelets. J Biol Chem 270:31377–31385

Nakamura F, Amieva MR, Hirota C, Mizuno Y, Furthmayr H (1996) Phosphorylation of 558T of moesin detected by site-specific antibodies in RAW264.7 macrophages. Biochem Biophys Res Commun 226:650–656

Nishida E (1985) Opposite effects of cofilin and profilin from porcine brain on rate of exchange of actin-bound adenosine 5′-triphosphate. Biochemistry 24:1160–1164

Nishiki T, Narumiya S, Morii N, Yamamoto M, Fujiwara M, Kamata Y, Sakaguchi G, Kozaki S (1990) ADP-ribosylation of the rho/rac proteins induces growth inhibition, neurite outgrowth and acetylcholine esterase in cultured PC-12 cells. Biochem Biophys Res Commun 167:265–272

Nishiyama T, Sasaki T, Takaishi K, Kato M, Yaku H, Araki K, Matsuura Y, Takai Y (1994) rac p21 is involved in insulin-induced membrane ruffling and rho p21 is involved in hepatocyte growth factor- and 12-O-tetradecanoylphorbol-13-acetate (TPA)-induced membrane ruffling in KB cells. Mol Cell Biol 14:2447–2456

Nobes C, Hall A (1994) Regulation and function of the Rho subfamily of small GTPases. Curr Opin Genet Dev 4:77–81

Noda M, Yasuda-Fukazawa C, Moriishi K, Kato T, Okuda T, Kurokawa K, Takuwa Y (1995) Involvement of rho in GTP gamma S-induced enhancement of phosphorylation of 20 kDa myosin light chain in vascular smooth muscle cells: inhibition of phosphatase activity. FEBS Lett 367:246–250

Nusrat A, Giry M, Turner JR, Colgan SP, Parkos CA, Carnes D, Lemichez E, Boquet P, Madara JL (1995) Rho protein regulates tight junctions and perijunctional actin organization in polarized epithelia. Proc Natl Acad Sci USA 92:10629–10633

Paterson HF, Self AJ, Garrett MD, Just I, Aktories K, Hall A (1990) Microinjection of recombinant p21rho induces rapid changes in cell morphology. J Cell Biol 111:1001–1007

Qiu RG, Chen J, McCormick F, Symons M (1995) A role for Rho in Ras transformation. Proc Natl Acad Sci USA 92:11781–11785

Reid T, Furuyashiki T, Ishizaki T, Watanabe G, Watanabe N, Fujisawa K, Morii N, Madaule P, Narumiya S (1996) Rhotekin, a new putative target for Rho bearing homology to a serine/threonine kinase, PKN, and rhophilin in the rho-binding domain. J Biol Chem 271:13556–13560

Ridley AJ, Hall A (1992) The small GTP-binding protein rho regulates the assembly of focal adhesions and actin stress fibers in response to growth factors. Cell 70:389–399

Ridley AJ, Comoglio PM, Hall A (1995) Regulation of scatter factor/hepatocyte growth factor responses by Ras, Rac, and Rho in MDCK cells. Mol Cell Biol 15:1110–1122

Schmalzing G, Richter HP, Hansen A, Schwarz W, Just I, Aktories K (1995) Involvement of the GTP binding protein Rho in constitutive endocytosis in *Xenopus laevis* oocytes. J Cell Biol 130:1319–1332

Sellers JR, Adelstein RS (1987) In: Boyer P, Erevs EG (eds) The enzymes. Academic Press, San Diego, pp 381–418

Settleman J, Albright CF, Foster LC, Weinberg RA (1992) Association between GTPase activators for Rho and Ras families. Nature 359:153–154

Stossel TP (1993) On the crawling of animal cells. Science 260:1086–1094

Strutt DI, Weber U, Mlodzik M (1997) The role of RhoA in tissue polarity and Frizzled signalling. Nature 387:292–295

Takaishi K, Sasaki T, Kato M, Yamochi W, Kuroda S, Nakamura T, Takeichi M, Takai Y (1994) Involvement of Rho p21 small GTP-binding protein and its regulator in the HGF-induced cell motility. Oncogene 9:273–279

Takeichi M (1990) Cadherins: a molecular family important in selective cell-cell adhesion. Annu Rev Biochem 59:237–252

Tanaka E, Sabry J (1995) Making the connection: cytoskeletal rearrangements during growth cone guidance. Cell 83:171–176

Tessier-Lavigne M, Goodman CS (1996) The molecular biology of axon guidance. Science 274:1123–1133

Tominaga T, Sugie K, Hirata M, Morii N, Fukata J, Uchida A, Imura H, Narumiya S (1993) Inhibition of PMA-induced, LFA-1-dependent lymphocyte aggregation by ADP ribosylation of the small molecular weight GTP binding protein, rho. J Cell Biol 120:1529–1537

Tsukita S, Tsukita S, Nagafuchi A, Yonemura S (1992) Molecular linkage between cadherins and actin filaments in cell-cell adherens junctions. Curr Opin Cell Biol 4:834–839

Tsukita S, Oishi K, Sato N, Sagara J, Kawai A, Tsukita S (1994) ERM family members as molecular linkers between the cell surface glycoprotein CD44 and actin-based cytoskeletons. J Cell Biol 126:391–401

Turunen O, Wahlstrom T, Vaheri A (1994) Ezrin has a COOH-terminal actin-binding site that is conserved in the ezrin protein family. J Cell Biol 126:1445–1453

Uehata M, Ishizaki T, Satoh H, Ono T, Kawahara T, Morishita T, Tamakawa H, Yamagami K, Inui J, Maekawa M, Narumiya S (1997) Calcium sensitization of smooth muscle mediated by a Rho-associated protein kinase in hypertension [see comments]. Nature 389:990–994

Van Aelst L, D'Souza-Schorey C (1997) Rho GTPases and signaling networks. Genes Dev 11:2295–2322

Vincent S, Settleman J (1997) The PRK2 kinase is a potential effector target of both Rho and Rac GTPases and regulates actin cytoskeletal organization. Mol Cell Biol 17:2247–2256

Watanabe G, Saito Y, Madaule P, Ishizaki T, Fujisawa K, Morii N, Mukai H, Ono Y, Kakizuka A, Narumiya S (1996) Protein kinase N (PKN) and PKN-related protein rhophilin as targets of small GTPase Rho. Science 271:645–648

Watanabe N, Madaule P, Reid T, Ishizaki T, Watanabe G, Kakizuka A, Saito Y, Nakao K, Jockusch BM, Narumiya S (1997) p140mDia, a mammalian homolog of *Drosophila* diaphanous, is a target protein for Rho small GTPase and is a ligand for profilin. EMBO J 16:3044–3056

Yoshioka K, Imamura F, Shinkai K, Miyoshi J, Ogawa H, Mukai M, Komagome R, Akedo H (1995) Participation of rhop21 in serum-dependent invasion by rat ascites hepatoma cells. FEBS Lett 372:25–28

Zigmond SH (1996) Signal transduction and actin filament organization. Curr Opin Cell Biol 8:66–73

Rnd Proteins: A New Family of Rho-Related Proteins That Interfere with the Assembly of Filamentous Actin Structures and Cell Adhesion

Pierre Chardin[1]

1 Introduction

Rho family proteins control actin organization. The Rnd proteins form a distinct branch of the Rho family: Rnd1 is expressed mostly in brain and liver, Rnd2 is highly expressed in testis and Rnd3 shows a ubiquitous, very low expression. In brain, Rnd1 is found in specialized neurons, mainly in the cortex, hippocampus and substantia nigra. Rnd1 exchanges GTP rapidly, has a low affinity for GDP, and lacks intrinsic GTPase activity, suggesting that in the cell it is constitutively in the GTP-bound form. Expression of Rnd1 in Swiss 3T3 fibroblasts inhibits the formation of actin stress fibers and induces a loss of focal adhesions and cell/cell contacts, leading to cell rounding (hence Rnd for "round"). In Swiss 3T3 fibroblasts and MDCK cells, Rnd1 localizes to adherens junctions. Thus, Rnd proteins are involved in rearrangements of the actin cytoskeleton and changes in cell adhesion that might play essential roles in cell migration and tumor invasion.

In mammalian cells, all members of the Rho family studied to date play important roles in regulating the dynamics of the actin cytoskeleton, and its reorganization in response to growth factors (see Chap. by Ridley). Like Ras, most members of the Rho family cycle between an inactive GDP-bound form and an active GTP-bound form. Three major regulators control their activity: RhoGDIs interact with the geranyl-geranylated form of these proteins to keep them in a "resting" cytosolic complex (Chap. by Zalcman et al.), exchange factors of the dbl family promote GDP → GTP exchange leading to their activation (Chap. by Stam and Collard), and GTPase activating proteins, GAPs, stimulate GTP hydrolysis and a return to the inactive GDP-bound form (see Chap. by Zalcman et al.).

In Swiss 3T3 cells, Rho regulates the assembly of focal adhesions and actin stress fibers, Rac regulates the formation of lamellipodia and membrane ruffles and Cdc42 controls the formation of filopodia (Chap. by Ridley). Several different "effectors" which interact with the GTP-bound form of Rho have been

Institut de Pharmacologie, CNRS UPR 411, 660 route des Lucioles, 06560 Valbonne, France
[1] *Present address*:
Cancer Center UCSF, 2340 Sutter Street, San Francisco, California 94115, USA

Progress in Molecular and Subcellular Biology, Vol. 22
Ph. Jeanteur (Ed.)

described, such as PKN or Rho associated kinases (ROK). One of these Rho stimulated kinases (p160ROKα) can phosphorylate a regulatory subunit of myosin light chain phosphatase, leading to its inhibition, and might also be able to phosphorylate myosin light chain (MLC) directly. Thus, activation of Rho would increase the phosphorylation of MLC and lead to the bundling of actin into stress fibers. Other results suggest that Rho regulates the interaction of actin with the plasma membrane by modifying the association of the ezrin/radixin/moesin (ERM) family proteins with their plasma membrane protein partners (Tsukita et al. 1997; Chap. by Kaibuchi).

We have recently cloned three new proteins that form a distinct branch of the Rho family. Rnd proteins are related to Rho in terms of sequence, but possess strikingly different functional properties. Rnd1 is expressed only in a few cell types; it seems to be constitutively in the GTP-bound form, associates with the plasma membrane at adherens junctions, and inhibits the formation of actin structures and focal adhesions.

2 Characterisation of Three New Members of the Rho Family

The mammalian Rho family comprises Rho (A, B and C isoforms), Rac (1, 2 and other isoforms), Cdc42Hs (Cdc42 and G25K isoforms), TC10, RhoG (Chap. by Fort), RhoD (Murphy et al. 1996) and RhoH/TTF (Dallery et al. 1997).

We used degenerated oligonucleotide mixes corresponding to the Rho "effector region" (loop α1-β2 and part of β2 in Rho structure, see Fig. 1) to clone bovine retina cDNAs for two new proteins (Rnd1 and Rnd2) related to Rho. The complete human cDNAs for Rnd1, Rnd2, and a third closely related member, Rnd3 were subsequently obtained (Fig. 1). Rnd3 is identical to the recently described RhoE (Foster et al. 1996), except that the published RhoE sequence lacks 15 N-terminal amino acids. The first five amino acids of this N-ter sequence (MKERRA) of Rnd3 are identical to those of Rnd1. In Rnd1 this sequence is essential for the intracellular localization and activity of the protein (Nobes et al. 1998).

Rnd proteins share 54–63% identity pairwise, around 45–49% identity with RhoA and slightly less identity with other Rho family members such as Rac or Cdc42, thus the three Rnd proteins form a new branch of the Rho family.

Rnd proteins display striking differences from other members of the Rho family in their size, charge and biochemical properties. Their expected molecular weights are higher (calculated: 25–28 kDa) due to N-terminal extensions for Rnd1 and Rnd3, and C-terminal extensions of about 30 a.a. for all three. Their apparent molecular mass on SDS-PAGE is about 31 kDa, whereas Rho appears as a protein of about 24 kDa. They also have very different isoelectric points: whereas most Rho family proteins are negatively charged at neutral

```
                                       **                  "effector"                 * *
Rnd1  mkerrapq----------pvvarcklvlvgdvqcgktamlqvlakdcypetyvptvfenytacleteeqrvelslwdtsgspyydnvrplcysdsdavl
Rnd2                  megqsgrckivvvgdaecgktallqvfakdaypgsyvptvfenytasfeidkrrielnmwdtsgssyydnvrplaypdsdavl
Rnd3  mkerrasqklssksimdpnqnvkckivvvgdsqcgktallhvfakdcfpenyvptvfenytasfeidtqrielslwdtsgspyydnvrplsypdsdavl
RhoA                   maairkklvivgdgacgktcllivfskdqfpevyvptvfenyvadievdgkqvelalwdtagqedydrlrplsypdtdvil
RhoB                   maairkklvvvgdgacgktcllivfskdefpevyvptvfenyvadievdgkqvelalwdtagqedydrlrplsypdtdvil
RhoC                   maairkklvivgdgacgktcllivfskdqfpevyvptvfenyiadievdgkqvelalwdtagqedydrlrplsypdtdvil
RhoD      mnasqvageeapqsghsvkvvlvgdggcgktslmmvfakgafpesysptvferynatlqmkgkpvhlqiwdtagqddydrlrplfypdanvll
Rac1                    mqaikcvvvgdgavgktcllisyttnafpgeyiptvfdnysanvmvdgkpvnlglwdtagqedydrlrplsypqtdvfl
RasH                    mteyklvvvgaggvgksaltiqliqnhfvdeydptiedsyrkqvvidgetclldildtagqeeysamrdqymrtgegfl
                             ---P1---        G1    P2                      --P3-
                        ===β1===      ///α1////         ====β2==== ====β3====      ////α2//// ===

Rnd1  lcfdisrpetvdsalkkwrteildy-cpstrvlligcktdlrtdlstlmelshqkqapisyeqgcaiakqlgaeiylegsaftseksihsifrtasmlcl
Rnd2  icfdisrpetldsvlkkwqgetqef-cpnakvvlvgckldmrtdlatlrelskqrlipvtheqgtvlakqvgavsyvecssrssersvrdvfhvatvasl
Rnd3  icfdisrpetldsvlkkwkgeiqef-cpntkmllvgcksdlrtdvstlvelsnhrqtpvsydqganmakqigaatyiecsalqsensvrdifhvatlacv
RhoA  mcfsidspdslenipekwtpevkhf-cpnvpiilvgnkkdlrndehtrrelakmkqepvkpeegrdmanrigafgymecsaktkdg-vrevfematraal
RhoB  mcfsvdspdslenipekwvpevkhf-cpnvpiilvankkdlrsdehvrtelarmkqepvrtddgramavriqaydylecsaktkeg-vrevfetatraal
RhoC  mcfsidspdslenipekwtpevkhf-cpnvpiilvgnkkdlrqdehtrrelakmkqepvrseegrdmanrisafgylecsaktkeg-vrevfematragl
RhoD  lcfdvtnpnsfdnvsnrwypevthf-ckgvpiivvgckidlrkdkvlvnnlrkkrlepvtyhrghdmarsvgavaylecsarlhdn-veavfqeaaeval
Rac1  icfslvspasfenvrakwypevrhh-cpntpiilvgtkldlrddkdtieklkekkltpitypqglamakeigavkylecsaltqrg-lktvfdeairavl
RasH  cvfainntksfedihqyreqikrvkdsddvpmvlvgnkcdlaa--------------rtvesrqaqdlarsygipyietsaktrqg-vedafytlvreir
                                        ---G2---    Rho Insert                   --G3---
      β4==   ///////α3////////       ==β5==  α3a  //α3b///        /////α4////  =β6=       //////α5/////

Rnd1    nkpsplpq kspvrslskrllhlpsrselisstfkkekakscsim
Rnd2    grghrqlrrtdsrrgmqrsaqlsgrpdrgnegeihkdrakscnlm
Rnd3  nktnknvkrnksqratkrishmpsrpelsavatdlrkdkaksctvm
RhoA                                 qarrgkkksgclvl
RhoB                              qkrygsqngcincckvl
RhoC                                qvrknkrrrgcpil
RhoD                            ssrrhnfwrritqncclat
Rac1                                   cpppvkkrkrkclll
RasH                        qhklrklnppdesgpgcmsckcvls
```

Fig. 1. Alignment of Rnd1, 2 and 3 with RhoA, B, C, RhoD, Rac1 and RasH. P1–3 and G1–3 indicate the regions binding the phosphate and the guanine moieties of GTP, respectively. The region considered as the effector binding site including the asparagine (FE*N*Y) ADP-ribosylated by the C3 transferase in Rho is indicated at the top. α-Helices and β-sheets (based on ras, rac and rho structures) are indicated below. The "Rho insert" (*α3b*) is a 14 amino acid helical insertion which is characteristic of Rho family proteins. The C-terminal parts are aligned on the CAAX sequence for prenylation. *Stars* over the sequence indicate the positions corresponding to ras activating mutations. Accession numbers for nucleotide sequences: Rnd1, Y07923; Rnd2, X95456; Rnd3, X95282

pH (pI 5.5 to 6.8, Huber et al. 1994), Rnd proteins are positively charged with pIs of 8.1–8.7.

The three guanine binding motifs (noted G1–3 on Fig. 1) are conserved in Rnd proteins, the two loops (noted P1, P3 on Fig. 1) and a highly conserved threonine (P2 on Fig. 1), involved in phosphate binding can also be recognized, for instance the three major residues that coordinate magnesium in the GTP bound form of Ras: T 17, T 35 and D 57 (Pai et al. 1990) are strictly conserved. However, three residues of the phosphate binding site that are important for the intrinsic GTPase activity differ in Rnd proteins. Ras glycine 12 is replaced by valine, alanine or serine in Rnd proteins, Ras glycine 13 is replaced by glutamine or glutamic acid, and Ras alanine 59 and glutamine 61 are both replaced by serine (indicated as *at the top of Fig. 1). Any of these substitutions in Ras decreases the GTPase rate and leads to constitutively active, oncogenic proteins. Furthermore, the presence of two substitutions at positions 12 and 59 (replaced by threonine) decreases the GTPase activity more than individual substitutions (John et al. 1988). Based on the effects of these substitutions on Ras, the prediction is that Rnd proteins have no intrinsic GTPase activity.

Rnd proteins are almost identical to Rho proteins in the "effector" region, including a conserved asparagine. In Rho proteins, this asparagine 41 is ADP-ribosylated by the C3 transferase, a clostridial toxin that inhibits Rho (Chardin et al. 1989; Chap. by Boquet et al.). Surprisingly, Rnd1, Rnd2 and Rnd3 proteins, either expressed in bacteria, or from rat tissues, are not ADP-ribosylated by C3. This result shows that other regions of Rho are also recognized by the C3 toxin and contribute to specificity (P. Chardin, in prep.).

Rnd proteins end with a "CAAX box" for prenylation, but unlike Rho, Rac or Cdc42, which have a C-terminal leucine specifying geranylgeranylation, Rnd proteins end with a methionine residue suggesting that they are farnesylated. Foster et al. (1996) have expressed an epitope-tagged version of RhoE/Rnd3 in COS cells, and have shown that this protein is indeed farnesylated. RhoA, RhoC, Rac and Cdc42 have C-terminal polybasic motifs, which are assumed to interact with the negatively charged heads of membrane phospholipids. These basic motifs are very close to the end of $\alpha 5$, which still makes contacts with the protein core, raising the possibility that membrane interactions directly affect the conformation of the core GTP binding domain in Rho, Rac or Cdc42. Rnd proteins also have polybasic motifs at the C-terminal end, but those are separated from the protein core by at least 30 residues. This "spacer" could form an additional small domain. Therefore, in Rnd proteins the GTP binding domain is probably more distant from the membrane. In Rnd1 this "spacer" is unusually rich in prolines, and it would be interesting to determine its structure and position relative to the core GTP binding domain.

3
Biochemical Properties of Rnd1: G-Proteins That Do Not Switch?

Rnd1 has been expressed in bacteria and purified in order to study its biochemical properties. In a physiological buffer containing Mg^{2+}, Rnd1 exchanges GTP rapidly (t1/2 of 1.4 min at 37 °C), whereas GDP binds very poorly to Rnd1 under the same conditions. The affinity for GDP is about 100 times lower than for GTP. Thus, Rnd1 exchanges GTP faster and has a much lower affinity for GDP than most other small G proteins (Nobes et al. 1998). Since in the cell the concentration of GTP is about ten times that of GDP, the biochemical properties of Rnd1 suggest that it will spontaneously bind GTP, with no need for an exchange factor.

Furthermore, Rnd1 has no detectable GTPase activity, even in the presence of RhoGAP (Nobes et al. 1998). It has also been shown that RhoE/Rnd3 does not hydrolyse GTP, even when p190 or CDC42 GAP are added, and a tagged version of RhoE expressed in COS cells is constitutively GTP-bound (Foster et al. 1996). These results raise the possibility that in vivo, Rnd proteins are constitutively in the GTP-bound form.

One could imagine that a GDI protein stabilizes the GDP-bound form of Rnd1, however, in contrast to Rho or Rac proteins which are mostly found in the cytosol, and are associated with a GDI, Rnd proteins from rat tissues are mostly found in the membrane fraction (see next paragraph), suggesting that they do not associate with a GDI. Furthermore, no interactions with RhoGDI-3 or LyGDI could be detected in the two-hybrid system (G. Zalcman and J. Camonis, unpubl.). This would be the first example of a G protein that does not switch between GDP- and GTP-bound forms. How could such a protein be regulated?

4
Expression in Tissues and Localization of Rnd1 in the Brain

Rnd1 is expressed mostly in the brain and liver, as a 1.9 kb mRNA. Rnd2 is abundant in testes, as a 1.5 kb mRNA, whereas Rnd3 is expressed at much lower levels in all tissue studied: lung, liver, heart, skeletal muscle, kidney, spleen, prostate, ovary, small intestine, and colon, as a 3.1 kb mRNA; Rnd3 expression is especially low in the brain, pancreas, thymus, testis, and peripheral bloods leukocytes. It could be that Rnd3 is expressed at higher levels in other cell types or at a specific stage during development.

The expression of Rnd1 could also be detected at the protein level. When rat brain and liver lysates are centrifuged at 100 000 g, the major portion of Rnd1 proteins is found in the pellet (P100) fraction, whereas most Rho appears in the supernatant (S100) fraction. In rat hepatocytes, the expression of the Rnd1 protein is similar to that found in total liver, suggesting that hepatocytes are the major liver cells expressing Rnd1.

Rnd1 appears to be mostly expressed in the cortex, and at similar levels in the occipital pole, frontal and temporal lobes. Lower levels of expression are observed in most other regions of the brain. The highest expression of Rnd1, observed either by in situ hybridization or by immunohistochemistry, is in the cerebral cortex, including the frontal, striate and primary olfactory cortex, and in the hippocampus, including the dentate gyrus, the cerebellar granular layer and pontine nuclei. A lower expression was found in the inferior and superior colliculus, substantia nigra, ventral tegmental area and several unidentified brain stem nuclei (Nobes et al. 1998). Expression of Rnd1 could be involved in actin cytoskeleton rearrangements occuring in these neurons (see next Sect.). Cytoskeletal plasticity is clearly of primary importance in neural cells, and it will be very interesting to characterize precisely the neurons that express Rnd1, and study the specific properties of their actin structures.

Rnd2 expression in testes of immature rats (6 days) is similar to that in adult rats. In testes, the differentiation of germ cells and maturation of spermatids is tightly coupled to their migration between the sertoli cells in seminiferous tubules. At places of germ cell/sertoli cell contacts, the rearrangements of cell junctions and associated cytoskeletal elements in sertoli cells controls germ cells differentiation (Russell 1993). The expression of Rnd2 could contribute to the control of these rearrangements of cellular junctions (see next Sect.).

5
Effects of Rnd1 Expression on the Formation of Actin Stress Fibers

Microinjection of Rnd1 expression vectors into the nucleus of quiescent serum-starved Swiss 3T3 fibroblasts does not induce the formation of actin stress fibers. Furthermore, when microinjected cells are stimulated with LPA for 30 min, cells expressing high levels of wild-type Rnd1 do not form actin stress fibers and lose focal adhesion (Fig. 2). After longer times, Rnd1 expressing cells round up, take on a dendritic appearance and detach from the glass coverslips. Rnd1 expression also blocks PDGF-stimulated lamellipodia protrusions (ruffling). When cells are grown in serum and form stress fibers, expression of Rnd1 promotes a rapid disappearance of stress fibers and focal adhesions nuclei (Nobes et al. 1998).

Mutant forms of Rnd1 (either a mutation corresponding to the N17 "dominant negative" mutation of Ras or to the A35 "effector site" mutation in Ras) do not inhibit LPA-induced stress fiber formation. Thus, the binding of GTP and a functional effector domain are both required for Rnd1 to block stress fiber formation. Expression of Rnd3 also blocks LPA-induced stress fiber formation, but expression of Rnd2 does not (Nobes et al. 1998).

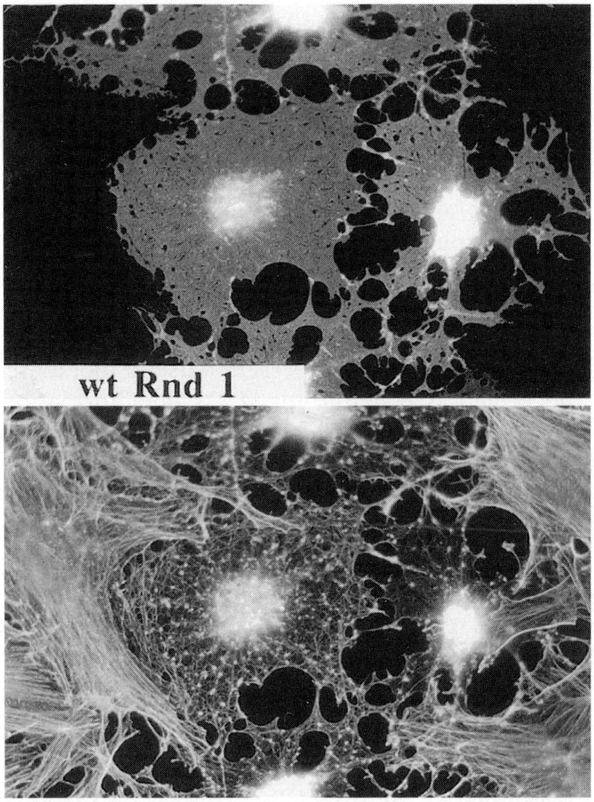

Fig. 2. Effects of Rnd1 expression on the formation of actin stress fibers. Serum-starved confluent Swiss 3T3 fibroblasts were injected with pRK5 vectors expressing Rnd1, and 2–3 hr later the cells were stimulated with LPA for 30 min, before fixation. Permeabilized cells were stained to show Rnd1 expression (*above*) and actin filaments (*below*)

Interestingly, a mutant Rnd1 deleted of its first six amino acids is unable to block stress fiber formation. This could explain the lack of effects on the cytoskeleton reported for RhoE, a N-terminal truncated version of Rnd3 (Foster et al. 1996). We found that at high levels of expression, even a CAAX box deleted version of Rnd1 could block stress fiber formation. It is possible, therefore, that the dramatic effects of Rnd1 on stress fiber formation is due to the high level of expression. At physiological levels of expression, the effects of Rnd1 on actin stress fibers are certainly more subtle and could be restricted to specific intracellular regions.

RhoE/Rnd3 was first found by the two hybrid system as a p190 (RhoGAP) binding protein (Foster et al. 1996). Clearly, a Rho protein constitutively bound to GTP could form a stable complex with RhoGAP and trap it, but this would presumably lead to increased Rho activity and stress fiber formation.

Thus, an interaction with p190 could hardly explain the effects of Rnd1 or Rnd3.

Since Rnd1 and Rho proteins have very similar sequences in the "effector region", but have antagonistic effects, it could be that Rnd1 bind the same effector as Rho proteins, without activating it. At least one of the targets responsible for the induction of actin stress fibers is most likely p160ROCK, that can phosphorylate and inhibit the regulatory subunit of myosin light chain phosphatase, and might also be able to phosphorylate myosin light chain (MLC) directly. Then, phosphorylated MLC induces the bundling of actin into stress fibers. Our results suggest that Rnd1, which is constitutively bound to GTP, could form a stable "unproductive" complex with the same effector as RhoA, and could thus represent a useful tool to help characterize which of the Rho targets is essential for stress fiber formation. However, using the two-hybrid system, no interaction of Rnd1 with citron, rhotekin, the rho binding domain of ROK, or PAK could be detected (G. Zalcman and J. Camonis, unpubl.). In addition, some of these interactions were tested by spotting Rho effector proteins on membranes and incubating them with labelled GTP bound Rnd1, also with negative results (N. Tapon, unpubl.). Since Rnd proteins are almost identical to Rho in the "effector domain" it strongly suggests that additional regions of Rho are also recognized by these effectors. In this respect it is interesting to note that Rnd and Rho differ in $\alpha 2$, the other major "switch" region.

6
Localization of Rnd1 in Swiss 3T3 Fibroblasts and Microinjected MDCK Cells

In Swiss 3T3 fibroblasts, Anti-Rnd1 antibodies give a dotted staining of the plasma membrane at points of cell-cell contacts, but no staining at places where there is no neighboring cell. This peripheral staining co-localizes with an anti-cadherin antibody, indicating that Rnd1 is localized in adherens junctions. Since high levels of wild-type Rnd1 induces cell rounding and loss of cell-cell contacts, it is difficult to localize the expressed protein. However, the "dominant-negative" and "effector site" mutants, that do not affect cell morphology, are predominantly localized at the cell periphery at points of cell-cell contact, when microinjected in confluent Swiss 3T3 fibroblasts (Nobes et al. 1998).

In MDCK cells, microinjection of wild type Rnd1 did not cause cell rounding and Rnd1 was localized at the cell periphery. This pattern closely resembles E-cadherin staining. Interestingly, the truncated form of wild type Rnd1, in which the N-terminal six amino acids are deleted, no longer localizes to adherens junctions; instead the expressed protein is present in the cytoplasm and nucleus. Deletion of the CAAX motif for prenylation leads to a predominant nuclear localization (Nobes et al. 1998).

These results suggest that a functional CAAX box is required for the binding of Rnd1 to membranes and that the N-terminal sequence provides an additional signal for the specific localization to adherens junctions. The N-terminal MKERRA... sequence is highly polar and mostly basic, and it might interact with the acidic head groups of phospholipids to stabilize membrane interaction or could bind directly to another component of adherens junctions.

It is not clear how Rnd1, that is predominantly localized in adherens junctions can affect actin stress fiber formation. It could bind one of the rho effectors without being able to activate it, and could thus trap this effector in a place where it can no longer be active. However one can not exclude the possibility that only a small amount of Rnd1 (corresponding to the low level of expression observed in most adult tissues) localizes in adherens junctions, whereas at higher levels of expression the putative "receptor" in adherens junctions would be saturated. Only the fraction of Rnd1 that does not localize in adherens junctions would then interfere with actin stress fiber formation. Alternatively, Rnd1 could interfere with Rho activation and impair the assembly of focal adhesions indirectly, as a result of the inhibition of stress fiber formation.

This localization of Rnd1 and its effects on cell morphology strongly suggest a role in the control of cell adhesion and motility. In support of this idea, the cDNA for a protein closely related to Rnd1 has recently been discovered in *Xenopus*, where it seems to affect cell adhesion early in development, and axis formation (K. Wuennenberg-Stapleton and K. Cho, pers. comm.).

7
A Role for Rnd Proteins in Transformation?

Rho family proteins play an important role in cell proliferation (Olson et al. 1995; Qiu et al. 1995a,b) and their exchange factors are also involved in transformation and invasion (Chap. Stam and Collard). Cells microinjected with activated Ras rapidly spread, with increased ruffling and actin stress fibers (Bar-Sagi and Feramisco 1986; Ridley et al. 1992), but display a completely different morphology later, with cell rounding and a fusiform appearance as seen in foci of Ras-transformed cells (Symons 1996). Several events contribute to cell rounding: for instance activation of the Ras/raf/MAP kinase pathway leads to the suppression of integrin activation, in the absence of de-novo protein synthesis (Hughes et al. 1997). Later, transformation by Ras also induces the expression and secretion of cathepsins D and L; matrix proteases involved in the acquisition of invasiveness. Whereas decreases in α-actinin, vinculin, or tropomyosin 2 or 3 could explain the loss of stress fibers in some transformed cells. Increased expression of a Rnd protein could also be one of the important mechanisms contributing to the rounding and loss of stress fibers in transformed cells. We have recently observed an increased expression of Rnd proteins in several transformed cell lines and are currently looking at increased expression of Rnd proteins in primary tumors. Invasiveness and malignancy frequently correlate with a decreased E-cadherin expression (Birchmeier and Berhens 1994), the co-localization of Rnd1 with cadherins suggests that both could be involved in the loss of cell-cell contacts.

Rnd2 maps to chromosome 17 q21, a region frequently rearranged in breast and ovarian tumors, the most frequent malignancies in women. More than one hundred kb of human genomic DNA, including the BRCA1 gene have been sequenced and Rnd2 is located right at the end of BRCA1, in the opposite orientation (Smith et al. 1996). Mutations in BRCA1 increase susceptibility to breast and ovarian cancers, but other genetic rearrangements of this region are also frequent. The human plakoglobin (γ-catenin) gene also localizes to 17 q21 and is subject to loss of heterozygosity in breast and ovarian cancers (Aberle et al. 1995). In cells, plakoglobin associates with E-cadherin and is predominantly localized in adherens junctions, as Rnd proteins. The fact that the two genes are present in the same chromosomal region, together with BRCA1, suggests that there might be a cluster of genes expressed in breast tissue, two of them coding for proteins regulating cell-cell contacts. We are currently looking at the expression of Rnd2 in breast tissue, and possible modifications of expression in breast tumors. The Rnd1 gene is located on chromosome 12 q12–q13; this region does not seem to be frequently rearranged in tumors. Rnd3 expression is increased in several transformed cell lines, (P. Chardin, in prep.).

8 Rnd Proteins as Targets of Farnesyl-Transferase Inhibitors

Based on the presence of a methionine as the C-terminal residue in the "CAAX box", Rnd proteins are predicted to be farnesylated instead of geranyl-geranylated as other Rho family proteins. It has been shown recently that RhoE/Rnd3 is indeed farnesylated (Foster et al. 1996). Interestingly, the farnesyltransferase inhibitor L-739, 749 induces cell flattening, actin stress fiber formation and inhibition of anchorage-independant growth after only 18 h of treatment (Prendergast et al. 1994), too fast to be explained by the loss of farnesylated Ras which is stable (half-life of 24 h). Since Rnd3 is expressed at higher levels in many transformed cells, where it could contribute to the loss of stress fibers and cell rounding, the inhibition of Rnd3 farnesylation would presumably induce opposite effects, namely stress fiber formation and cell spreading, as seen when cells are treated with L737–749. Rnd proteins are thus potentially important targets for these drugs.

9
Conclusions

Cdc42, Rac and Rho control dynamic rearrangements of the cytoskeleton that can occur in the time scale of minutes. They can be rapidly activated by exchange factors and inactivated by GAPs, both being tightly coupled to regulatory signals. We suggest that Rnd proteins, which may constitutively be in the GTP-bound form, are regulated on a longer time-scale (hours/days). In adults, the expression of Rnd proteins appears to be restricted to a few specialized cells, where they could be used to induce a sustained increase in the flexibility of the actin cytoskeleton, to modify cell-cell contacts and increase cell motility. The expression of Rnd proteins seems to be increased in many transformed cells where they could contribute to the loss of actin stress fibers, rounded morphology, increased motility and invasiveness.

Acknowledgements. Cellular effects and localization of Rnd1 have been studied by Kate Nobes in Alan Hall's laboratory. Sonia Paris has studied the biochemical properties of Rnd1 and Inger Lauritzen its localization in rat brain. M.-G. Mattei has determined the chromosomal localization. Many thanks to Gerard Zalcman, Jacques Camonis Nicolas Tapon and Katrin Wuennenberg-Stapleton for sharing unpublished results. PC is supported by INSERM and the work in Alan Hall's laboratory was supported by a short-term EMBO fellowship.

References

Aberle H, Bierkamp C, Torchard D et al. (1995) The human plakoglobin gene localizes on chromosome 17 q21 and is subjected to loss of heterozygosity in breast and ovarian cancers. Proc Natl Acad Sci USA 92:6384–6388

Adamson P, Paterson HF, Hall A (1992) Intracellular localization of the p21Rho proteins. J Cell Biol 119:617–627

Aktories K, Just I (1995) In vitro ADP-ribosylation of Rho by bacterial ADP-ribosyl transferases. Methods in enzymology, 256, part B. Academic Press, New York, pp 184–195

Bar-Sagi D, Feramisco JR (1986) Induction of membrane ruffling and fluid-phase pinocytosis in quiescent fibroblasts by Ras proteins. Science 233:1061–1068

Ben-Ze'ev A (1997) Cytoskeletal and adhesion proteins as tumor suppressors. Curr Opin Cell Biol 9:99–108

Birchmeier W, Berhens J (1994) Cadherin expression in carcinomas: role in the formation of cell junctions and the prevention of invasiveness. Biochim Biophys Acta 1198:11–26

Bokoch GM, Bohl BP, Chuang TH (1994) Guanine nucleotide exchange regulates membrane translocation of rac/Rho GTP-binding proteins. J Biol Chem 269:31674–31679

Chardin P, Boquet P, Madaule P, Popoff MR, Rubin EJ, Gill DM (1989) The mammalian G protein RhoC is ADP-ribosylated by clostridium botulinum exoenzyme C3 and affects actin microfilaments in Vero cells. EMBO J 8:1087–1092

Dallery-Prudhomme E, Roumier C, Denis C, Preudhomme C, Kerckaert JP, Galiegue-Zouitina S (1997) Genomic structure and assignment of the RhoH/TTF small GTPase gene (ARHH) to 4p13 by in situ hybridization. Genomics 43:89–94

Der CJ, Finkel T, Cooper GM (1996) Biological and biochemical properties of human H-ras genes mutated at codon 61. Cell 44:167–176

Foster R, Hu K-Q, Lu Y, Nolan KM, Thissen J, Settleman J (1996) Identification of a novel human Rho protein with unusual properties: GTPase deficiency and in vivo farnesylation. Mol Cell Biol 16:2689–2699

Habets GGM, Scholte EHM, Zuydgeest D, Van der Kammen RA, Stam JC, Berns A, Collard JG (1994) Identification of an invasion-inducing gene, Tiam-1, that encodes a protein with homology to GDP-GTP exchangers for Rho-like proteins. Cell 77:539–547

Hirshberg M, Stockley RW, Dodson G, Webb MR (1997) The crystal structure of human racl, a member of the rho-family complexed with a GTP analogue. Nat Str Biol 4:147–151

Huber LA, Ullrich O, Takai Y et al. (1994) Mapping of Ras-related GTP-binding proteins by GTP overlay following two-dimensional gel electrophoresis. Proc Natl Acad Sci USA 91:7874–7878

Hughes PE, Renshaw MW, Pfaff M et al. (1997) Suppression of integrin activation: a novel function of a Ras/raf-initiated MAP Kinase pathway. Cell 88:521–530

John J, Frech M, Wittinghofer A (1988) Biochemical properties of H-Ras encoded p21 mutants and mechanism of the autophosphorylation reaction J Biol Chem 263:11792–11799

Kimura K, Ito M, Amano M et al. (1996) Regulation of myosin phosphatase by Rho and Rho-associated kinase. Science 273:245–248

Murphy C, Saffrich R, Grummt M, Gournier H, Rybin V, Rubino M, Auvinen P, Lutcke A, Parton RG, Zerial M (1996) Endosome dynamics regulated by a Rho protein. Nature 384:427–432

Nobes CD, Hall A (1995) Rho, rac, and Cdc42 GTPases regulate the assembly of multimolecular focal complexes associated with actin stress fibers, lamellipodia, and filopodia. Cell 81:53–62

Nobes CD, Lauritzen I, Mattei M-G, Paris S, Hall A, Chardin P (1998) A new member of the Rho family, Rnd1, promotes disassembly of actin filament structures and loss of cell adhesion. J Cell Biol 141:187–197

Olson MF, Ashworth A, Hall A (1995) An essential role for Rho, rac, and Cdc42 GTPases in cell cycle progression through G1. Science 269:1270–1272

Pai E, Krengel U, Petsko GA et al. (1990) Refined crystal structure of the triphosphate conformation of H-ras p21 at 1.35 Å resolution: implications for the mechanism of GTP hydrolysis. EMBO J 9:2351–2359

Prendergast GC, Davide JP, deSolms SJ et al. (1994) Farnesyltransferase inhibition causes morphological reversion of Ras-transformed cells by a complex mechanism that involves regulation of the actin cytoskeleton. Mol Cell Biol 14:4193–4202

Qiu R.-G, Chen J, Kirn D, McCormick F, Symons M (1995a) An essential role for rac in Ras transformation. Nature 374:457–459

Qiu R-G, Chen J, McCormick F, Symons M (1995b) A role for Rho in Ras transformation. Proc Natl Acad Sci USA 92:11781–11785

Ridley AJ, Hall A (1992) The small GTP-binding protein Rho regulates the assembly of focal adhesions and actin stress fibers in response to growth factors. Cell 70:389–399

Ridley AJ, Patterson HF, Johnston CL, Diekmann D, Hall A (1992) The small GTP-binding protein rac regulates growth factor-induced membrane ruffling. Cell 70:401–410

Russell LD (1993) In: Russell LD, Griswold MD (eds) The sertoli cell Cache River Press, Florida

Smith TM, Lee MK, Szabo CI, Jerome N et al. (1996) Complete genomic sequence and analysis of 117 kb of human DNA containing the gene BRCA1. Genome Res 6:1029–1049

Symons M (1996) Rho family GTPases: the cytoskeleton and beyond. Trends Biochem Sci 21:178–181

Tsukita S, Yonemura S, Tsukita S (1997) ERM family: from cytoskeleton to signal transduction. Curr Opin Cell Biol 9:70–75

Van Aelst L, D'Souza-Schorey C (1997) Rho GTPases and signaling networks. Genes Dev 11:2295–2322

Wei Y, Zhang Y, Derewanda U, Liu X, Minor W, Nakamoto RK, Somlyo AV, Somlyo AP, Derewanda ZS (1997) Crystal structure of RhoA-GDP and its functional implications. Nat Str Biol 4:699–703

The DH Protein Family, Exchange Factors for Rho-Like GTPases

Jord C. Stam and John G. Collard

1
Introduction

1.1
Small GTPases

Members of the Ras superfamily of GTPases play critical roles in regulating a wide range of cellular processes, including growth, differentiation, various forms of vesicle transport, nuclear transport and actin cytoskeleton regulation (Bourne et al. 1990). These small GTPases (about 21 kDa) act as molecular switches that are active when they are GTP-bound, and inactive when GTP is converted to GDP, due to their intrinsic GTPase activity (Boguski and McCormick 1993). Cycling between the active and the inactive conformation is influenced by three classes of regulatory proteins (Fig. 1). GTPase activating proteins (GAPs) stimulate the intrinsic GTPase activity of small GTPases, thereby accelerating the return to the inactive conformation. Guanine nucleotide exchange factors (GEFs) catalyze the release of bound GDP which is rapidly exchanged by GTP, due to the high intracellular ratio of free GTP/GDP in vivo. GEFs, which are also termed guanine nucleotide dissociation stimulators (GDSs) or guanine nucleotide release factors (GRFs), induce the active conformation. Guanine nucleotide dissociation inhibitors (GDIs) bind GTPases and stabilize their conformation, either in the GDP-bound- or in the GTP-bound state (Boguski and McCormick 1993). In addition, GDIs may extract GTPases from membranes, due to binding of GDI to the carboxyl-terminal isoprene group of the GTPases (Gosser et al. 1997). The intracellular localization of the regulators of small GTPases may also determine the activity of these proteins.

The Ras superfamily of proteins consists of about 60 proteins, which are divided into seven subfamilies: Ras, Rho, Rab, ARF, Sar1, Ran and Rad/Gem, on the basis of amino acid sequence similarities. The members of these subfamilies are also related with respect to their functions (Bourne et al. 1990; Boguski and McCormick 1993). The genes encoding the oncogenic form of two

Division of Cell Biology, The Netherlands Cancer Institute, Plesmanlaan 121, 1066 CX Amsterdam, The Netherlands

Progress in Molecular and Subcellular Biology, Vol. 22
Ph. Jeanteur (Ed.)

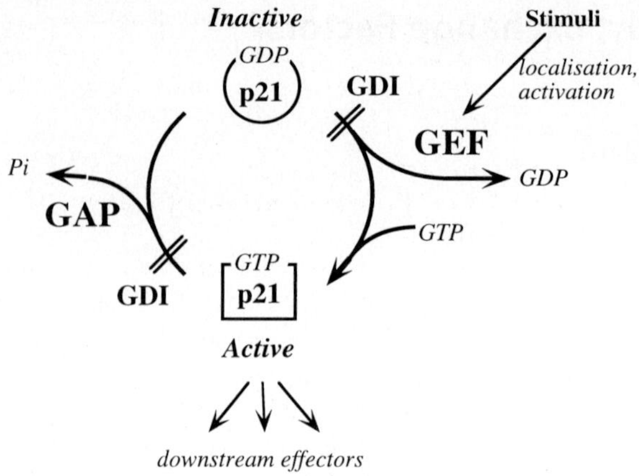

Fig. 1. The regulation of the activity of small GTPases, as explained in the text. Extracellular or intracellular stimuli may change the localization and/or activity of a GEF, resulting in activation of the small GTPase, and downstream signaling. Regulation by GAPs or GDIs might be similarly subject to stimuli

Ras proteins, were originally recovered from transforming viruses. To date, Ras genes are well established (proto-) oncogenes, which are mutated in about 30% of human tumors. The transforming capacity of Ras proteins is due to mutations in single amino acids (amino acids 12, 13, or 61), which keep Ras in the active conformation (Barbacid 1987; Bos 1989, 1995). The extensively investigated Ras proteins are valuable prototypes for the other members of the Ras GTPase superfamily, which are similarly activated by corresponding mutations.

Ras proteins play a central role in signaling pathways in which extracellular stimuli lead to transcriptional activation. Ligand activated tyrosine kinase receptors recruit Sos, a GEF for Ras, to the membrane, resulting in Ras activation and signaling through a protein kinase cascade, which ultimately leads to transcriptional activation. This pathway is subject to many regulatory inputs, and in addition branches to other signaling pathways (Marshall 1996). The small GTPases of the other subfamilies are assumed to mediate signaling pathways in a similar fashion. Various data show cross-talk between different GTPases, which may result in coordination of specific tasks mediated by these GTPases during complex cellular processes (Chant and Stowers 1995).

1.2 Rho-Like GTPases

The Rho-like GTPases comprise a subfamily of the Ras superfamily, consisting of at least 13 members: RhoA, -B, -C, -D, -E, and -G, Rac1, 2, and 3, TC10, Cdc42Hs1 and 2, and TTF. They are about 30% homologous in sequence to Ras and 50–55% to each other (Hall 1994). Whereas Rho was serendipitously discovered by *ho*mology to *R*as, other members were isolated in functional assays: Rac as a toxin substrate and Cdc42 as a genetic mutant involved in cell division in *S. cerevisiae.* The common action of this class of GTPases is the regulation of the actin cytoskeleton (Hall 1994). Injection of constitutively active V14RhoA in serum-starved Swiss 3T3 cells, resulted in formation of stress fibers, whereas active V12Rac1 induced membrane ruffling, and V12Cdc42 induced formation of microspikes. The extracellular ligands LPA, PDGF, and bradykinin, respectively, induced the same phenotypes, suggesting that they activate the corresponding GTPases. This was confirmed by the use of C3 toxin, which specifically inactivates Rho, and the use of dominant negative constructs of Rac and Cdc42 (Ridley and Hall 1992; Ridley et al. 1992; Kozma et al. 1995; Nobes and Hall 1995; Nobes et al. 1995). Using this model system it was also shown that in a hierachical order, Cdc42 activation may lead to activation of Rac, which in turn may activate Rho: the concept of the GTPase cascade (Nobes and Hall 1995).

Whereas the regulation of the actin cytoskeleton in Swiss 3T3 cells proved very revealing, the functions of the Rho-like GTPases and their regulators are diverse and cell type dependent. For instance, Rac2 mediates NADPH superoxide complex activity in neutrophils (Segal and Abo 1993; Bokoch 1995), Cdc42 regulates cell polarity in yeast cells and mediates cytolytic responses in cytotoxic T-lymphocytes (Stowers et al. 1995; Chant 1996). RhoA has been implicated in cytokinesis as well as in neurite retraction in neuronal cell types (Kishi et al. 1993; Jalink et al. 1994; Drechsel et al. 1997). Independent of the cytoskeletal rearrangements, Rho-like GTPases have also been implicated in transcriptional activation. Constitutively active V12Cdc42 and V12Rac1, but not V14RhoA, activate c-Jun amino-terminal kinase (JNK) in COS cells. All three GTPases activate serum response factor (SRF)-dependent signaling pathways (Coso et al. 1995; Hill et al. 1995; Minden et al. 1995; Zhang et al. 1995b). Again, responses may be cell type-dependent, since in HEK293 cells, RhoA was shown to activate JNK (Teramoto et al. 1996).

For each of the GTPases, a number of putative effectors have been identified (Tapon and Hall 1997), which may be involved in the different cellular responses. These can be as diverse as regulating cell morphology, cell motility, tumor cell invasion, endocytosis, secretion, and the establishment of intercellular junctions, as well as coordinating aspects of cell cycle progression. Furthermore, it becomes clear that many of these processes require the combined activities of various members of the Rho-like GTPases, often in combination with Ras. Most strikingly, Ras-induced oncogenic transformation requires the activity of Rho-like GTPases, and Ras can activate Rac (Ridley et al. 1992; Symons 1995; Van Aelst and D'Souza-Schorey 1997). The involvement of Rho-like GTPases in many cellular processes has been established by using constitutively active and dominant negative mutants of the GTPases. The particular GEFs involved are often unknown. Since dominant negative forms of GTPases are assumed to act by sequestering the corresponding GEF (Feig

1994; Quilliam et al. 1995) this implies that processes which can be inhibited by dominant negative GTPases are GEF-dependent. Indeed, for the signaling pathways studied in more detail, the major form of regulation is activation of GEFs (Feig 1994). It should be noted that the GTPases are small molecules which are switched on and off, at specific intracellular locations. The GEFs, on the other hand, are usually large in size and contain multiple domains which allow them to interact with different cellular components and to integrate various signals. Therefore, GEFs are likely to be the key targets of signal transduction pathways activating Rho-like GTPases. In this review we will summarize our knowledge of DH proteins (Table 1), the GEFs that activate Rho-like GTPases.

2
Identification of DH Proteins, GEFs for Rho-Like GTPases

Two lines of research led to the identification of DH proteins. Genetic clues were obtained from a collection of *c*ell *d*ivision *c*ycle (CDC) mutants of the yeast *S. cerevisiae* (Hartwell 1971). The Cdc24 protein and the small GTPase Cdc42 similarly regulated the development of normal cell shape and the selection of nonrandom budding sites in yeast (Sloat et al. 1981; Adams et al. 1990). Defects in Cdc24 could be complemented by overexpression of wild type Cdc42 (Bender and Pringle 1989), providing genetic evidence that Cdc24 is an upstream regulator of the GTPase Cdc42. Biochemical evidence came from studies on Dbl, an oncogene which was isolated as a transforming gene, after transfection of NIH3T3 cells with DNA from a human *D*iffuse *B* cell *l*ymphoma (Dbl; Eva and Aaroson 1985). Dbl and Cdc24 appeared to contain a region with sequence similarity, now known as the Dbl homology (DH) domain (Ron et al. 1991). In vitro experiments showed that Dbl (Hart et al. 1991), and Cdc24 (Zheng et al. 1994), both act as a GEF for Cdc42. The DH domain of about 240 amino acids was also present in the Vav, BCR and Ect2 proteins. Mutational analysis showed that the DH domain is indeed responsible for the GEF activity of Dbl (Hart et al. 1994), as well as for its transforming capacity.

The BCR (*b*reak point *c*luster *r*egion) gene was found to be disrupted in a reciprocal translocation event with the ABL (Abelson) proto-oncogene, which yields the Philadelphia chromosome, implicated in the pathogenesis of leukemias. Whereas the transforming activity is explained by the formation of fusion proteins with an activated ABL tyrosine kinase activity, the BCR DH domain might play a modulatory role (Heisterkamp et al. 1985; Hermans et al. 1987; Clark et al. 1988; Lugo et al. 1990). Similar to Dbl, the Vav and the Ect2 genes have been identified as transforming genes by transfection of genomic DNAs or cDNA libraries into NIH 3T3 cells (Katzav et al. 1989; Miki et al. 1993). This method appeared very successful for isolating genes encoding DH-containing proteins. The transforming DH-proteins Dbs, Lbc, Lfc, Lsc, Net1, Ost, and Tim were similarly identified, making the genes encoding the DH

Table 1. DH Proteins, GEFs for the Rho family of GTPases

	Name	Etymology	Organism[a]	Accession no.[b] / Swiss Prot no[c]	Size in aa[d]	Notes[e]
1.	Abr	Active BCR-related	H	U01147	859	Identified with GAP-specific probes
2.	Bcr	Break point cluster region	H	Y00661 / P11274	1271	Involved in formation of Philadelphia chromosome
3.	CDC24	Cell division control 24 protein	S.c	M16809 / P11433	854[f]	= CLS4, calcium regulatory protein
4.	Dbl	Diffuse B-cell lymphoma	H	X12556 / P10911	925	(Proto-)oncogene; MCF2 is a truncated oncogenic form of Dbl
5.	Dbs	Dbl's big sister	H	S76838	1149	(Proto-)oncogene; DNA from murine 32D hemopoietic cell line cDNA library
6.	Ect2		H	L11316	738	(Proto-)oncogene; DNA from keratinocyte cDNA library
7.	Fgdl	Faciogenital displasia	H	U11690	961	
8.	Lbc	Lymphoid blast crisis	H	U03634	424	(Proto-)oncogene; DNA from human myeloid leukemia
9.	Lfc	Lbc's first cousin	M	U28495	573	(Proto-)oncogene isolated along with Dbs; binds Rac, independent of status of the nucleotide bound
10.	Lsc	Lbc's second cousin	M	U58203	919	(Proto-)oncogene; isolated along with Dbs
11.	Net1	Neuro epithelioma transforming	H	U02081	601	(Proto-)oncogene; the oncogene is 460 amino acids
12.	Ost	Osteosarcoma	R	Z35654 :E	872	(Proto-)oncogene
13.	P-CIP 10a	PAM COOH-terminal interactor protein 10a	R	U88157	1899	= "Kalirin" (Alam et al. 1997); interacts with *p*eptidylglycine *a*lpha-amidating *m*onooxygenase, a peptide-processing enzyme (Alam et al. 1996)
14.	P-CIP 10b		R	U88156	1919	
15.	Ras-GRF	Ras-guanine-nucleotide releasing factor	R	X67241 :E/ P28818	1244	Is RasGEF, activated by heterotrimeric G-proteins
16.	Ras-GRF2		M	U67326	1189	(Fam et al. 1997)
17.	p115-RhoGEF	Rho-guanine nucleotide exchange factor	H	U64105	912	Binds nucleotide depleted RhoA; also (proto-)oncogene

	Name	Etymology	Organism[a]	Accession no.[b] / Swiss Prot no[c]	Size in aa[d]	Notes[e]
18.	RhoGEF-RIP2	-Rho-interacting protein 2	M	U73199	1693	Induces neurite retraction in neuronal N1E-115 cells
19.	Rom1	*RHO1 m*ulticopy suppressor	S.c	/ P53046	1155	(N-terminal-) sequence not complete; homologous to ROM2 (42% identity, with 78% identity in DH)
20.	Rom2		S.c	/ P51862	1356	ROM 1–5 and PKC isolated as multicopy suppressors of RHO1 in yeast
21.	Scd1	Shape and conjugation deficiency	S.p	U12538 / P40995	834	Selected together with Ras1 and Scd2 from cells that did not mate (Chang et al. 1994)
22.	SIF2	Still life type 2	D	D86546	2044	SIF1 is a splice variant (differnt N-terminus)
23.	mSos1	Son of sevenless	M	Z11574 :E	1336	Is RasGEF; dSos binds the sevenless receptor in *Drosophila*
24.	mSos2		M	Z11664 :E	1297	mSos1 67% identical to mSos2
25.	Tiam1	T-lymphoma invasion and metastasis	M	U05245	1591	Also (proto-)oncogene
26.	Tim	Transforming immortalised mammary	H	U02082	519	(Proto-)oncogene; DNA from mammary epithelial cell line
27.	Trio	Three catalytic domains	H	U42390	2861	Binds LAR transmembrane tyrosine phos phatase
28.	unc-89	Uncoordinated movement	C.el.		6632	
29.	Vav	Sixth letter of the Hebrew alphabet	M	X64361 :E / P27870	845	(Proto-)oncogene; expressed in hematopoietic cell types
30.	Vav2		M	U37017 / Q60992	868	Over 50% identical to Vav (Henske et al. 1995), widely expressed

[a] H, human; M, mouse; D, *Drosophila*; S.c., *Saccharomyces cerevisiae*; S.p., *Schizosaccharomyces pombe*; R, rat; C.el., *Caenorhabditis elegans*, homologues from other species have been identified for many DH proteins.
[b] Genbank no., unless indicated E, EMBL accession no.
[c] Swiss Prot no. is indicated when available because of the extensive annotations in this library.
[d] Size indicated in aa, amino acids; the apparent Mr on PAGE gels is usually somewhat larger (possibly due to modifications like phosphorylation).
[e] Comments in addition to the text.
[f] Previously reported to be 736 amino acids.

family of proteins the largest family of (proto-)oncogenes (Chan et al. 1994, 1996; Horii et al. 1994; Toksoz and Williams 1994; Whitehead et al. 1995a,b, 1996; Glaven et al. 1996). In the NIH3T3 transformation assay, transformed cells are scored as foci, groups of cells with aberrant cell morphology. The DH proteins activate Rho-like GTPases, which are known to regulate the actin cytoskeleton and cell morphology. This may contribute to their frequent isolation with this method.

Other DH proteins have been identified due to their implication in other biological processes. Tiam1 was identified by proviral insertion mutagenesis, due to its capacity to induce invasiveness in T-lymphoma cells (Habets et al. 1994). The FGD1 gene was cloned by positional cloning, in search of the human gene causing *f*aciogenital *d*isplasia. This disease, also known as the Aarskog-Scott syndrome, causes developmental aberrations like impaired growth, facial abnormalities and urogenital malformations (Pasteris et al. 1994). In *Drosophila*, a screen for genes involved in locomotion, led to the identification of Sif. This protein, which is highly homologous to Tiam1, is implicated in synapse formation (Sone et al. 1997). Two GEFs, p115-RhoGEF and RhoGEF (RIP2) were isolated as binding partners of RhoA, using biochemical methods and yeast two-hybrid screening (Hart et al. 1996; Gebbink et al. 1997).

Other DH domain-containing proteins were identified due to the presence of additional catalytic domains, suggesting multifunctionality and cross-talk. Among these, Trio was isolated as a protein binding to the phosphatase LAR. Trio harbors two DH domains, one specific for Rac and one for Rho (Debant et al. 1996). Sos and RasGRF are GEFs for Ras and contain, in addition to the RasGEF domain (CDC25-domain), a DH domain (Bowtell et al. 1992; Shou et al. 1992). Bcr and Abr (active Bcr related) contain an additional Rho-GAP domain, whereas Bcr also contains a kinase domain (Hariharan and Adams 1987; Diekmann et al. 1991; Maru and Witte 1991; Heisterkamp et al. 1993; Tan et al. 1993). *C. elegans* genetics led to the identification of the unc-89 gene, a gene involved in worm motility and muscle structure. The gene encodes a giant modular protein (6632 amino acids) composed of Ig- and signal transduction domains, among which is a DH domain (Benian et al. 1996).

An increasing number of DH proteins are expected to be identified. Genome sequencing projects from human and other species have already identified additional DH proteins (Nagase et al. 1997), which are not mentioned here since no further data are available.

3 Structure and Function of DH Proteins

3.1 Catalytic Function

Several groups have investigated the substrate specificity of DH proteins in vitro, using recombinant GEFs, purified from bacteria or baculoviruses

Table 2. Specificities of DH proteins for activating Rho-like GTPases

Name[a]	Specificity in vitro[b] +	Specificity in vitro[b] −	Specificity in vivo[c] (fibroblasts)
Abr	Cdc42, Rac1,2, RhoA	Rap1A, H-Ras	
Bcr	Cdc42, Rac1,2, RhoA	Rap1A, H-Ras	
Cdc24	Cdc42Sc	Rho1, Rsr1	
Dbl	Cdc42Hs, RhoA, Rac1 lipid modified	Rab3A, Rac1, Rap1B, K-Ras, TC10	CDC42, Rac, Rho
Dbs	Cdc42, RhoA	Rac1, H-Ras	
Ect2		Cdc42Mm, Rac1, RhoA	
Fgd1	Cdc42Hs		CDC42
Lbc	RhoA	Cdc42, Rac1, H-Ras, TC10, RhoG	Rho
Lfc	RhoA	Cdc42, Rac1, H-Ras	
Lsc	RhoA	Cdc42, Rac1, H-Ras	
Ost	Cdc42, RhoA	Rac1, 2, Rap 1A, H-Ras, TC10, RhoB, C, G	
RasGRF		Cdc42Hs, Rac2, RalA, H-Ras, RhoA	
RhoGEF RIP2	RhoA (2-hyb)[d]	Rac1, Cdc42Hs (2-hyb)	
Tiam1	Cdc42Hs, Rac1, RhoA		Rac
Trio	D1: Rac1 D2: RhoA	D1: Cdc42, Ras, RhoA D2: Cdc42, Rac1, Ras	
Vav	CDC42, Rac, Rho		CDC42, Rac, Rho

[a] Only DH proteins that have been tested are indicated.
[b] Specificity in vitro: +, indicates GEF activity; −, indicates Rho-like GTPases that were tested and were negative.
[c] Specificity in vivo: induction of actin rearrangements specific for Rho-like GTPase members.
[d] Determined by two-hybrid analysis in yeast.

(Table 2). GEF activity is demonstrated both by stimulation of [^{3}H]GDP dissociation and by exchange of bound GDP by [^{35}S]GTPγS. Dbl as well as Cdc24, were shown to bind Cdc42 and to catalyze the exchange of GDP for GTP (Hart et al. 1991; Zheng et al. 1994, 1995a). Dbl can also bind to other Rho-like GTPases, but not Ras family members, and also shows exchange activity towards RhoA (Miki et al. 1993; Yaku et al. 1994). Some DH proteins appeared highly specific in vitro. Cdc24 activates Cdc42Sc, while Lbc, Lfc and Lsc activate RhoA only (Zheng et al. 1995b; Glaven et al. 1996). Other DH proteins are capable of activating different GTPases: Dbl and Ost activate both Cdc42 and RhoA, whereas Abr, Bcr and Tiam1 can activate RhoA, Rac1 and Cdc42 (Miki et al. 1993; Horii et al. 1994; Chuang et al. 1995; Michiels et al. 1995).

For Ras, the protein regions involved in interactions with regulators and downstream effectors have previously been established. Studies in which analogous point mutations were introduced in RhoA and Cdc42, showed that multiple sites are involved in binding and catalysis by the DH proteins Lbc and Dbl. Mutation of a conserved tyrosine in the putative effector region of the

GTPases impairs binding, whereas mutations in other amino acids impair only catalysis by DH proteins (Li and Zheng 1997).

In extension to these findings, GEF binding to Rho-like GTPases and exchange activity do not always coincide. Ect2 binds to RhoC, RhoA and Rac1, but does not display exchange activity towards these proteins (Miki et al. 1993). Since Ect2 induces morphological transformation similar to the other Rho GEFs, Ect2 might activate a Rho-like GTPase which has not been tested, or identified, yet. The Ost oncogene product shows GEF activity towards Cdc42 and RhoA, but not Rac1. However, Ost binds GTP-bound Rac1, raising the possibility that Ost is a downstream effector of Rac (Horii et al. 1994).

To study activity and specificity of DH proteins in vivo, the specific effects of distinct Rho-like GTPases on actin regulation in fibroblasts (Ridley and Hall 1992; Ridley et al. 1992; Kozma et al. 1995; Nobes and Hall 1995) have been exploited. For instance, similar to V12Rac1, Tiam1 induces membrane ruffling in fibroblasts. This effect is inhibited by dominant negative N17Rac1, but not by dominant negative N19RhoA or N17Cdc42, indicating that Tiam1 activates Rac in vivo (Michiels et al. 1995; Van Leeuwen et al. 1995). FGD1 specifically induced the formation of microspikes mediated by Cdc42, a specificity that was confirmed in vitro (Olson et al. 1996; Zheng et al. 1996a). Vav was previously reported to activate Ras (Gulbins et al. 1993, 1994a,b,c), but this was not supported by later studies (Bustelo et al. 1994; Khosravi-Far et al. 1994; Crespo et al. 1996, 1997; Han et al. 1997). In fibroblasts, Vav activates RhoA, Rac1 and Cdc42 (Olson et al. 1996). Vav is, however, only expressed in haematopoietic cells and Vav may act as a specific activator of Rac in lymphoid cells (Crespo et al. 1996, 1997). In general, GEF specificities that have been determined in vivo indicate a more restricted specificity.

3.2 Functional Domains in DH Proteins

3.2.1 *DH and PH Domains*

Based on sequence homologies, the DH domain has been estimated to be about 180 amino acids in size. Within this domain, three regions have been identified with higher conservation; conserved regions (CR) 1–3, which are expected to be critical for the Rho GEF function. An alignment of all known DH domains has recently been presented (Whitehead et al. 1997). In all cases analyzed, mutations in the DH domain abolish GEF activity in vitro, was well as the phenotypes induced by the DH protein in vivo (e.g. transforming activity or induction of cytoskeletal rearrangements).

The DH domain is invariably flanked at its C-terminus by a pleckstrin homology (PH) domain of about 100 amino acids. The relevance of this domain was shown by deletion studies, in which deletions in the PH domains of Dbl, Ost, Dbs and Lfc resulted in loss of transforming capacity (Ron et al. 1989; Horii et al. 1994; Whitehead et al. 1995a,b; Hart et al. 1996). PH domains have been identified in more than 100 proteins, many of which have important functions in signal transduction pathways. The sequence homology between PH domains is very low, but the 3-D structure of various PH domains appeared well conserved and lent structural support for their reported binding properties to phosphatidyl inositol (PI)-phospholipids (Lemmon et al. 1996). Recently, different PH domains were shown to have distinct binding affinities for specific PI species, e.g. for PIP2, or PIP3. This tentatively provides a mechanism allowing proteins with a PH domain to respond to lipid signaling (Lemmon et al. 1996; Klarlund et al. 1997). In addition, many PH domains can bind to Gβγ subunits of heterotrimeric G proteins. Binding of ligands to G protein-coupled receptors induces dissociation of the Gα subunit from the Gβγ subunits, making each available for signaling interactions. Both PI-lipids and Gβγ subunits are membrane-bound, suggesting that PH domains may mediate signal-dependent membrane localization (Shaw 1996). In agreement with this, the transforming capacity of a mutant of Lfc, which carries a deletion in the PH domain, could be restored by the addition of a membrane-localization signal (the RasCAAX motif), suggesting that the membrane localization of Lfc was required for its transforming capacity (Whitehead et al. 1995b). In contrast, the RasCAAX motif could not replace the PH domain of Dbl. This PH domain is suggested to mediate targeting to cytoskeletal components (Zheng et al. 1996b).

A few DH proteins, Tiam1, RasGRF and FGD1, contain a second PH domain. Interestingly, for both RasGRF and Tiam1 and N-terminal PH domain, but not the DH-flanking C-terminal PH (Phc) domain, is required for localization at the plasma membrane (Buchsbaum et al. 1996; Michiels et al. 1997). A large deletion in the PHc domain did not interfere with membrane localization of Tiam1 in fibroblasts, but membrane ruffling was impaired (Michiels et al. 1997). This suggests that the PHc of Tiam1 may have an alternative function for the activity of the protein and that part of the DH-flanking PH domain is required for GEF activity. In agreement with this, sequence identity between Sif (Sone et al. 1997) and Tiam1 (Habets et al. 1994) is high in the DH domain and extends into the N-terminal half of the adjacent PH domain, after which sequence identity drops from 57 to 30%. The high sequence conservation suggests a functional requirement. This notion may also apply to other DH domains, many of which were reported to be overlapping partly with the flanking PH domain (Fig. 2).

Further analysis of the N-terminal PH domain of Tiam1 and RasGRF showed that their function is dependent on cooperative interactions with adjacent domains (Buchsbaum et al. 1996; Stam et al. 1997). In RasGRF, the PH domain together with a coiled-coil sequence and an IQ domain, are required for Ca^{2+}-mediated RasGRF activation. In Tiam1, a functional unit PHn-CC-Ex, encompassing the N-terminal PH domain, a coiled-coil sequence and an adjacent extending domain, mediates receptor-stimulated membrane

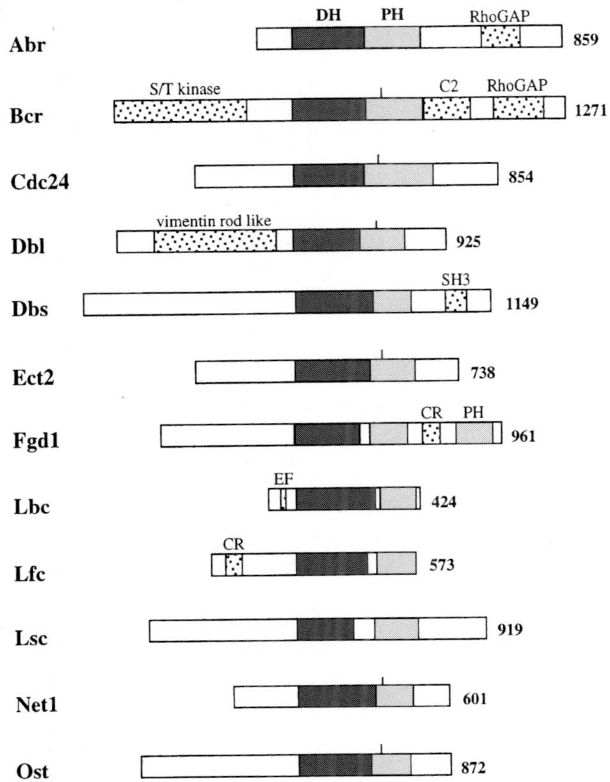

Domains / motifs

and their putative functions

Ac	rich in acidic amino acids, protein/protein interaction
C2	involved in calcium-dependent lipid binding
CC	coiled coil
CR	cysteine rich, zinc butterfly motif, binds diacyl glycerol (DAG)and phorbol ester (PE)
Cdc25	RasGEF motif
CH	calponin homology, putative actin binding
DH	Dbl homology
EF	EF-hand, calcium binding
HLH	leucine-rich helix-loop-helix, may mediate heterodimerization
Ig-like	immune globulin-like
IQ	calmodulin binding
LZ	leucine zipper
opa	repeated Q residues, presence correlated with differential expression
PH	pleckstrin homology
PDZ	also called DHR or GLGF domains, function in sub-membrane proteins
PEST	rich in amino acids P, E, S and T, regulates rapid degradation
PHn-CC-Ex	functional region, regulates membrane-localisation in Tiam1
REM	Ras exchanger motif
RhoGAP	GTPase activating protein for Rho-like GTPases
SH2	Src homology 2, binds phosphotyrosine residues
SH3	Src homology 3, binds proline-rich motifs
SP-D1	spectrin-like
S/T kinase	serine / threonine kinase
vimentin rod like	coiled coil region

Fig. 2. Schematic representation of the DH proteins, drawn to scale. Not shown are DH proteins which are close homologues to the indicated proteins and contain similar domains: RasGRF2, Vav2, mSos2 (Bowtell et al. 1992; Schuebel et al. 1996; Fam et al. 1997). The size of the DH domain is taken from original articles and often overlaps with the adjacent PH domain (indicated with a *vertical bar*); alternative allignments (Whitehead et al. 1997) indicate shorter DH domains that do not overlap (see text)

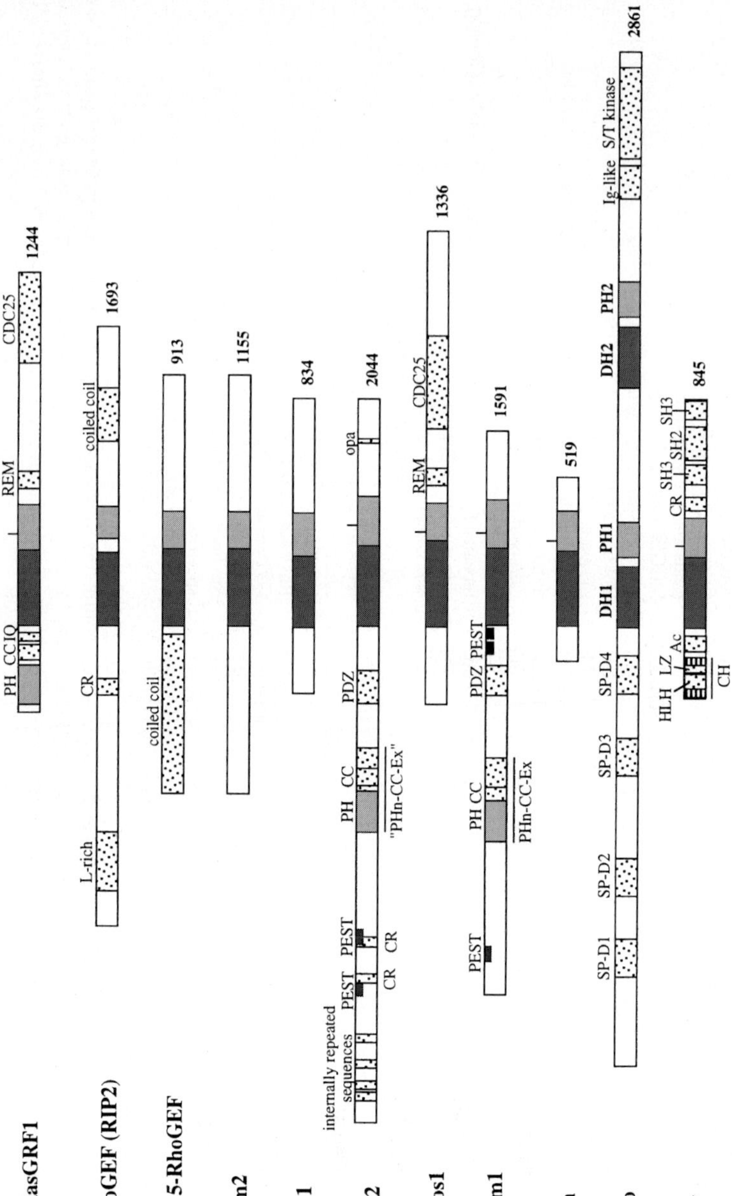

Fig. 2. *Continued*

translocation. A Tiam1 fragment, containing only the N-terminal PH domain, binds to PIP3 (Rameh et al. 1997), but does not localize at the membrane. For membrane localization, the entire PHn-CC-Ex domain is required. This PHn-CC-Ex domain also mediates the association of Tiam1 with a Triton X-100 - insoluble fraction, suggesting that membrane localization requires binding to protein complexes as well as PH domain-mediated lipid binding (Stam et al. 1997). For Sos, on the other hand, the isolated PH domain by itself is sufficient to mediate serum-dependent membrane localization (Chen et al. 1997).

3.2.2
Other Domains in DH Proteins

In addition to the DH-PH combination shared by all DH proteins, many DH proteins carry additional protein motifs (Fig. 2). These can be distinguished in regulatory domains and domains with a known catalytic function, suggesting multifunctionality. Information concerning the functions and interactions between the respective domains is very limited. Interestingly, the transforming capacity of DH proteins is generally stimulated by truncation of other domains, suggesting that these negatively regulate the catalytic activity of the DH protein. For instance, the transforming capacity of Dbl is strongly enhanced by deletion of 300 amino acids in the N-terminal portion of proto-Dbl, which bears similarity to the coiled-coil forming rod domain of the intermediate filament protein vimentin, and is most likely involved in similar protein interactions (Ron et al. 1991).

Figure 2 displays the domains that have been identified in DH proteins, and indicates their putative functions. Various domains are shared by several DH proteins. A cysteine-rich zinc butterfly motif is found in Lfc, Vav, FGD1 and RhoGEF (RIP2), and is thought to be involved in lipid interactions. Deletion of this domain abolished transforming activity of Vav (Coppola et al. 1991), but not of Lfc (Whitehead et al. 1995b). Furthermore, Vav contains a calponin homology region, which is possibly involved in actin binding, whereas Tiam1 and Dbl were also suggested to be associated with cytoskeletal components (Castresana and Saraste 1995; Zheng et al. 1996b; Stam et al. 1997).

Several DH proteins might be Ca^{2+}-regulated. A yeast Cdc24 mutant had previously been identified as a Ca^{2+}-regulatory mutant, CLS4 (Miyamoto et al. 1987) and Lbc and Bcr contain an EF-hand motif and C2 domain, respectively, which might bind intracellular Ca^{2+} (Toksoz and Williams 1994; Romero and Fischer 1996). Furthermore, among many other domains, Trio harbors a serine/threonine kinase domain, most similar to calcium/calmodulin-dependent kinases, such as smooth muscle myosin light chain kinase (Debant et al. 1996).

Among the proteins with multiple catalytic domains, Abr and Bcr are particularly intriguing, since they contain a GEF as well as a GAP domain, which are active on Cdc42 and Rac (Chuang et al. 1995). Dbs and Vav contain a SH3 domain, whereas FGD1 contains a 250 amino acid proline-rich region (22%), in which two sites may bind to SH3 domains (Pasteris et al. 1994). For some DH proteins there are indications that regulation of their turnover rate may be important. Tiam1 and Sif share several domains, among which PEST domains (single letter code for amino acids), which may regulate (high) protein turnover, and RasGRF2 harbors a cyclin destruction box (Habets et al. 1994; Fam et al. 1997; Sone et al. 1997). Whereas the half-life of full length Dbl is 1 h, the truncated onco-Dbl is much more stable (half-life 5–6 h; Ron et al. 1989).

In conclusion, the DH-PH combination may mediate membrane-localization and catalytic activity and the additional domains may regulate the catalytic activity, intracellular localization, (cross-talk-) interactions with other cellular components, and/or protein stability. The relations between the activity of multiple catalytic domains in some DH proteins are intriguing, but remain to be elucidated.

4 Cellular Functions of DH Proteins

4.1 Oncogenic Transformation

The role of BCR in human cancer is well established, but is not directly related to DH domain activity. Reciprocal translocation between chromosome 9 and 22, yielding the diagnostic Philadelphia chromosome, causes leukemia due to fusion of BCR with ABL. A 185 kDa BCR-ABL fusion protein lacking the BCR DH domain, shows activated ABL tyrosine kinase activity and is associated with acute lymphatic leukemia (ALL). A 210-kDa fusion protein which contains the DH domain, is associated with chronic myeloid leukemia (CML; Chan et al. 1987; Hermans et al. 1987; Clark et al. 1988; Witte 1988; Lugo et al. 1990). The DH domain is not required for BCR-ABL mediated oncogenicity, but it seems to modulate the transforming capacity of the fusion protein (Lugo et al. 1990) and to contribute to the pathology of the disease.

In search of oncogenes, most DH proteins were identified by transfer of DNA from various tumors into NIH3T3 cells, a method which frequently resulted in the isolation of mutated Ras genes as well. Cells that were morphologically transformed by DH proteins produced tumors in nude mice, after subcutaneous injection. NIH3T3 cells that have been transformed by Dbl, Dbs, Ect2, Lbc, Lfc, Net1, Ost, Tim, Vav or p115-RhoGEF form similar foci, consisting of rounded, densely-packed cells (Eva and Aaronson 1985; Katzav et al. 1989; Chan et al. 1994, 1996; Horii et al. 1994; Toksoz and Williams 1994; Whitehead et al. 1995a,b, 1996; Glaven et al. 1996; Hart et al. 1996). This phenotype is clearly different from the spindle shaped, highly refractile phenotype induced by Ras or Src in these cells. Other DH proteins like Tiam1 and Vav2 induced a flat epithelial-like morphology in NIH3T3 cells (Van Leeuwen et al. 1995; Schuebel et al. 1996). The Tiam1-transfected cells also produced tumors in nude mice (Van Leeuwen et al. 1995). Thus, transforming activity might be a general capacity of activated DH proteins. Indeed, constitutively active V14RhoA, V12Rac1 and V12Cdc42 induce an oncogenic phenotype (Perona et al. 1993; Khosravi-Far et al. 1995; Qiu et al. 1995a,b, 1997; Van Leeuwen et al. 1995; Lin et al. 1997). The oncogenic activity of V14RhoA is relatively weak compared with V12Rac in NIH3T3 cells (Van Leeuwen et al. 1995), and is absent in Rat1 cells and in another NIH3T3 cell line (Avraham and Weinberg 1989; Qiu et al. 1995a). It should be noted that oncogenicity of DH proteins is also cell type-dependent, being most effective in NIH3T3 cells (Whitehead et al. 1995a, 1997).

The actual involvement of oncogenic DH proteins or Rho-like GTPases in human tumorigenesis has yet to be established. In marked contrast to Ras, no mutations in Rho-like GTPases have been identified in human tumors so far (Moscow et al. 1994). Moreover, there are no indications that the DH proteins isolated by the focus forming assay, had contributed to the tumorigenesis of the tumors from which the DNA was derived. The invasion-inducing Tiam1 gene is expressed in many tumor cell types, including B- and T-lymphomas, neuroblastomas, melanomas and carcinomas (Habets et al. 1995). Whether the expression levels or activating mutations correlate with progression to metastatic disease also remains to be elucidated. Transgenic expression of oncogenic Dbl, driven by various promoters, did not result in spontaneous tumor formation (Eva et al. 1991). However, in mice deficient in the tumor suppressor gene p53, Dbl expression in neuroepithelial tissues increased tumor incidence. This may suggest that DH proteins play a role in tumorigenesis in cooperation with other genes (Colucci et al. 1995).

It is likely that reorganization of the actin cytoskeleton contributes to the transforming capacity of DH-proteins. Several studies have suggested that the actin cytoskeleton is an important component in the maintenance of a transformed phenotype (Symons 1995). Furthermore, the transcriptional activation induced by Rho-like GTPases has been assumed to contribute to cellular transformation. Oncogenic transformation is correlated with the activation of mitogen activated protein kinase (MAPK) cascade pathways. Whereas Ras activates the ERK2 MAPK pathway, Rho-like GTPases and their GEFs specifically activate related kinase cascades, i.e. the SAPK or JNK pathway (Coso et al. 1995; Minden et al. 1995; Zhang et al. 1995b). In addition, active RhoA, Rac1 and Cdc42 have been implicated in activation of the nuclear transcription factor-KB (Perona et al. 1997) and in transcription regulation through the serum response factor (Hill et al. 1995). In analogy to activation of ERK2, transcription regulation by Rho-like GTPases might also be important in tumorigenesis. However, recent experiments suggest that JNK-activation is not correlated with tumorigenesis (Joneson et al. 1996; Lamarche et al. 1996; Michiels et al. 1997; Westwick et al. 1997).

The oncogenic transforming capacity of DH proteins was proposed to be due to the induction of anchorage-independent growth. Normal adherent cells require adhesion to extracellular matrix (ECM) components, as well as growth factor stimulation, for proliferation (Fig. 3). Tumor cells are thought to escape both these requirements due to constitutive activation of the signaling pathways involved (Varner and Cheresh 1996). The role of growth factors in proliferation has focused on the activation of the Ras/MAPK pathway. Adhesion is mediated by integrins and is accompanied by integrin clustering, which results in the formation of intracellular signaling complexes that contribute to growth regulation (Clark and Brugge 1995; Burridge and Chrzanowska 1996). It has been proposed that Rho-like GTPases and their GEFs may induce anchorage-

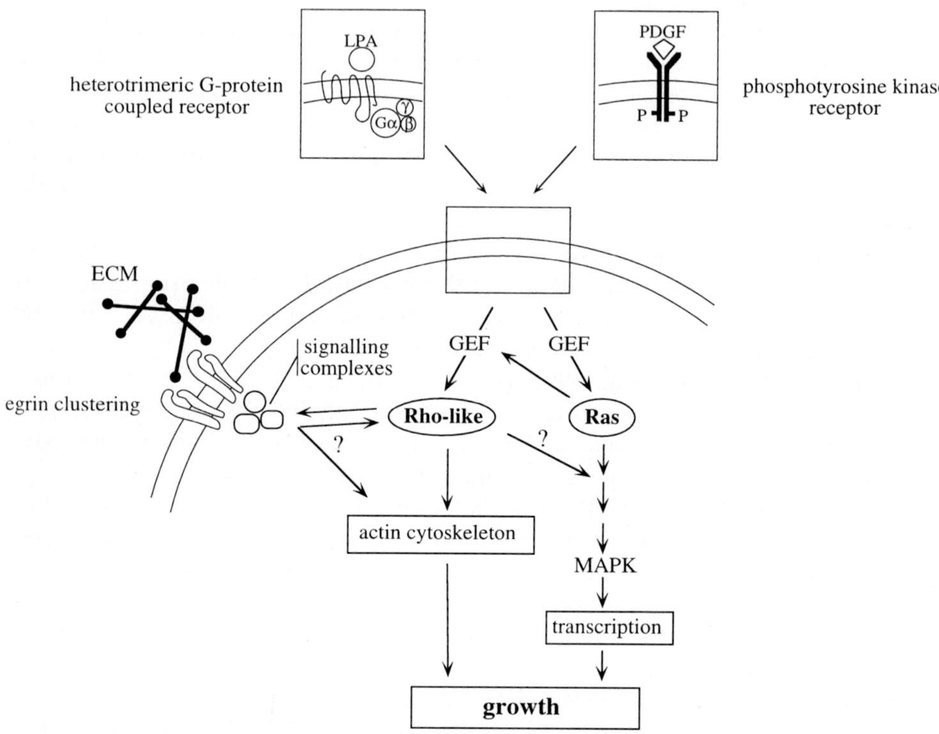

Fig. 3. Rho-like GTPases in signaling pathways regulating growth. Phosphotyrosine kinase receptors as well as heterotrimeric G-protein coupled receptors activate both Ras and Rho-like GTPases. Normal growth may require activation of both pathways. Rho-like GTPases induce formation of adhesive contacts (Hotchin and Hall 1996; Van Leeuwen et al. 1997), in which integrins are clustered. Integrins bind extracellular matrix (ECM) components and ECM binding is normally required for (anchorage-dependent) growth. At sites where integrins cluster, many signaling proteins have been described to be assembled into complexes. ECM-induced signaling contributes to actin-cytoskeleton regulation (Defilippi et al. 1997; Van Leeuwen et al. 1997) and to MAPK activation (Schlaepfer et al. 1994; Renshaw et al. 1996). Active forms of the Rho-like GTPases Rac, Rho and Cdc42, as well as DH proteins have been described as inducing anchorage-independent growth (Khosravi-Far et al. 1995, 1996 Qiu et al. 1995a,b, 1997; Schwartz et al. 1996; Lin et al. 1997; suggesting a model in which these proteins act downstream of integrins in regulation of growth. (Schwartz et al. 1996)

independent growth in soft agar, by constitutive activation of the integrin pathway (Schwartz et al. 1996). In this model, Rho-like GTPases are considered to act downstream of integrins in integrin-mediated growth control. It should be noted that Rho-like GTPases can also act upstream of integrins. In Swiss 3T3 cells, adhesion to ECM components is not sufficient for clustering of integrins, which requires active Rho-like GTPases. Integrin clustering is a prerequisite for the formation of "focal complexes" in which signaling complexes are brought together (Hotchin and Hall 1995; Nobes and Hall 1995). Further research should resolve the connection between these models, which is not yet clear (Fig. 3). Whatever the precise mechanism, Dbl and Lbc transformed NIH3T3 cells were reported to display anchorage-independent but serum-dependent growth (Schwartz et al. 1996). This was also found for V12Cdc42-transformed Rat1 cells, whereas V12Rac1-transformed Rat1 cells showed anchorage-independent, and also serum-independent growth (Qiu et al. 1995a, 1997).

4.2 Invasion

A search for genes involved in the process of tumor cell invasion and metastasis led to the identification of Tiam1 (Habets et al. 1994). Non-invasive T-lymphoma cells were mutagenized by retroviral insertion and invasive variants were selected by their capacity to infiltrate into fibroblast monolayers in vitro. Forty percent of the independently generated invasive clones carried a proviral insertion in the *Tiam1* gene, resulting in truncation of the Tiam1 protein. The invasive T-lymphoma variants also produced metastases when injected intravenously into immune-deficient mice (Habets et al. 1994). Tiam1 was established to be a GEF for the Rho-like GTPase Rac. Accordingly, constitutively active V12Rac1 renders T-lymphoma cells invasive (Michiels et al. 1995).

Subsequent analyses showed that V12Cdc42, but not V14RhoA, also induces invasion of T-lymphoma cells, similarly to V12Rac1 (Stam et al. submitted). Although V14RhoA does not induce invasiveness, endogenous RhoA activity is required for invasion since C3 transferase, which inactivates Rho, inhibits invasion. The invasive capacity of T-lymphoma cells is also dependent on the presence of serum, which can be substituted by the serum component lysophosphatidic acid (LPA) or by sphingosine 1-phosphate (S1P). Both lipids are implicated in Rho activation. T-lymphoma cells expressing both V12Rac1 and V14RhoA, however, still required LPA or S1P for invasion. Further analysis indicated that actin-driven invasion of T-lymphoma cells can be induced by Tiam1/Rac1- or Cdc42-activated pathways, in cooperation with active RhoA and a RhoA-independent PLC activation (Stam et al. 1998).

Invasion and metastasis of carcinoma cells is often associated with reduced E-cadherin-mediated cell-cell adhesion (Behrens et al. 1989; Frixen et al. 1991;

Vleminckx et al. 1991; Takeichi 1993; Birchmeier and Behrens 1994; Shiozaki et al. 1996), and mutations in E-cadherin, β-, or α-catenin have been identified in human tumors and tumor cell lines (Hirano et al. 1992; Becker et al. 1994; Berx et al. 1995; Morin et al. 1997). Interestingly, in epithelial Madin-Darby Canine Kidney (MDCK) cells, Tiam1 localized to adherens junctions (Hordijk et al. 1997). Ectopic expression of Tiam1 or V12Rac inhibited hepatocyte growth factor-induced scattering by increasing E-cadherin-mediated cell-cell adhesion, accompanied by actin polymerization at cell-cell contacts. In Ras-transformed MDCK cells, expression of Tiam1 or V12Rac restored E-cadherin-mediated adhesion, resulting in phenotypic reversion and loss of invasiveness (Hordijk et al. 1997). These results identify Tiam1 as a protein that can modulate E-cadherin-mediated adhesion, and are consistent with studies in keratinocytes and *Drosophila* epithelium, which have implicated Rac in the assembly of adherens junctions (Eaton et al. 1995; Braga et al. 1997). As discussed above, both Tiam1 and RacV12 induce invasion of T-lymphoma cells (Habets et al. 1994; Michiels et al. 1995). This effect appears to be cell-type specific and in epithelial cells Tiam1 and RacV12 rather suppress invasion. These opposing effects may ultimately be the result of the same Tiam1/Rac-regulated phenomenon; an increase in adhesive properties of cells. Tiam1 and Rac1, similar to RhoA, promote integrin-mediated adhesion in different cell types (Tapon and Hall 1997; Van Leeuwen et al. 1997). In lymphoid cells, this might increase integrin-mediated heterotypic cell-cell- and cell-matrix interactions, which could promote invasion of this cell type. In Ras-transformed epithelial cells, Tiam1/Rac-signaling reverts the Ras phenotype and inhibits invasion by increasing E-cadherin-mediated homotypic cell-cell adhesion (Hordijk et al. 1997).

4.3 Neuronal Development

Differentiation-dependent expression of Tiam1 in the developing brain also suggests a role for this GEF in the control of neuronal morphology (Ehler et al. 1997). In agreement with this, Tiam1 induces cell spreading and neurite outgrowth in a neuroblastoma cell line (Van Leeuwen et al. 1997). Tiam1 specifically induced the formation of adhesive contacts at the cell periphery when plated on laminin, and neurite formation was stimulated only on a laminin substrate, not on fibronectin. This suggests that reciprocal interactions between integrins and Rho-like GTPases mediate neuronal differentiation (Van Leeuwen et al. 1997). Interestingly, Rac and Rho seem to have opposing roles in neurite outgrowth. V14RhoA inhibited Tiam1-induced neurite formation (Van Leeuwen et al. 1997), in agreement with a model in which Rac and Cdc42 induce neurite outgrowth (Luo et al. 1994, 1996; Kozma et al. 1997) and Rho antagonizes this process by inducing neurite retraction (Jalink et al. 1994). The balance between the activities of these GTPases may determine neuronal outgrowth and morphology (Kozma et al. 1997; Van Leeuwen et al. 1997).

Recently, the still life protein (Sif) was identified by a functional screen in *Drosophila* (Sone et al. 1997). Mutant flies with impaired motor activities carried a mutation in Sif, which represents a *Drosophila* homologue of Tiam1. The Sif protein localized to synaptic junctions and is thought to regulate (motor-)neuron synapse formation (Sone et al. 1997). In this respect it may be noted that Tiam1 and Sif contain a PDZ domain, which is found in many synaptic proteins, and may play a role in neuronal cells (Sheng 1996; Ponting et al. 1997).

4.4 Development

DH proteins have also been implicated in human growth and development. Patients with the Aarskog-Scott syndrome display impaired growth, facial abnormalities, and urogenital malformations. Positional cloning identified the faciogenital displasia FGD1 gene, which is truncated in affected individuals, such that the encoded protein lacks a functional DH domain (Pasteris et al. 1994). FGD1 was shown to be a GEF specific for Cdc42, by its effects on the actin cytoskeleton in Swiss 3T3 cells and by its GEF activity towards Cdc42 in vitro (Olson et al. 1996; Zheng et al. 1996a). The FGD1 gene is expressed in most tissues examined.

The requirement for Vav in developing embryos was established in knock-out mice, which died very early during development (Zmuidzinas et al. 1995). Crosses of Vav (+/−) heterozygous mice yielded apparently normal Vav (−/−) E3.5 embryos, but no post implantation embryos (>E7.5). Further analysis showed that Vav (−/−) blastocysts do not hatch, a process in which normal embryos attach to ECM plates and spread. Vav expression in normal embryos is high in trophoblast-derived cells which invade the maternal tissue (Zmuidzinas et al. 1995). Since it is known that Vav may activate Rac, like Tiam1, one of the functions of Vav could be to regulate invasion of trophoblast cells (Olson et al. 1996).

In mature tissues, Vav is expressed exclusively in hematopoietic cells, suggesting an important role of Vav in hematopoiesis. However, Vav (−/−) ES cells differentiated normally into erythroid, macrophage and mast cells (Zhang et al. 1994; Zmuidzinas et al. 1995). The role of Vav in lymphoid cells was investigated in chimeric mice, which appeared to be able to develop Vav (−/−) T- and B-lymphocytes. However, these cells were severely compromised in their ability to proliferate in response to antigen receptor activation (Fischer et al. 1995; Tarakhovsky et al. 1995; Zhang et al. 1995a). These and other studies established a role for Vav in T- and B-cell receptor signaling (Romero and Fischer 1996).

4.5 Modulation of RasGEF Activity

Activation of Ras is mediated by the catalytic (Cdc25-) domain of the Ras exchange proteins Sos and RasGRF. Tyrosine kinase receptor-induced Ras activation through Sos has been well established and involves binding of Sos to Grb2 and recruitment to the plasma membrane. However, it appeared that the combination of the Grb2 binding domain with the catalytic domain did not activate Ras in vivo. Instead, the additional DH-PH combination was required to enhance Sos activity in vivo (McCollam et al. 1995). The mechanisms involved are unknown, but considering the relatively low sequence similarity with other DH proteins it has been suggested that the stimulation of Sos activity is not due to Rho GEF activity, but presumably due to binding to membrane components. Indeed, the isolated PH domain of Sos was shown to mediate signal-induced membrane translocation (Chen et al. 1997).

A different Ras pathway requires activation of RasGRF, which also harbors a DH-PH domain. In brain neurons, Ras can be activated by elevated calcium levels. This results in binding of calmodulin to the IQ domain of p140 RasGRF, modulation of interactions with other components, and increased Ras activation mediated by the catalytic RasGEF domain (Farnsworth et al. 1995). For the DH domain of RasGRF, no exchange activity on Rho-like GTPases could be established. However, mutations analogous to those that block the exchange activity of DH domains in other GEFs, as well as mutations in the PH domain, prevent RasGRF activation by Ca^{2+} in vivo (Freshney et al. 1997). A deletion of the DH-PH domains yielded a constitutively active RasGEF, which was no longer Ca^{2+}-regulated (Buchsbaum et al. 1996), suggesting that the DH-PH domain participates in the IQ-mediated Ca^{2+} regulation of RasGEF activity.

5 DH Proteins in Signaling Pathways

Analogous to Ras activation, it is likely that GEFs are involved in the activation of Rho-like GTPases in response to various stimuli and by cross-talk mechanisms. Whereas the role of Rho-like proteins in an increasing number of signaling pathways is being revealed, little is known about the GEFs involved. Only recently are data emerging that demonstrate the involvement of specific DH proteins in signaling pathways and indicate how GEFs may be activated.

5.1 Activation of GEF Activity by Phosphorylation

Vav is known to be phosphorylated on tyrosine in response to various stimuli. Recently, tyrosine phosphorylation by Lck was shown to increase GEF activity of Vav, in vitro as well as in vivo (Crespo et al. 1997; Han et al. 1997). In COS cells Vav activates Rac more strongly than RhoA or Cdc42. Moreover, dominant negative N17Rac1, but not N19RhoA, N17Cdc42 or N17Ras, down-regulate Vav signaling (Crespo et al. 1996). This suggests that Vav specifically activates Rac in vivo. A wide variety of stimuli in hematopoietic cells can activate non-receptor PTKs, which have been shown to phosphorylate Vav, in addition to other substrates. Lck, Fyn, ZAP-70, Syk, Tec, JAK2 and Tyk-2 have all been implicated in phosphorylating Vav (Matsuguchi et al. 1995; Deckert et al. 1996; Miyazato et al. 1996; Crespo et al. 1997; Gotoh et al. 1997; Han et al. 1997; Raab et al. 1997; Uddin et al. 1997). Through its SH2 domain, Vav may bind to several of the other phosphorylated substrates, such as Cbl and SLP-76 (Tuosto et al. 1996; Wu et al. 1996; Marengere et al. 1997; Raab et al. 1997), as well as to the kinases themselves (Wu et al. 1997). In addition, the resulting tyrosine-phosphorylation of Vav allows other signaling molecules, like the regulatory p85 subunit of PI3K (Shigematsu et al. 1997) to bind to Vav. By reversible in vivo cross-linking studies in mast cells, it was found that Vav constitutively associates with Grb2, Raf-1 and MAPK to form an active multimeric signaling complex that can be recruited and activated upon receptor stimulation (Song et al. 1996). This provides a biochemical link with MAPK activation as well. Thus, in addition to activating Rac, which may regulate cytoskeletal changes and transcriptional activation, Vav may convey signals to many different signaling pathways. Most DH proteins are phosphoproteins, mostly phosphorylated on serine/threonine. Similar phosphorylation events as found for Vav might regulate the activity of other GEFs. Tiam1 was found to be phosphorylated specifically on threonine residues in response to various growth factors like LPA, PDGF, bradykinin, and bombesin. The biological role of this phosphorylation, which appeared to be PKC-dependent, remains to be established (Fleming et al. 1997).

5.2
Activation of Rho-like GTPases by Membrane Localization of GEFs

Many signaling pathways have been described in which particular ligands activate distinct Rho-like GTPases. The possible mechanisms by which the GEF involved can be activated, are illustrated for Rac (Fig. 4). PDGF induces Rac-mediated membrane ruffling in Swiss 3T3 cells. Mutational analysis of the PDGF receptor and the inhibition of ruffling by wortmannin indicated that activation of PI3K is involved, which is one of the downstream effectors of the PDGF receptor (Ridley et al. 1992; Kotani et al. 1994; Wennstrom et al. 1994). Indeed, activated PI3K was shown to activate Rac (Hawkins et al. 1995). Interestingly, PI3K also represents a downstream effector of Ras, that is required for Ras-induced cell transformation as well as Ras-induced membrane ruffling (Rodriguez et al. 1994; Rodriguez-Viciana et al. 1997). PI3K catalyses the formation of PIP3, which might bind to the PH domain of the GEFs involved.

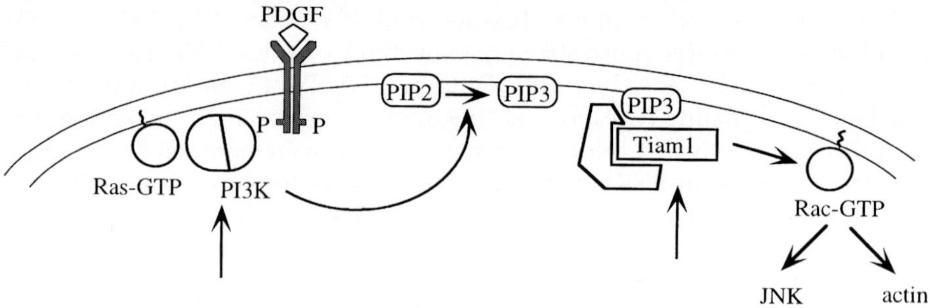

Fig. 4. Activation of Rac mediated by recruitment of a GEF to the plasma membrane. Activated growth factor receptors as well as active Ras recruit PI3K to the membrane, resulting in PI3K mediated formation of PIP3. The N-terminal PH-domain of Tiam1 can bind PIP3 (Rameh et al. 1997). This PH domain is part of a larger protein interaction domain and the entire domain is required for plasma membrane localization of Tiam1 (Stam et al. 1997). The model proposes that formation of PIP3 may provide a membrane-localizing stimulus, which is effective only when Tiam1 also interacts with proteins. Membrane localization of Tiam1 results in activation of membrane-bound Rac, leading to stimulation of JNK activity and F-actin polymerization at the plasma membrane

In agreement with this, it was found that the N-terminal PH domain of the Rac-GEF Tiam1 binds PIP3 (Rameh et al. 1997) and regulates translocation of Tiam1 to the plasma membrane (Michiels et al. 1997; Stam et al. 1997). However, whether Tiam1, or a related GEF, indeed mediates PDGF-induced signaling pathways remains to be established.

Recently, yeast genetics unraveled a possibly similar activation pathway in *S. cerevisiae*. The phosphatidyl inositol kinase homologue Tor2 is involved in cell cycle-dependent polarized distribution of the actin cytoskeleton. It appeared that Tor2 activates Rom2, which is a GEF for Rho1 and Rho2. Deletion of the PH domain in Rom2 abolished activation, suggesting that Tor2-induced lipid products bind to the PH domain, thereby activating Rom2 (Schmidt et al. 1997).

By analogy, other pathways involve the activation of heterotrimeric G-proteins, like the bradykinin and LPA receptors, which leads to the dissociation of the α and $\beta\gamma$ subunits. The $\beta\gamma$ subunits are membrane-localized and are candidate binding partners for PH domains of DH proteins (Shaw 1996). This provides another mechanism to localize DH proteins to the plasma membrane in response to receptor stimulation. Indeed, it has been shown that regulation of polarized cell growth in yeast involves Cdc24p interaction with $\beta\gamma$ subunits in response to pheromone-receptor activation (Zhao et al. 1995).

Thus, in addition to proposed models in which regulated membrane translocation of Rho-like GTPases determines their activity (Bokoch 1995; Hawkins et al. 1995), these data underscore the importance of membrane translocation of GEFs.

5.3 Signal Mediation by DH Proteins in Protein Complexes

Yeast genetics has proven to be a useful tool for determining functional interactions between proteins. The Ras-like Rsr1 gene was identified as a multicopy suppressor of a mutation in Cdc24 (Bender and Pringle 1989). Rsr1 (also known as Bud1) is a Ras GTPase protein that directly interacts with GEF Cdc24, which leads to activation of the GTPase Cdc42, providing one of the first examples of a GTPase cascade (Zheng et al. 1994, 1995a). A similar cascade controls morphogenesis in fission yeast. Scd1, a *Schizosaccharomyces pombe* homologue of Cdc24, interacts with Cdc42sp and acts functionally downstream of both Ras1 and Scd2. In addition, two-hybrid and in vitro biochemical studies suggested that all 4 components act cooperatively to form a complex (Chang et al. 1994). In screens designed to identify genes required for efficient mating, the formin Bni1p was isolated from *S. cerevisiae*, which interacted with GTP-bound Cdc42. This suggests complete unraveling of a Rho-like GTPase mediated pathway, in which pheromone receptor activation leads to establishment of cell polarity, since Bni1p also interacts with actin and the actin associated proteins profilin and Bud6p. Bni1p may thus represent a Cdc42 effector regulating the actin cytoskeleton (Evangelista et al. 1997). Through a 75 amino acid C-terminal region, Cdc24p also binds Bem1p, which is also involved in bud-site assembly (Peterson et al. 1994). It is very likely that many other DH proteins are also involved in the formation of complexes containing various signaling proteins.

5.4 Role of DH Proteins in Ras Transformation

Proto-oncogenic forms of Vav as well as Dbl, which by themselves are poorly oncogenic, can cooperate with proto-oncogenic Ras to induce cell transformation (Bustelo et al. 1994). Subsequent studies indicate that Rac, Cdc42, and Rho are required for Ras transformation (Khosravi-Far et al. 1995; Prendergast et al. 1995; Qiu et al. 1995a, b, 1997). V12Rac was shown to cooperate with Raf in induction of transformation. This led to the suggestion that mitogenic signaling bifurcates at the level of Ras to signal to Rac, as well as to Raf which activates MAPK through a kinase cascade (Khosravi-Far et al. 1995; Qiu et al. 1995a). It is likely that Rac is activated via PI3K, which is an effector of Ras (Rodriguez-Viciana et al. 1997). How Cdc42 and Rho are involved in Ras-induced transformation is less clear. The transforming activity of Lsc, a GEF specific for Rho, is reduced by dominant negative N19RhoA as well as by N17Cdc42 and N17Rac1 (Whitehead et al. 1996). This implies that a linear GTPase cascade model, as proposed for actin rearrangement-induction in fibroblasts, cannot explain transforming activity. It is possible that Rho, Rac and Cdc42 activity are required for Ras transformation independently of each other. As mentioned earlier, all three Rho-like GTPases may be important for the induction of anchorage-independent growth.

It should be noted that the role of Rho-like GTPases in Ras-transformation has been established in fibroblasts. In Ras-transformed epithelial MDCK cells, Tiam1 and V12Rac suppress the Ras phenotype and revert fibroblastoid Ras-transformed cells into epithelial cells. This suggests that Rac may counteract Ras in transformation of epithelial cells (Hordijk et al. 1997). The cooperation of Rho-like GTPases and their activators with Ras in cell transformation may thus not be valid for all cell types.

6 Concluding Remarks

In the last couple of years, Rho-like GTPases have been implicated as control molecules in a large number of cellular processes, by using dominant negative and constitutively active forms of Rho-like GTPases. To understand how these processes are regulated will require the identification of the DH proteins involved and their mode of activation. In vivo as well as in vivo assays established the specificities of DH proteins towards Rho-like GTPases. However, the actual in vivo function of these DH proteins is more difficult to assess. While for each Rho-like GTPase several DH proteins have been identified, tissue- and cell type-specific expression patterns may give some clues as to which DH protein is involved in which particular processes that are mediated by Rho-like GTPases. Furthermore, the generation of dominant negative forms of DH proteins could help determine the involvement of specific DH proteins in particular cellular signaling pathways.

As shown for Tiam1, DH proteins may have different functions in different cell types. Whereas Tiam1 induces invasion of T-lymphoma cells and oncogenicity in fibroblasts, the protein reverts the Ras-phenotype and inhibits invasion of MDCK cells. These effects are all correlated with changes in cytoskeletal organization, underscoring the importance of cell type-dependent actions of DH proteins on the actin cytoskeleton.

Also, little is known at present of the relative contribution of GDI and GAP proteins in regulating the activity of Rho-like GTPases in specific signal transduction pathways, such as cytoskeletal rearrangement and transcription activation. Notwithstanding the questions remaining, the last couple of years have yielded compelling evidence demonstrating the importance of DH proteins in the regulation of many cellular processes mediated by Rho-like GTPases.

Acknowledgements. We thank Frits Michiels, Frank van Leeuwen and Eva Sander for critical reading of the manuscript. The work in John Collard's lab is supported by grants from the Dutch Cancer Society and the Netherlands Organization for Scientific Research.

References

Adams AE, Johnson DI, Longnecker RM, Sloat BF, Pringle JR (1990) CDC42 and CDC43, two additional genes involved in budding and the establishment of cell polarity in the yeast *Saccharomyces cerevisiae*. J Cell Biol 111:131–142

Alam MR, Johnson RC, Darlington DN, Hand TA, Mains RE, Eipper BA (1996) Novel proteins that interact with the COOH-terminal cytosolic routing determinants of an integral membrane peptide-processing enzyme. J Biol Chem 271:28636–28640

Alam MR, Johnson RC, Darlington DN, Hand TA, Mains RE, Eipper BA (1997) Kalirin, a cytosolic protein with spectrin-like and GDP/GTP exchange factor-like domains that interacts with peptidylglycine alpha-amidating monooxygenase, an integral membrane peptide-processing enzyme. J Biol Chem 272:12667–12675

Avraham H, Weinberg RA (1989) Characterization and expression of the human rhoH12 gene product. Mol Cell Biol 9:2058–2066

Barbacid M (1987) Ras genes. Annu Rev Biochem 56:779–827

Becker KF, Atkinson MJ, Reich U, Becker I, Nekarda H, Siewert JR, Hofler H (1994) E-cadherin gene mutations provide clues to diffuse type gastric carcinomas. Cancer Res 54:3845–3852

Behrens J, Mareel MM, Roy van FM, Birchmeier W (1989) Dissecting tumor cell invasion: epithelial cells acquire invasive properties after the loss of uvomorulin-mediated cell-cell adhesion. J Cell Biol 108:2435–2447

Bender A, Pringle JR (1989) Multicopy suppression of the cdc24 budding defect in yeast by CDC42 and three newly identified genes including the ras-related gene RSR1. Proc Natl Acad Sci USA 86:9976–9980

Benian GM, Tinley TL, Tang X, Borodovsky M (1996) The *Caenorhabditis elegans* gene unc-89, required fpr muscle M-line assembly, encodes a giant modular protein composed of Ig and signal transduction domains. J Cell Biol 132:835–848

Berx G, Cleton-Jansen AM, Nollet F, Leeuw de WJ, Vijver van de M, Cornelisse C, Roy van F (1995) E-cadherin is a tumour/invasion suppressor gene mutated in human lobular breast cancers. EMBO J 14:6107–6115

Birchmeier W, Behrens J (1994) Cadherin expression in carcinomas: role in the formation of cell junctions and the prevention of invasiveness. Biochim Biophys Acta 1198:11–26

Boguski MS, McCormick F (1993) Proteins regulating Ras and its relatives. Nature 366:643–654

Bokoch GM (1995) Regulation of the human phagocyte respiratory burst by small GTP-binding proteins. Trends Cell Biol 5:109–113

Bos JL (1989) Ras oncogenes in human cancer: a review. Cancer Res 49:4682–4689

Bos JL (1995) p21ras: an oncoprotein functioning in growth factor-induced signal transduction. Eur J Cancer 31A:1051–1054

Bourne HR, Sanders DA, McCormick F (1990) The GTPase superfamily: a conserved switch for diverse cell functions. Nature 348:125–132

Bowtell D, Fu P, Simon M, Senior P (1992) Identification of murine homologues of the *Drosophila* son of sevenless gene: potential activators of ras. Proc Natl Acad Sci USA 89:6511–6515

Braga VM, Machesky LM, Hall A, Hotchin NA (1997) The small GTPases Rho and Rac are required for the establishment of cadherin-dependent cell-cell contacts. J Cell Biol 137:1421–1431

Buchsbaum R, Telliez JB, Goonesekera S, Feig LA (1996) The N-terminal pleckstrin, coiled-coil, and IQ domains of the exchange factor Ras-GRF act cooperatively to facilitate activation by calcium. Mol Cell Biol 16:4888–4896

Burridge K, Chrzanowska WM (1996) Focal adhesions, contractility, and signaling. Annu Rev Cell Dev Biol 12:463–518

Bustelo XR, Suen KL, Leftheris K, Meyers CA, Barbacid M (1994) Vav cooperates with Ras to transform rodent fibroblasts but is not a Ras GDP/GTP exchange factor. Oncogene 9:2405–2413

Castresana J, Saraste M (1995) Does Vav bind to F-actin through a CH domain? FEBS Lett 374:149–151
Chan AM, McGovern ES, Catalano G, Fleming TP, Miki T (1994) Expression cDNA cloning of a novel oncogene with sequence similarity to regulators of small GTP-binding proteins. Oncogene 9:1057–1063
Chan AM, Takai S, Yamada T, Miki T (1996) Isolation of a novel oncogene, NET1, from neuroepithelioma cells by expression cDNA cloning. Oncogene 12:1259–1266
Chan LC, Karhi KK, Rayter SI, Heisterkamp N, Eridani S, Powles R, Lawler SD, Groffen J, Foulkes JG, Greaves MF (1987) A novel abl protein expressed in Philadelphia chromosome positive acute lymphoblastic leukaemia. Nature 325:635–637
Chang EC, Barr M, Wang Y, Jung V, Xu HP, Wigler MH (1994) Cooperative interaction of *S. pombe* proteins required for mating and morphogenesis. Cell 79:131–141
Chant J (1996) Generation of cell polarity in yeast. Curr Opin Cell Biol 8:557–565
Chant J, Stowers L (1995) GTPase cascades choreographing cellular behavior: movement, morphogenesis, and more. Cell 81:1–4
Chen RH, Corbalan-Garcia S, Bar-Sagi D (1997) The role of the PH domain in the signal-dependent membrane targeting of Sos. EMBO J 16:1351–1359
Chuang TH, Xu X, Kaartinen V, Heisterkamp N, Groffen J, Bokoch GM (1995) Abr and Bcr are multifunctional regulators of the Rho GTP-binding protein family. Proc Natl Acad Sci USA 92:10282–10286
Clark EA, Brugge JS (1995) Integrins and signal transduction pathways: the road taken. Science 268:233–239
Clark SS, McLaughlin J, Timmons M, Pendergast AM, Ben NY, Dow LW, Crist W, Rovera G, Smith SD, Witte ON (1988) Expression of a distinctive BCR-ABL oncogene in Ph1-positive acute lymphocytic leukemia (ALL). Science 239:775–777
Colucci D, Amato GL, Santelli G, D'Alessio A, Chiappetta G, Mineo A, Manzo G, Vecchio G, De Franciscis V (1995) Dbl expression driven by the neuron specific enolase promoter induces tumor formation in transgenic mice with a p53(+/−) genetic background. Biochem Biophys Res Commun 216:762–770
Coppola J, Bryant S, Koda T, Conway D, Barbacid M (1991) Mechanism of activation of the vav protooncogene. Cell Growth Differ 2:95–105
Coso OA, Chiariello M, Yu JC, Teramoto H, Crespo P, Xu N, Miki T, Gutkind JS (1995) The small GTP-binding proteins Rac1 and Cdc42 regulate the activity of the JNK/SAPK signaling pathway. Cell 81:1137–1146
Crespo P, Bustelo XR, Aaronson DS, Coso OA, Lopez Barahona M, Barbacid M, Gutkind JS (1996) Rac-1 dependent stimulation of the JNK/SAPK signaling pathway by Vav. Oncogene 13:455–460
Crespo P, Schuebel KE, Ostrom AA, Gutkind JS, Bustelo XR (1997) Phosphotyrosine-dependent activation of Rac-1 GDP/GTP exchange by the vav proto-oncogene product. Nature 385:169–172
Debant A, Serra Pages C, Seipel K, O'Brien S, Tang M, Park SH (1996) The multidomain protein Trio binds the LAR transmembrane tyrosine phosphatase, contains a protein kinase domain, and has separate rac-specific and rho-specific guanine nucleotide exchange factor domains. Proc Natl Acad Sci USA 93:5466–5471
Deckert M, Tartare, Deckert S, Couture C, Mustelin T, Altman A (1996) Functional and physical interactions of Syk family kinases with the Vav proto-oncogene product. Immunity 5:591–604
Defilippi P, Olivo C, Tarone G, Mancini P, Torrisi MR, Eva A (1997) Actin cytoskeleton polymerization in Dbl-transformed NIH3T3 fibroblasts is dependent on cell adhesion to specific extracellular matrix proteins. Oncogene 14:1933–1943
Diekmann D, Brill S, Garrett MD, Totty N, Hsuan J, Monfries C, Hall C, Lim L, Hall A (1991) Bcr encodes a GTPase-activating protein for p21rac. Nature 351:400–402
Drechsel DN, Hyman AA, Hall A, Glotzer M (1997) A requirement for Rho and Cdc42 during cytokinesis in *Xenopus* embryos. Curr Biol 7:12–23

Eaton S, Auvinen P, Luo L, Jan YN, Simons K (1995) CDC42 and Rac1 control different actin-dependent processes in the *Drosophila* wing disc epithelium. J Cell Biol 131:151–164
Ehler E, Van Leeuwen F, Collard JG, Salinas PC (1997) Expression of Tiam-1 in the developing brain suggests a role for the Tiam-1-Rac signaling pathway in cell migration and neurite outgrowth. Mol Cell Neurosci 9:1–12
Eva A, Aaronson SA (1985) Isolation of a new human oncogene from the diffuse B-cell lymphoma. Nature 316:273–275
Eva A, Graziani G, Zannini M, Merin LM, Khillan JS, Overbeek PA (1991) Dominant dysplasia of the lens in transgenic mice expressing the dbl oncogene. New Biol 3:158–168
Evangelista M, Blundell K, Longtine MS, Chow CJ, Adames N, Pringle JR, Peter M, Boone C (1997) Bni1p, a yeast formin linking cdc42p and the actin cytoskeleton during polarized morphogenesis. Science 276:118–122
Fam NP, Fan WT, Wang Z, Zhang LJ, Chen H, Moran MF (1997) Cloning and characterization of Ras-GRF2, a novel guanine nucleotide exchange factor for Ras. Mol Cell Biol 17:1396–1406
Farnsworth CL, Freshney NW, Rosen LB, Ghosh A, Greenberg ME, Feig LA (1995) Calcium activation of Ras mediated by neuronal exchange factor Ras-GRF. Nature 376:524–527
Feig LA (1994) Guanine-nucleotide exchange factors: a family of positive regulators of Ras and related GTPases. Curr Opin Cell Biol 6:204–211
Fischer KD, Zmuldzinas A, Gardner S, Barbacid M, Bernstein A, Guidos C (1995) Defective T-cell receptor signalling and positive selection of Vav-deficient CD4+ CD8+ thymocytes. Nature 374:474–477
Fleming IN, Elliot CM, Collard JG, Exton JH (1997) Lysophosphatydic acid induces threonine phosphorylation of Tiam1 in Swiss 3T3 fibroblasts via activation of protein kinase C. J Biol Chem 272:33105–33110
Freshney NW, Goonesekera SD, Feig LA (1997) Activation of the exchange factor Ras-GRF by calcium requires an intact Dbl homology domain. Febs Lett 407:111–115
Frixen UH, Behrens J, Sachs M, Eberle G, Voss B, Warda A, Lochner D, Birchmeier W (1991) E-cadherin-mediated cell-cell adhesion prevents inasiveness of human carcinoma cells. J Cell Biol 113:173–185
Gebbink MF, Kranenburg O, Poland M, Van Horck FP, Houssa B, Moolenaar WH (1997) Identification of a novel, putative Rho-specific GDP/GTP exchange factor and a RhoA-binding protein: control of neuronal morphology. J Cell Biol 137:1603–1613
Glaven JA, Whitehead IP, Nomanbhoy T, Kay R, Cerione RA (1996) Lfc and Lsc oncoproteins represent two new guanine nucleotide exchange factors for the Rho GTP-binding protein. J Biol Chem 271:27374–27381
Gosser YQ, Nomanbhoy TK, Aghazadeh B, Manor D, Combs C, Cerione RA, Rosen MK (1997) C-terminal binding domain of Rho GDP-dissociation inhibitor directs N-terminal inhibitory peptide to GTPases. Nature 387:814–819
Gotoh A, Takahira H, Geahlen RL, Broxmeyer HE (1997) Cross-linking of integrins induces tyrosine phosphorylation of the proto-oncogene product Vav and the protein tyrosine kinase Syk in human factor-dependent myeloid cells. Cell Growth Differ 8:721–729
Gulbins E, Coggeshall KM, Baier G, Katzav S, Burn P, Altman A (1993) Tyrosine kinase-stimulated guanine nucleotide exchange activity of Vav in T cell activation. Science 260:822–825
Gulbins E, Coggeshall KM, Baier G, Telford D, Langlet C, Baier BG, Bonnefoy BN, Burn P, Wittinghofer A, Altman A (1994a) Direct stimulation of Vav guanine nucleotide exchange activity for Ras by phorbol esters and diglycerides. Mol Cell Biol 14:4749–4758
Gulbins E, Coggeshall KM, Langlet C, Baier G, Bonnefoy, Berard N, Burn P, Wittinghofer A, Katzav S, Altman A (1994b) Activation of Ras in vitro and in intact fibroblasts by the Vav guanine nucleotide exchange protein. Mol Cell Biol 14:906–913
Gulbins E, Langlet C, Baier G, Bonnefoy Berard N, Herbert E, Altman A, Coggeshall KM (1994c) Tyrosine phosphorylation and activation of Vav GTP/GDP exchange activity in antigen receptor-triggered B cells. J Immunol 152:2123–2129

Habets GG, Scholtes EH, Zuydgeest D, Kammen van der RA, Stam JC, Berns A, Collard JG (1994) Identification of an invasion-inducing gene, Tiam-1, that encodes a protein with homology to GDP-GTP exchangers for Rho-like proteins. Cell 77:537–549

Habets GG, Kammen van der RA, Stam JC, Michiels F, Collard JG (1995) Sequence of the human invasion-inducing TIAM1 gene, its conservation in evolution and its expression in tumor cell lines of different tissue origin. Oncogene 10:1371–1376

Hall A (1994) Small GTP-binding proteins and the regulation of the actin cytoskeleton. Ann Rev Cell Biol 10:31–54

Han J, Das B, Wei W, Van AL, Mosteller RD, Khosravi Far R, Weswick JK, Der CJ, Broek D (1997) Lck regulates Vav activation of members of the Rho family of GTPases. Mol Cell Biol 17:1346–1353

Hariharan IK, Adams JM (1987) cDNA sequence for human Bcr, the gene that translocates to the Abl oncogene in chronic myeloid leukaemia. EMBO J 6:115–119

Hart MJ, Eva A, Evans T, Aaronson SA, Cerione RA (1991) Catalysis of guanine nucleotide exchange on the CDC42Hs protein by the dbl oncogene product. Nature 354:311–314

Hart MJ, Eva A, Zangrilli D, Aaronson SA, Evans T, Cerione RA, Zheng Y (1994) Cellular transformation and guanine nucleotide exchange activity are catalyzed by a common domain on the dbl oncogene product. J Biol Chem 269:62–65

Hart MJ, Sharma S, elMasry N, Qiu RG, McCabe P, Polakis P, Bollag G (1996) Identification of a novel guanine nucleotide exchange factor for the Rho GTPase. J Biol Chem 271:25452–25458

Hartwell LH (1971) Genetic control of the cell division cycle in yeast. IV. Genes controlling bud emergence and cytokinesis. Exp Cell Res 69:265–276

Hawkins PT, Eguinoa A, Qiu RG, Stokoe D, Cooke FT, Walters R, Wennstrom S, Claesson WL, Evans T, Symons M (1995) PDGF stimulates an increase in GTP-Rac via activation of phosphoinositide 3-kinase. Curr Biol 5:393–403

Heisterkamp N, Stam K, Groffen J, De Klein A, Grosveld G (1985) Structural organization of the bcr gene and its role in the Ph′ translocation. Nature 315:758–761

Heisterkamp N, Kaartinen V, Van Soest S, Bokoch GM, Groffen J (1993) Human ABR encodes a protein with GAPrac activity and homology to the DBL nucleotide exchange factor domain. J Biol Chem 268:16903–16906

Henske EP, Short MP, Jozwiak S, Bovey CM, Ramlakhan S, Haines JL, Kwiatkowski DJ (1995) Identification of VAV2 on 9q34 and its exclusion as the tuberoussclerosis gene TSC1. Ann Hum Genet 59:25–37

Hermans A, Heisterkamp N, Von Linden M, Van Baal S, Meijer D, Van der Plas D, Wiedemann LM, Groffen J, Bootsma D, Grosveld G (1987) Unique fusion of bcr and c-abl genes in Philadelphia chromosome positive acute lymphoblastic leukemia. Cell 51:33–40

Hill CS, Wynne J, Treisman R (1995) The Rho family GTPases RhoA, Rac1, and CDC42Hs regulate transcriptional activation by SRF. Cell 81:1159–1170

Hirano S, Kimoto N, Shimoyama Y, Hirohashi S, Takeichi M (1992) Identification of a neural alpha-catenin as a key regulator of cadherin function and multicellular organization. Cell 70:293–301

Hordijk PL, Ten Klooster JP, Van der Kammen RA, Michiels F, Oomen LCJM, Collard JG (1997) Inhibition of invasion of epithelial cells by Tiam1-Rac signalling. Science 278:1484–1486

Horii Y, Beeler JF, Sakaguchi K, Tachibana M, Miki T (1994) A novel oncogene, ost, encodes a guanine nucleotide exchange factor that potentially links Rho and Rac signaling pathways. EMBO J 13:4776–4786

Hotchin NA, Hall A (1995) The assembly of integrin adhesion complexes requires both extracellular matrix and intracellular rho/rac GTPases. J Cell Biol 131:1857–1865

Hotchin NA, Hall A (1996) Regulation of the actin cytoskeleton, integrins and cell growth by the Rho family of small GTPases. Cancer Surv 27:311–322

Jalink K, Van Corven EJ, Hengeveld T, Morii N, Narumiya S, Moolenaar WH (1994) Inhibition of lysophosphatidate- and thrombin-induced neurite retraction and neuronal cell rounding by ADP ribosylation of the small GTP-binding protein Rho. J Cell Biol 126:801–810

Joneson T, McDonough M, Bar-Sagi D, Van Aelst L (1996) RAC regulation of actin polymerization and proliferation by a pathway distinct from Jun kinase. Science 274:1374–1376

Katzav S, Martin Zanca D, Barbacid M (1989) vav, a novel human oncogene derived from a locus ubiquitously expressed in hematopoietic cells. EMBO J 8:2283–2290

Khosravi-Far R, Chrzanowska-Wodnicka M, Solski PA, Eva A, Burridge K, Der CJ (1994) Dbl and Vav mediate transformation via mitogen-activated protein kinase pathways that are distinct from those activated by oncogenic Ras. Mol Cell Biol 14:6848–6857

Khosravi-Far R, Solski PA, Clark GJ, Kinch MS, Der CJ (1995) Activation of Rac1, RhoA, and mitogen-activated protein kinases is required for Ras transformation. Mol Cell Biol 15:6443–6453

Khosravi-Far R, White MA, Westwick JK, Solski PA, Chrzanowska-Wodnicka M, Van Aelst L, Wigler MH, Der CJ (1996) Oncogenic Ras activation of Raf/mitogen-activated protein kinase-independent pathways is sufficient to cause tumorigenic transformation. Mol Cell Biol 16:3923–3933

Kishi K, Sasaki T, Kuroda S, Itoh T, Takai Y (1993) Regulation of cytoplasmic division of *Xenopus* embryo by rho p21 and its inhibitory GDP/GTP exchange protein (rho GDI). J Cell Biol 120:1187–1195

Klarlund JK, Guilherme A, Holik JJ, Virbasius JV, Chawla A, Czech MP (1997) Signaling by phosphoinositide-3,4,5-trisphosphate through proteins containing pleckstrin and Sec7 homology domains. Science 275:1927–1930

Kotani K, Yonezawa K, Hara K, Ueda H, Kitamura Y, Sakaue H, Ando A, Chavanieu A, Calas B, Grigorescu F (1994) Involvement of phosphoinositide 3-kinase in insulin- or IGF-1-induced membrane ruffling. EMBO J 13:2313–2321

Kozma R, Ahmed S, Best A, Lim L (1995) The Ras-related protein Cdc42Hs and bradykinin promote formation of peripheral actin microspikes and filopodia in Swiss 3T3 fibroblasts. Mol Cell Biol 15:1942–1952

Kozma R, Sarner S, Ahmed S, Lim L (1997) Rho family GTPases and neuronal growth cone remodelling: relationship between increased complexity induced by Cdc42Hs, Rac1, and acetylcholine and collapse induced by RhoA and lysophosphatidic acid. Mol Cell Biol 17:1201–1211

Lamarche N, Tapon N, Stowers L, Burbelo PD, Aspenstrom P, Bridges T, Chant J, Hall A (1996) Rac and Cdc42 induce actin polymerization and G1 cell cycle progression independently of p65PAK and the JNK/SAPK MAP kinase cascade. Cell 87:519–529

Lemmon MA, Ferguson KM, Schlessinger J (1996) PH domains: diverse sequences with a common fold recruit signaling molecules to the cell surface. Cell 85:621–624

Li R, Zheng Y (1997) Residues of the Rho family GTPases Rho and Cdc42 that specify sensitivity to Dbl-like guanine nucleotide exchange factors. J Biol Chem 272:4671–4679

Lin R, Bagrodia S, Cerione R, Manor D (1997) A novel Cdc42Hs mutant induces cellular transformation. Curr Biol 7:794–797

Lugo TG, Pendergast AM, Muller AJ, Witte ON (1990) Tyrosine kinase activity and transformation potency of bcr-abl oncogene products. Science 247:1079–1082

Luo L, Liao YJ, Jan LY, Jan YN (1994) Distinct morphogenetic functions of similar small GTPases: *Drosophila* Drac1 is involved in axonal outgrowth and myoblast fusion. Genes Dev 8:1787–1802

Luo L, Hensch TK, Ackerman L, Barbel S, Jan LY, Jan YN (1996) Differential effects of the Rac GTPase on Purkinje cell axons and dendritic trunks and spines. Nature 379:837–840

Marengere LE, Mirtsos C, Kozieradzki I, Veillette A, Mak TW, Penninger JM (1997) Proto-oncoprotein Vav interacts with c-Cbl in activated thymocytes and peripheral T cells. J Immunol 159:70–76

Marshall CJ (1996) Ras effectors. Curr Opin Cell Biol 8:197–204

Maru Y, Witte ON (1991) The BCR gene encodes a novel serine/threonine kinase activity within a single exon. Cell 67:459–468

Matsuguchi T, Inhorn RC, Carlesso N, Xu G, Druker B, Griffin JD (1995) Tyrosine phosphorylation of p95Vav in myeloid cells is regulated by GM-CSF, IL-3 and steel factor and is constitutively increased by p210BCR/ABL. EMBO J 14:257–265

McCollam L, Bonfini L, Karlovich CA, Conway BR, Kozma LM, Banerjee U, Czech MP (1995) Functional roles for the pleckstrin and Dbl homology regions in the Ras exchange factor Son-of-sevenless. J Biol Chem 270:15954–15957

Michiels F, Habets GG, Stam JC, Kammen van der RA, Collard JG (1995) A role for Rac in Tiam1-induced membrane ruffling and invasion. Nature 375:338–340

Michiels F, Stam JC, Hordijk PL, Kammen van der RA, Ruuls-Van Stalle L, Feltkamp CA, Collard JG (1997) Regulated membrane localization of Tiam1, mediated by the N-terminal pleckstrin homology domain, is required for Rac-dependent membrane ruffling and JNK activation. J Cell Biol 137:387–398

Miki T, Smith CL, Long JE, Eva A, Fleming TP (1993) Oncogene ect2 is related to regulators of small GTP-binding proteins. Nature 362:462–465

Minden A, Lin A, Claret FX, Abo A, Karin M (1995) Selective activation of the JNK signaling cascade and c-Jun transcriptional activity by the small GTPases Rac and Cdc42Hs. Cell 81:1147–1157

Miyamoto S, Ohya Y, Ohsumi Y, Anraku Y (1987) Nucleotide sequence of the CLS4 (CDC24) gene of *Saccharomyces cerevisiae.* Gene 54:125–132

Miyazato A, Yamashita Y, Hatake K, Miura Y, Ozawa K, Mano H (1996) Tec protein tyrosine kinase is involved in the signaling mechanism of granulocyte colony-stimulating factor receptor. Cell Growth Differ 7:1135–1139

Morin PJ, Sparks AB, Korinek V, Barker N, Clevers H, Vogelstein B, Kinzler KW (1997) Activation of beta-catenin-Tcf signaling in colon cancer by mutations in beta-catenin or APC. Science 275:1787–1790

Moscow JA, He R, Gnarra JR, Knutsen T, Weng Y, Zhao WP, Whang PJ, Linehan WM, Cowan KH (1994) Examination of human tumors for rhoA mutations. Oncogene 9:189–194

Nagase T, Ishikawa K, Nakajima D, Ohira M, Seki N, Miyajima N, Tanaka A, Kotani H, Nomura N, Ohara O (1997) Prediction of the coding sequences of unidentified human genes. VII. The complete sequences of 100 new cDNA clones from brain which can code for large proteins in vitro. DNA Res 4:141–150

Nobes CD, Hall A (1995) Rho, rac, and cdc42 GTPases regulate the assembly of multimolecular focal complexes associated with actin stress fibers, lamellipodia, and filopodia. Cell 81:53–62

Nobes CD, Hawkins P, Stephens L, Hall A (1995) Activation of the small GTP-binding proteins rho and rac by growth factor receptors. J Cell Sci :225–233

Olson MF, Pasteris NG, Gorski JL, Hall A (1996) Faciogenital dysplasia protein (FGD1) and Vav, two related proteins required for normal embryonic development, are upstream regulators of Rho GTPases. Curr Biol 6:1628–1633

Pasteris NG, Cadle A, Logie LJ, Porteous ME, Schwartz CE, Stevenson RE, Glover TW, Wilroy RS, Gorski JL (1994) Isolation and characterization of the faciogenital dysplasia (Aarskog-Scott syndrome) gene: a putative Rho/Rac guanine nucleotide exchange factor. Cell 79:669–678

Perona R, Esteve P, Jimenez B, Ballestero RP, Ramon Y, Cajal S, Lacal JC (1993) Tumorigenic activity of rho genes from *Aplysia californica.* Oncogene 8:1285–1292

Perona R, Montaner S, Saniger L, Sanchez PI, Bravo R, Lacal JC (1997) Activation of the nuclear factor-kappaB by Rho, CDC42, and Rac-1 proteins. Genes Dev 11:463–475

Peterson J, Zheng Y, Bender L, Myers A, Cerione R, Bender A (1994) Interactions between the bud emergence proteins Bem1p and Bem2p and Rho-type GTPases in yeast. J Cell Biol 127:1395–1406

Ponting CP, Philips C, Davis KE, Blake DJ (1997) PDZ domains: targeting signalling molecules to sub-membranous sites. Bioessays 19:469–479

Prendergast GC, Khosravi FR, Solski PA, Kurzawa H, Lebowitz PF, Der CJ (1995) Critical role of Rho in cell transformation by oncogenic Ras. Oncogene 10:2289–2296

Qiu RG, Chen J, Kirn D, McCormick F, Symons M (1995a) An essential role for Rac in Ras transformation. Nature 374:457–459

Qiu RG, Chen J, McCormick F, Symons M (1995b) A role for Rho in Ras transformation. Proc Natl Acad Sci USA 92:11781–11785

Qiu RG, Abo A, McCormick F, Symons M (1997) Cdc42 regulates anchorage-independent growth and is necessary for Ras transformation. Mol Cell Biol 17:3449–3458

Quilliam LA, Khosravi FR, Huff SY, Der CJ (1995) Guanine nucleotide exchange factors: activators of the Ras superfamily of proteins. Bioessays 17:395–404

Raab M, da SA, Findell PR, Rudd CE (1997) Regulation of Vav-SLP-76 binding by ZAP-70 and its relevance to TCR zeta/CD3 induction of interleukin-2. Immunity 6:155–164

Rameh LE, Arvidsson A, Carraway KLI, Couvillon AD, Rathbun G, Crompton A, VanRentterghem B, Czech MP, Ravichandran KS, Burakoff SJ, Wang D, Cantley LC (1997) A comparative analysis of the phosphoinositide binding specificity of pleckstrin homology domains. J Biol Chem 272:22059–22066

Renshaw MW, Toksoz D, Schwartz MA (1996) Involvement of the small GTPase rho in integrin-mediated activation of mitogen-activated protein kinase. J Biol Chem 271:21691–21694

Ridley AJ, Hall A (1992) The small GTP-binding protein rho regulates the assembly of focal adhesions and actin stress fibers in response to growth factors. Cell 70:389–399

Ridley AJ, Paterson HF, Johnston CL, Diekmann D, Hall A (1992) The small GTP-binding protein rac regulates growth factor-induced membrane ruffling. Cell 70:401–410

Rodriguez VP, Warne PH, Dhand R, Vanhaesebroeck B, Gout I, Fry MJ, Waterfield MD, Downward J (1994) Phosphatidylinositol-3-OH kinase as a direct target of Ras. Nature 370: 527–532

Rodriguez-Viciana P, Warne PH, Khwaja A, Marte BM, Pappin D, Das P, Waterfield MD, Ridley A, Downward J (1997) Role of phosphoinositide 3-OH kinase in cell transformation and control of the actin cytoskeleton by Ras. Cell 89:457–467

Romero F, Fischer S (1996) Structure and function of Vav. Cell Signal 8:545–553

Ron D, Graziani G, Aaronson SA, Eva A (1989) The N-terminal region of proto-dbl down regulates its transforming activity. Oncogene 4:1067–1072

Ron D, Zannini M, Lewis M, Wickner RB, Hunt LT, Graziani G, Tronick SR, Aaronson SA, Eva A (1991) A region of proto-dbl essential for its transforming activity shows sequence similarity to a yeast cell cycle gene, CDC24, and the human breakpoint cluster gene, bcr. New Biol 3:372–379

Schlaepfer DD, Hanks SK, Hunter T, Van der Geer P (1994) Integrin-mediated signal transduction linked to Ras pathway by GRB2 binding to focal adhesion kinase. Nature 372:786–791

Schmidt A, Bickle M, Beck T, Hall MN (1997) The yeast phosphatidylinositol kinase homology TOR2 activates RHO1 and RHO2 via the exchange factor ROM2. Cell 88:531–542

Schuebel KE, Bustelo XR, Nielsen DA, Song BJ, Barbacid M, Goldman D, Lee IJ (1996) Isolation and characterization of murine vav2, a member of the vav family of proto-oncogenes. Oncogene 13:363–371

Schwartz MA, Toksoz D, Khosravi-Far R (1996) Transformation by Rho exchange factor oncogenes is mediated by activation of an integrin-dependent pathway. EMBO J 15:6525–6530

Segal AW, Abo A (1993) The biochemical basis of the NADPH oxidase of phagocytes. Trends Biochem Sci 18:43–47

Shaw G (1996) The pleckstrin homology domain: an intriguing multifunctional protein module. Bioessays 18:35–46

Sheng M (1996) PDZs and receptor/channel clustering: rounding up the latest suspects. Neuron 17:575–578

Shigematsu H, Iwasaki H, Otsuka T, Ohno Y, Arima F, Niho Y (1997) Role of the vav proto-oncogene product (Vav) in erythropoietin-mediated cell proliferation and phosphatidylinositol 3-kinase activity. J Biol Chem 272:14334–14340

Shiozaki H, Oka H, Inoue M, Tamura S, Monden M (1996) E-cadherin mediated adhesion system in cancer cells. Cancer 77:1605–1613

Shou C, Farnsworth CL, Neel BG, Feig LA (1992) Molecular cloning of cDNAs encoding a guanine-nucleotide-releasing factor for Ras p21. Nature 358:351–354

Sloat BF, Adams A, Pringle JR (1981) Roles of the CDC24 gene product in cellular morphogenesis during the *Saccharomyces cerevisiae* cell cycle. J Cell Biol 89:395–405

Sone M, Hoshino M, Suzuki E, Kuroda S, Kaibuchi K, Nakagoshi H, Siago K, Nabeshima Y, Hama C (1997) Still life, a protein in synaptic terminals of *Drosophila* homologous to GDP-GTP exchangers. Science 275:543–547

Song JS, Gomez J, Stancato LF, Rivera J (1996) Association of a p95 Vav-containing signaling complex with the Fc epsilon RI gamma chain in the RBL-2H3 mast cell line. Evidence for a constitutive in vivo association of Vav with Grb2, Raf-1, and ERK2 in an active complex. J Biol Chem 271:26962–26970

Stam JC, Sander EE, Michiels F, Van Leeuwen FN, Kain HET, Van der Kammen RA, Collard JG (1997) Targeting of Tiam1 to the plasma membrane requires the cooperative function of the N-terminal pleckstrin homology domain and an adjacent protein interaction domain. J Biol Chem 272:28447–28454

Stam JC, Michiels F, Van der Kammen RA, Moolenaar WH, Collard JG (1998) Invasion of T-lymphoma cells: cooperation between Rho family GTPases and lysophospholipid receptor signaling. EMBO J 17:4066–4074

Stowers L, Yelon D, Berg LJ, Chant J (1995) Regulation of the polarization of T cells toward antigen-presenting cells by Ras-related GTPase CDC42. Proc Natl Acad Sci USA 92:5027–5031

Symons M (1995) The Rac and Rho pathways as a source of drug targets for Ras-mediated malignancies. Curr Opin Biotech 6:668–674

Takeichi M (1993) Cadherins in cancer: implications for invasion and metastasis. Curr Opin Cell Biol 5:806–811

Tan EC, Leung T, Manser E, Lim L (1993) The human active breakpoint cluster region-related gene encodes a brain protein with homology to guanine nucleotide exchange proteins and GTPase-activating proteins. J Biol Chem 268:27291–27298

Tapon N, Hall A (1997) Rho, Rac and Cdc42 GTPases regulate the organization of the actin cytoskeleton. Curr Opin Cell Biol 9:86–92

Tarakhovsky A, Turner M, Schaal S, Mee PJ, Duddy LP, Rajewsky K, Tybulewicz VL (1995) Defective antigen receptor-mediated proliferation of B and T cells in the absence of Vav. Nature 374:467–470

Teramoto H, Crespo P, Coso OA, Igishi T, Xu N, Gutkind JS (1996) The small GTP-binding protein rho activates c-Jun N-terminal kinases/stress-activated protein kinases in human kidney 293T cells. Evidence for a Pak-independent signaling pathway. J Biol Chem 271:25731–25734

Toksoz D, Williams DA (1994) Novel human oncogene lbc detected by transfection with distinct homology regions to signal transduction products. Oncogene 9:621–628

Tuosto L, Michel F, Acuto O (1996) p95vav associates with tyrosine-phosphorylated SLP-76 in antigen-stimulated T cells. J Exp Med 184:1161–1166

Uddin S, Sweet M, Colamonici OR, Krolewski JJ, Platanias LC (1997) The vav proto-oncogene product (p95vav) interacts with the Tyk-2 protein tyrosine kinase. FEBS Lett 403:31–34

Van Aelst L, D'Souza-Schorey C (1997) Rho GTPases and signalling networks. Genes Dev 11:2295–2322

Van Leeuwen FN, van der Kammen RA, Habets GG, Collard JG (1995) Oncogenic activity of Tiam1 and Rac1 in NIH3T3 cells. Oncogene 11:2215–2221

Van Leeuwen FN, Kain HET, Van der Kammen RA, Michiels F, Kranenburg O, Collard JG (1997) Tiam1 affects neuronal morphology; opposing roles for Rac and Rho. J Cell Biol 139:797–807

Varner JA, Cheresh DA (1996) Integrins and cancer. Curr Opin Cell Biol 8:724–730

Vleminckx K, Vakaet LJ, Mareel M, Fiers W, van RF (1991) Genetic manipulation of E-cadherin expression by epithelial tumor cells reveals an invasion suppressor role. Cell 66:107–119

Wennstrom S, Hawkins P, Cooke F, Hara K, Yonezawa K, Kasuga M, Jackson T, Claesson-Welsh L, Stephens L (1994) Activation of phosphoinositide 3-kinase is required for PDGF-stimulated membrane ruffling. Curr Biol 4:385–393

Westwick JK, Lambert QT, Clark GJ, Symons M, Van Aelst L, Pestell RG, Der CJ (1997) Rac regulation of transformation, gene expression, and actin organization by multiple, PAK-independent pathways. Mol Cell Biol 17:1324–1335

Whitehead I, Kirk H, Kay R (1995a) Retroviral transduction and oncogenic selection of a cDNA encoding Dbs, a homolog of the Dbl guanine nucleotide exchange factor. Oncogene 10:713–721

Whitehead I, Kirk H, Tognon C, Trigo-Gonzalez G, Kay R (1995b) Expression cloning of lfc, a novel oncogene with structural similarities to guanine nucleotide exchange factors and to the regulatory region of protein kinase C. J Biol Chem 270:18388–18395

Whitehead IP, Khosravi-Far R, Kirk H, Trigo-Gonzalez G, Der CJ Kay R (1996) Expression cloning of lsc, a novel oncogene with structural similarities to the Dbl family of guanine nucleotide exchange factors. J Biol Chem 271:18643–18650

Whitehead IP, Campbell S, Rossman KL, Der CJ (1997) Dbl family proteins. Biochim Biophys Acta 1332:F1–F23

Witte ON (1988) Closely related BCR/ABL oncogenes are associated with the distinctive clinical biologies of Philadelphia chromosome positive chronic myelogenous and acute lymphocytic leukemia. Curr Top Microbiol Immunol 141:42–49

Wu J, Motto DG, Koretzky GA, Weiss A (1996) Vav and SLP-76 interact and functionally cooperate in IL-2 gene activation. Immunity 4:593–602

Wu J, Zhao Q, Kurosaki T, Weiss A (1997) The Vav binding site (Y315) in ZAP-70 is critical for antigen receptor-mediated signal transduction. J Exp Med 185:1877–1882

Yaku H, Sasaki T, Takai Y (1994) The Dbl oncogene product as a GDP/GTP exchange protein for the Rho family: its properties in comparison with those of Smg GDS. Biochem Biophys Res Commun 198:811–817

Zhang R, Tsai FY, Orkin SH (1994) Hematopoietic development of vav-/- mouse embryonic stem cells. Proc Natl Acad Sci USA 91:12755–12759

Zhang R, Alt FW, Davidson L, Orkin SH, Swat W (1995a) Defective signalling through the T- and B-cell antigen receptors in lymphoid cells lacking the vav proto-oncogene. Nature 374:470–473

Zhang S, Han J, Sells MA, Chernoff J, Knaus UG, Ulevitch RJ, Bokoch GM (1995b) Rho family GTPases regulate p38 mitogen-activated protein kinase through the downstream mediator Pak1. J Biol Chem 270:23934–23936

Zhao ZS, T. L, Manser E, Lim L (1995) Pheromone signalling in *Saccharomyces cerevisiae* requires the small GTP-binding protein Cdc42p and its activator CDC24. Mol Cell Biol 15:5246–5257

Zheng Y, Cerione R, Bender A (1994) Control of the yeast bud-site assembly GTPase Cdc42. Catalysis of guanine nucleotide exchange by Cdc24 and stimulation of GTPase activity by Bem3. J Biol Chem 269:2369–2372

Zheng Y, Bender A, Cerione RA (1995a) Interactions among proteins involved in bud-site selection and bud-site assembly in *Saccharomyces cerevisiae*. J Biol Chem 270:626–630

Zheng Y, Olson MF, Hall A, Cerione RA, Toksoz D (1995b) Direct involvement of the small GTP-binding protein Rho in lbc oncogene function. J Biol Chem 270:9031–9034

Zheng Y, Fischer DJ, Santos MF, Tigyi G, Pasteris NG, Gorski JL, Xu Y (1996a) The faciogenital dysplasia gene product FGD1 functions as a Cdc42Hs-specific guanine-nucleotide exchange factor. J Biol Chem 271:33169–33172

Zheng Y, Zangrilli D, Cerione RA, Eva A (1996b) The pleckstrin homology domain mediates transformation by oncogenic dbl through specific intracellular targeting. J Biol Chem 271:19017–19020

Zmuidzinas A, Fischer KD, Lira SA, Forrester L, Bryant S, Bernstein A, Barbacid M (1995) The vav proto-oncogene is required early in embryogenesis but not for hematopoietic development in vitro. EMBO J 14:1–11

RhoGAPs and RhoGDIs, (His)stories of Two Families

Gérard Zalcman[1], Olivier Dorseuil[2], Juan Antonio Garcia-Ranea[3], Gérard Gacon[2], and Jacques Camonis[1]

As all GTPases, Rho proteins are found bound either to GDP or GTP. In the former state, they are supposedly inactive, in the latter state, they take an "active" conformation that allows them to interact with their so-called effectors.

Identifying the molecular partners of each Rho GTPase and unravelling the molecular mechanisms controlling their activity are a major challenge to understand the various cellular functions under their control. The first part of this challenge was and remains very fruitful: Rho proteins are the GTPases that have the largest number of partners identified, partners of the GDP-bound state as well as partners of the GTP-bound state. However, the question of the respective biological function of most of these partners has not yet received an answer.

In this chapter, we will review two families of Rho partners. The GTPases activating proteins (GAP) are in charge of accelerating the inactivation of Rho by accelerating their intrinsic GTPase activity. A peculiar aspect of Rho-GAPs is whether Rho-GAPs are solely negative regulators or can be Rho effectors – a question reminiscent of ancient questions about Ras-GAPs. Two cases will illustrate a negative answer to this question.

RhoGDI are proteins largely abundant in cells that sequestrate Rho from membranes by solubilizing them in the cytosol. Thus Rho exist in three states: a transient GTP-bound state where it meets effectors and/or GAPs, a transient GDP-bound state and a GDP-bound GDI-complexed state. It is unclear whether a GTP-bound GDI-complexed state exists in vivo. Most of Rho is found complexed to a GDI and the accepted paradigm was that GDIs keep Rhos out of reach of their site of action by solubilizing them. The finding that the third Rho-GDI identified is not in the cytosol but rather bound to what could be the cytoskeleton makes scientists just more confused.

[1] Institut Curie, INSERM U-248, 26 rue d'Ulm, 75248 Paris cedex 05, France
[2] INSERM U-257, I.C.G.M., 26 rue du Fbg Saint Jacques, 75014 Paris, France
[3] Protein Design Group, CNB-CSIC, Cantoblanco, Madrid, Spain

Progress in Molecular and Subcellular Biology, Vol. 22
Ph. Jeanteur (Ed.)

1 RhoGDI (GDP Dissociation Inhibitors): Who Needs Them?

The interconversion of Rho proteins between their GTP-bound active conformations and their GDP-bound conformations is tightly regulated by three classes of proteins. Guanine nucleotide exchange factors (GEFs), most of them sharing the Dbl homology (DH) domain, stimulate the replacement of GDP by GTP and upregulate Rho proteins [1]. GTPase-activating proteins (GAPs) accelerate the intrinsic GTP hydrolytic activity, and thereby downregulate Rho proteins [2]. The guanine nucleotide dissociation inhibitors (GDIs) inhibiting the GDP to GTP exchange reaction are often considered as downregulators of Rho proteins [3].

1.1 A Family of RhoGDIs

Three GDIs have so far been reported to act on Rho proteins:

1. The ubiquitously expressed Rho-GDI (renamed later Rho-GDI-1 and more recently RhoGDIα) was first purified from bovine brain in 1990 [3, 4]. The cDNA and genomic human sequence organization were characterized later[1]. The RhoGDI-1 gene exhibits four exons and four introns (Mulheron et al., unpubl.). The human RhoGDI-1 locus has been assigned to 17q25.3-25-3.
2. Human D4/Ly-GDI[2], also named Rho-GDI-2 and Rho-GDIβ, was characterized as a RhoGDI specifically expressed in hematopoietic tissues [5, 6], whereas a murine homologue was shown to exhibit the same tissue specificity [2].
3. More recently, we have identified a third RhoGDI acting upon RhoB and RhoG but not Rac that we named RhoGDI-3[3] [8]. Subsequently, a human homologue of RhoGDI-3, RhoGDIγ[4], was characterized [9].

At the nucleic level, human RhoGDI-1 cDNA shares 68% and 66% identity with RhoGDI-2 (LyGDI) and RhoGDI-3 (RhoGDIγ), respectively. At the protein level, the three proteins share roughly 50% identical amino acids. The main sequence difference between them is found in the N-terminus where RhoGDI-3 has an amphipatic leucine-rich extension (Fig. 1). A yeast RhoGDI protein has also been purified and was shown to be 36% identical to human RhoGDI-1 and RhoGDI-2, demonstrating an evolutionary conservation of this family [10].

[1] Maeda et al. *Accession No. D13989*, Mulheron et al. *Accession No. X63863.*
[2] Lelias et al. *Accession No. L20688*, Scherle et al. *Accession No. LO7916.*
[3] Zalcman et al. *Accession No. L42463.*
[4] Adra et al. *Accession No. U82532.*

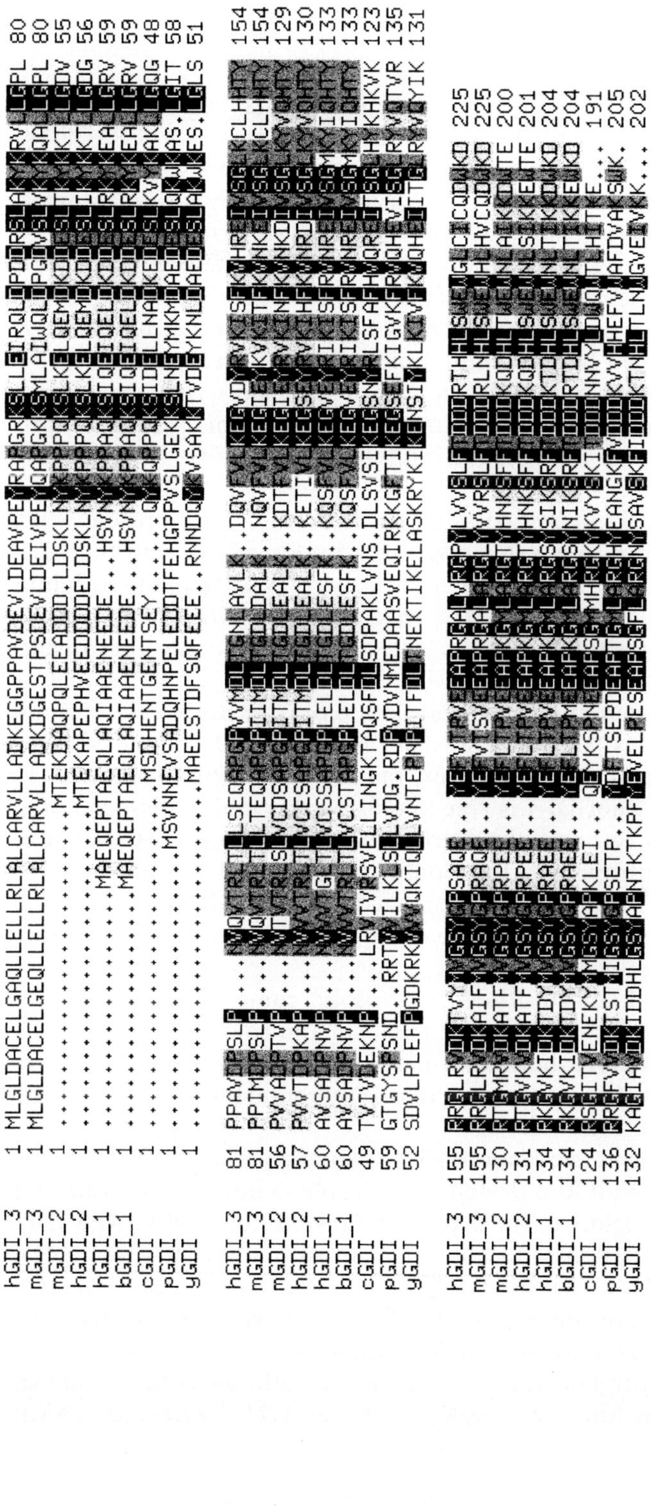

Fig. 1. Alignment of mammalian, worm and yeast GDIs (*h* human; *m* mouse; *b* bovine; *c C. elegans*; *p* pig; *y* yeast)

1.2 Biochemical Properties

Human RhoGDI-1 is composed of 204 amino acids with a theoretical Mr of 23.4kDa but it migrates as a 27–28kDa protein on SDS-PAGE [4]. Evidence for the existence of alternative splice transcripts came from the identification of two related cDNAs in a cDNA library of transformed human amnion cells [11]. These two cDNAs probably arose by alternative splicing since they diverge from RhoGDI-1 at the amino terminus, with a sequence identical to the reported sequence of the second intron of genomic RhoGDI-1. At least one of these transcripts has been detected by Northern blot in SV40-transformed human embryonic lung fibroblasts, using a probe specific for the intron sequence [11]. The role of such transcripts encompassing different N-terminal peptides remains unclear, since this part of the protein was recently shown to mediate the inhibitory effect on GDP dissociation [12].

RhoGDI-1 was initially identified as a bovine brain cytosol protein which forms a complex in vitro in a 1:1 molar ratio, preferentially with the GDP-bound form of RhoB. Its first characterised property was its ability to inhibit the dissociation of GDP from RhoB [4], hence the name GDI, which might be confusing about the function since, in fact, many GTPase effectors display some GDI activity when they interact with their GTPase [12]. Subsequently, RhoGDI-1 was also shown to inhibit dissociation of GDP RhoA [4], Rac1 and Rac2 [14] and CDC42 Hs [15, 16]. RhoGDI-1 binds the GDP bound forms of these GTPases with a tenfold higher affinity than for the GTP-bound forms [17, 15, 18]. Consequently, RhoGDI-1 is able to only weakly inhibit dissociation of GTPγS from CDC42Hs, Rac and Rho. Furthermore, the RhoA-RhoGDI-1 complex purified from neutrophile cytosol only contains GDP-bound RhoA [19, 20]. This RhoA-RhoGDI complex is stabilized by RhoGDI phosphorylation [21], possibly on a residue located on the inner wall of the hydrophobic pocket interacting with the isoprene [12].

The existence of a RhoGDP-RhoGDI cytosolic complex fits well with a model where Rho proteins cycle between cytosol and some membranes [22–24] and where the active GTP-bound form of Rho is membrane-localized, and interacts there with its effectors. The inactive GDP-bound form appears mainly cytosolic, complexed to RhoGDI.

Rho signalling is terminated upon GTP spontaneous or GAP-stimulated hydrolysis: the GDP-bound Rho protein is then able to interact with RhoGDI-1 and this interaction results in the cytosolic translocation of Rho-RhoGDI-1 complexes [23, 15]. Consistent with this model, in vitro, RhoGDI-1 is unable to extract from membranes GTP-bound Rho that was exogenously bound to these membranes [23], presumably because GTP-membrane Rho is engaged in an effector interaction and not accessible to RhoGDI-1.

However, in NK lymphocytes, an alternative pathway either to down-regulate or re-localise RhoA signalling, bypasses GTP hydrolysis: cAMP-dependent phosphorylation of serine 188 of membrane-localized GTP-bound RhoA enables phosphorylated GTP-RhoA to be removed from membranes by RhoGDI-1. It seems that RhoA phosphorylation favours the naturally weak interaction of RhoGDI-1 with RhoA-GTP [24]. The half-life of this phosphorylated RhoA-GTP::RhoGDI-1 complex in cytosol is not known.

As the Rac-RhoGDI-1 complex purified from neutrophil was shown to be able, in vitro, to stimulate NAPDH oxidase activity [22], it seems likely that Rac in this complex is in the active GTP-bound conformation.

Post-translational isoprenylation of the C-terminal CAAX region of Rho proteins was shown to be absolutely necessary for the high-affinity interaction with RhoGDI-1 [14, 15, 25, 26], indicating that RhoGDI-1 could bind to the carboxy-terminal isoprene group of these proteins. Indeed, Rho proteins are mainly post-translationally geranyl-geranylated, with the notable exception of RhoB proteins that are found either farnesylated or geranyl-geranylated, and then additionally palmitoyled. However, a low affinity interaction has been detected in the two-hybrid system between non isoprenylated RhoG (deleted in its CAAX sequence) and RhoGDI-2 (P. Fort, pers. comm.), indicating that a low-affinity GDI::Rho interaction could involve other regions than the C-terminal isoprene.

Functional properties of RhoGDI result from the following biochemical properties:

1. The strong interaction between RhoGDI-1 and its GDP-bound substrate does not only result in GDP dissociation inhibition, but also prevents the subsequent binding of GTP, maintaining the GTPase in its inactive stage [4].
2. Because of the strength of this interaction, when RhoA-GDP is complexed to RhoGDI-1, at least in vitro, the GDP dissociation catalyzed by purified smgGDS or Dbl is inhibited [27, 28]. In other words, Rho-exchange factors are unable, by themselves, to activate the Rho proteins complexed with RhoGDI-1.
3. In the same way, spontaneous or GAP-stimulated GTP hydrolysis (by purified RhoGAP or Bcr), are blocked by in vitro prebinding of RhoGDI-1 to Rac- or CDC42Hs-GTP [15, 16, 34], leading to the hypothesis that RhoGDI-1 could serve as a chaperone-like protein, to transport activated Rac through the cytosol to its site of action [25]. However, the physiological relevance of this activity stills awaits further in vivo demonstration.
4. Rho-GDP inactivation by *EDIN* exoenzyme-mediated ADP-ribosylation was found to be inhibited by prebinding to RhoGDI-1 [27], presumably because RhoGDI-1 complexed to RhoA sterically hinders ADP-ribosyl-transferase interaction with the effector site. On the other hand, in vitro ADP-ribosylation of RhoA has no effect on its ability to bind RhoGDI-1 [25].

1.3
From Structure to Function: Insights Deduced from RhoGDI-1 3-D Structure

The spatial structure of RhoGDI-1 was recently determined by a combination of X-ray crystallography and NMR spectroscopy [12, 30] leading to the following picture. The 60 N-terminal residues of RhoGDI-1 are flexible and unstructured in solution. Deletion of the first 58 amino acids by trypsin proteolysis gives a stable protease-resistant and folded domain of about 16–17 kDa. The 2.5-Å crystal structure of this folded domain revealed that it consists of nine β strands in two antiparallel sheets and one short 3_{10} helix. Seven β strands in two sheets form the top and the bottom wall of a β sandwich structure, which is significantly homologous to an immunoglobulin-like fold. Several hydrophobic residues line the inner walls of this actual pocket, these residues being conserved in RhoGDI-1, 2 and 3. This pocket is considered to interact with the isoprene group of Rho proteins, since interaction of RhoGDI-1 with a farnesylated peptide causes resonance changes in residues in and near this pocket. Deletions of increasing size in the N-terminal part of RhoGDI-1 lowers affinity of RhoGDI-1 to Cdc42, but the folded domain by itself still binds the GTPase with a nanomolar affinity. On the other hand, the N-terminal region by itself shows a low affinity (within a millimolar range) toward Cdc42Hs. However, both domains are required to partition Cdc42 from membranes since combined interactions of the RhoGDI N- and C-termini provide sufficient force to extract GTPases from membranes [12].

N-terminal deletants of RhoGDI-1 were also tested for GDI activity upon Cdc42: only full-length RhoGDI-1 and a 22 amino acid N-terminal deletant were able to inhibit GDP dissociation, showing that the residues 23–42 are required for inhibitory activity. It is proposed that the C-terminal folded domain contributes binding energy whereas the N-terminal peptide mediates the inhibitory effects. Since RhoGDIs do not bind other members of the Ras superfamily that are also isoprenylated, other protein-protein contacts seem to be necessary to get a specific interaction. This is indeed the case, since interaction with non-isoprenylated Cdc42 or Rac, although weak, still resulted in additional regions of resonance change in the N-terminal region of RhoGDI-1. Thus, the model is that the GTPase isoprene group is buried inside the C-terminal RhoGDI-1 hydrophobic pocket and this interaction directs the GDN N-terminal extremity to also contact the bound GTPase; this latter contact impairs GDP dissociation. Besides the C-terminal isoprene, an additional Rho region interacting with RhoGDI-1 could be the Rho effector domain. Indeed, the inhibitory RhoGDI peptide contains the $_{32}$-Gln-Lys-Ser-Ile-$_{35}$ sequence that is reminiscent of the Ras-binding region of Raf (Gln-Arg-Thr-Val) [30]. Alternatively, the Rho insert region, which is well-exposed at the surface of Rho molecules, could represent a good candidate for an additional RhoGDI interacting region. In fact, when the Rho insert is removed from CDC42HS, RhoGDI was still able to extract it from membranes, but was unable to inhibit GDP dissociation [31]. So the Rho insert isn't involved in Rho::RhoGDI interaction per se, but it is required for the execution of some of the RhoGDI effects upon Rho.

1.4
In Vivo Functional RhoGDI as a Specific Tool for Gaining Insight into Rho Function

Recombinant GST-RhoGDI-1 produced in E. coli has the same inhibitory activity in an in vitro GDP-GTP exchange reaction with RhoA or Rac as does RhoGDI-1 purified from bovine brain [25, 27]. GST-RhoGDI-1 fusion proteins were therefore micro-injected into cells as a means of inhibiting cellular endogenous Rho proteins, and to identify the physiological functions of Rho proteins. Thus, RhoGDI-1 was successively shown:

1. To inhibit Swiss 3T3 cell spontaneous motility [32] or keratinocytes HGF-induced motility [33]. This inhibition was shown to be mediated by RhoA inhibition, since micro-injection of ADP-ribosyl-transferase C3 in cells led to the same effect by specifically inactivating endogenous Rho proteins. Moreover, co-microinjection of RhoGDI-1 and RhoA-GTP but not Rac1-GTP led to normal cell motility, overexpression of RhoA-GTP overwhelming the RhoGDI-induced trapping of RhoA-GDP.
2. To inhibit sperm-induced cytoplasmic division of a *Xenopus* embryo and to impair the actin-myosin furrow formation. The involvement of RhoA in these effects was also established by the use of C3 exoenzyme or by the co-microinjection of RhoA-GTP [34].
3. To inhibit insulin-, HGF- or TPA-induced membrane ruffling in KB cells. Insulin-induced ruffling was shown to involve Rac1, whereas HGF- or TPA-induced ruffling was shown to involve RhoA [35].
4. To inhibit translocation to membrane ruffling areas of overexpressed myc-tagged RhoA, to inhibit cell-cell adhesion and to inhibit cleavage furrow during cell cytokinesis [36].
5. To inhibit exocytosis in mast cells [37].

All these RhoGDI inhibitory effects confirmed the role of Rho proteins in regulating various actin-myosin dependent cell functions. Moreover, overexpression of RhoGDI-1 induced cell rounding-up, loss of cell-cell contacts, and detachment of cells from culture dishes [11] – as does inactivation of Rho proteins by C3 botulinum toxin – confirming that inhibition of Rho function by RhoGDI could impair actin-regulated cell adhesion.

1.5
Cellular Localisation

Subcellular fractionation led to the detection of RhoGDI-1 in the cytosolic fraction, complexed with Rho [38, 20, 4]. Furthermore, HGF- or TPA-induced translocation of RhoA from cytosol to membrane ruffling area were inhibited by microinjection of RhoGDI-1, arguing for the existence of a cytosolic RhoA-RhoGDI-1 complex where RhoA is impaired in its interaction with its membrane-located effector(s) [36].

A more complex view of the physiological function and cellular localisation of RhoGDI-1 came recently with the finding that RhoGDI-1 co-immunoprecipitates with CD44/moesin complexes in BHK cells [39]. The ERM proteins (ezrin, radixin, moesin) are the cross-linkers between filamentous actin and the plasma membrane. They interact with the cytoplasmic domain of a cell-surface integral membrane glycoprotein, CD44, in the presence of phospho-inositides such as 4,5-PIP_2. The occurrence, in vivo, of RhoGDI-1 in the CD44/moesin complex would indicate that a portion of RhoGDI-1 cellular pool could be recruited to the sub-membrane area, in association with RhoA. The direct interaction between ERM proteins and RhoGDI was recently shown to release Rho proteins from the inhibitory effect of RhoGDI [4], authorising the GDP/GTP exchange reaction catalysed by Dbl-like exchange factors. RhoGTP, in turn, would promote the enhancement of the CD44/ERM complex formation detected in the insoluble fraction of BHK cell lysates. One signalling pathway linking Rho to the actin/ERM/CD44 complex could involve the Rho-induced PIP-5 kinase up-regulation, leading to the elevation of 4,5-PIP_2 in the plasma membrane. Alternatively, Rho protein could activate ser/thr kinases to phosphorylate ERM proteins, inducing their membrane association. Anyway, a Rho-induced conformational change in moesin could contribute to its role in stress fibre formation and cortical actin polymerization [41].

1.6
A Hematopoietic RhoGDI, RhoGDI-2 (Ly/D4-RhoGDI) Could Play a Role During Apoptosis

RhoGDI-2 encodes a protein of 201 amino acids that was found to interact with RhoA [33, 36], CDC42Hs, Rac1 [7, 42], and RhoG (S. Doucet-Brutin and P. Fort, pers. comm). However, RhoGDI-2 was 10–20 times less potent than RhoGDI-1 in inhibiting a GDP/GTP exchange reaction tested with CDC42Hs, and this accounts for a weaker binding affinity [7, 42]. A single Asp $\rightarrow$ Ile change at position 174 near the carboxy terminus of RhoGDI-2 (corresponding to Ile 177 in RhoGDI-1) increases RhoGDI-2 activity upon CDC42Hs to levels closer but still not identical to RhoGDI-1 activity [42]. Ile 177 of RhoGDI-1 lines one side of the hydrophobic pocket, able to interact with Rho C-terminal isoprene group [3, 44]. However, an increase in binding affinity is not enough to account for differences in inhibitory activity on Rho proteins sharing the same post-translational modification. Moreover, the ability of RhoGDI-2 to interact with non-isoprenylated RhoG protein (P. Fort, pers. comm., and personal data) confirms that other protein-protein contacts (identified as

N-terminal residues 23–63 in RhoGDI-1), although weak, are necessary for specific interactions.

The most striking specificity of RhoGDI-2 relies on its regulation during hematopoietic differentiation. When embryonic stem (ES) cells, cultured in semisolid conditions, differentiate in erythroid, granulocytic or megakaryocytic lineage colonies, RhoGDI-2 transcript is first detected on day 10 of differentiation, with an increased expression by day 12. In contrast, RhoGDI-2 transcript is downregulated in hematopoietic cell lines induced to undergo terminal differentiation into macrophages by PMA (U937), whereas HL60 cells induced to differentiate into neutrophils by DMSO do not show any change in RhoGDI-2 mRNA content [5].

At the proteic level, RhoGDI-2 was found to be phosphorylated on threonine, 5 min after stimulation of Jurkat T cells with PMA which mimics activation of T cells through the antigen receptor [6].

When Jurkat T cells were treated with the agonist anti-Fas monoclonal antibody that activates Fas receptor, they underwent apoptosis. Overexpression of IL-1-β-converting enzyme (ICE)-like cysteine proteases results in such apoptosis. During purification of ICE activity, two polypeptides of 19 and 23 kDa co-purified with the ICE activity [43, 44]. N-terminal microsequencing revealed the 19 and 23 kDa proteins were truncated forms of RhoGDI-2 beginning at residues 56 and 19, respectively. Inspection of the protein sequence revealed that RhoGDI-2 (but neither RhoGDI-1 nor RhoGDI-3) contains two potential ICE protease cleavage sites, DELD^{19} S and LLGD^{55} G. The existence of an N-terminal deleted RhoGDI-2 of 19 kDa (IEF 1120) had been previously reported in human keratinocytes indicating that proteolytic cleavage of RhoGDI-2 is not specific to hematopoietic tissues [11].

ICE itself was shown to cleave RhoGDI-2 at asparagine 55 [43], whereas the ICE-related protease CPP32 (apopain) was found to efficiently cleave RhoGDI-2 at asparagine 19 [44]. The larger proteolytic fragment of 23 kDa could retain some inhibitory activity since the 22–63 region is unaffected, but the ICE cleavage product of 19 kDa was shown to be biochemically inactive [43, 12]. Thus, proteolytic cleavage of flexible RhoGDI N-termini could represent a general mechanism for regulating their activity, leading to fragments with different inhibitory properties and/or with dominant negative properties. Whether such cleavages occur in vivo for RhoGDI-1 or RhoGDI-3 remains to be established.

ICE also plays an important pro-inflammatory role by converting pro-IL-1β to its mature form, which in turn induces leukocyte accumulation and proliferation and prostaglandin synthesis at the inflammatory sites. Therefore, a role in the inflammatory reaction for Rho activation, by proteolytic release of the RhoGDI-2 inhibitory effect, has been speculated, since it has been suggested that release of Rac from RhoGDI-2, could induce NAPDH oxidase activation, with production of toxic oxygen [43].

As RhoGDI-1 is expressed in all tissues where RhoGDI-2 has been detected, their respective functions in the same cell remains elusive. Targeted disruption of RhoGDI-2 did not help to answer this question since embryonic stem cells with disruption of both alleles of the RhoGDI-2 gene undergo normal hematopoietic differentiation and exhibit normal phagocytosis [45]. Only a slight reduction of the capacity of RhoGDI-2$^{-/-}$ macrophages to produce superoxide was observed, somehow involving RhoGDI-2 in the modulation of NAPDH oxidase.

1.7 A Third RhoGDI Exhibits Catalytic Specificity and Cytoskeleton Localisation

In a two-hybrid screen planned to identify proteins interacting with RhoB, we have identified a mouse cDNA encoding for a new RhoGDI that we named RhoGDI-3 [8]. Murine RhoGDI-3 differs drastically from RhoGDI-1 and RhoGDI-2 at its N-terminus in that it contains a 21 amino acid extension compared with RhoGDI-1 and -2. This neo N-terminus is predicted to be structured in three helix segments, highly amphipatic. The human homologue also encodes such an amphipatic N-terminal extension [9]. RhoGDI-3 mRNA expression in murine tissues was found to be restricted to brain, lung and testis, contrasting with the ubiquitous RhoGDI-1 and the hematopoietic RhoGDI-2. Two human RhoGDI-3 mRNAs were highly expressed in brain and pancreas, and detected at only very low levels in testis and small intestine [9]. Human RhoGDI-3 gene was localized on chromosome 16p13.3 (locus HS14G4) and its genomic structure is made of 6 exons and 5 introns, without a TATA box or CAAT box in the promoter region (personal data).

In the two-hybrid system, RhoGDI-3 interacts with full-length RhoB and RhoG; these interactions occur as well with the wild-type allele as with the N19 and V14 alleles, suggesting that Rho-GDI-3 is able to interact with Rho bound to GTP as well as GDP. RhoGDI-3 was unable to interact with Rho with deleted C-terminal CAAX boxes. Thus this GDI-3 is a bona fide GDI that interacts with post-translationally isoprenylated Rho proteins. On the other hand, RhoGDI-3 was unable to interact with Rac1 or Rac2 in the two-hybrid system. Human RhoGDI-3 was shown to interact in vitro with RhoA and CDC42Hs, but still unable to be retained by GST-isoprenylated Rac fusion proteins [7]. We showed an in vitro inhibitory effect of RhoGDI-3 on the dissociation of GDP and GTP RhoB protein overexpressed in HeLa cells and partially purified by immunoprecipitation. Moreover, RhoGDI-3, as RhoGDI-1, was able to release the GDP-bound form (but not the GTP-bound form) of RhoB from crude membranes of cells overexpressing RhoB. RhoGDI-3 also inhibits the dissociation of GDP from CDC42 expressed in insect cells, but with a 20–30 times weaker GDI activity than RhoGDI-1 [7]. With antibodies raised against a conserved region of RhoGDI-1, -2, -3, we were able to detect exogenously expressed RhoGDI-3 (migrating at a higher apparent molecular weight than

RhoGDI-1) in the insoluble fraction of HeLa cells. RhoGDI-3 was not extracted from this particulate fraction by detergents (CHAPS, TritonX100), suggesting its association with a cytoskeletal subcellular fraction. Using immunofluorescence, exogenously expressed Rho-GDI-3 is seen clustered around the nucleus, maybe in a vesicular intracellular compartment [7]. Thus, biochemical, fractionation and immunolocalisation data led to the hypothesis that, in vivo, RhoGDI-3 preferentially acts on RhoB which was previously localised in the very same vesicular perinuclear compartment. Lastly, overexpression of human RhoGDI-3 in hamster cells led to the loss of actin stress fibres with rounding up of the cells, a feature also encountered after Rho inactivation by the C3 ADP-ribosyl transferase. Once again, the role of the N-terminal extension of RhoGDI-3 appears to be of major importance. Cerione et al. mentioned that an N-terminal peptide is essential for perinuclear localisation [3], and we have found that, in the two-hybrid system, an N-terminal deleted RhoGDI-3 loses its capacity to interact with RhoG, but retains its ability to interact with RhoB.

2
GTPase Activating Proteins (GAPs) for Rho-Like GTPases: A Catalytic Module Found Within a Plethora of Multifunctional Proteins

Following biochemical identification and cloning of p50-RhoGAP, the first GAP for Rho proteins, a large family of proteins harbouring GAP activity for Rho GTPases (Rho-GAPs) has been discovered in the past few years. More than 25 members have already been described from yeast to human, and constitute a large class of Rho-regulatory/interacting factors (Table 1). All these Rho-GAPs, each with its own specificity for peculiar GTPases, share a conserved RhoGAP domain, the "GAP catalytic domain" which mediates binding and catalytic activity with Rho proteins (Fig. 2). In addition to their RhoGAP catalytic domain, most of these Rho-GAPs bear other structures involved in signalling, including Src-homology 2 or 3 (SH2 or SH3), pleckstrin-homology (PH), Dbl-homology (DH) and ser/thr kinase domains, as well as SH3-binding sites or zinc finger-like sequences.

In contrast to large heterotrimeric G proteins, Ras-like GTPases have, in general, a low intrinsic activity for GTP hydrolysis in vitro and their in vivo conversion into the GDP-bound form depends largely on GAP activity. This suggests that the biological function of Rho-GAPs is to down-regulate Rho functions. This was shown to be true in different cellular studies where various RhoGAP domains were able to inhibit Rho-dependent events. However, Rho-GAPs interact with the GTP-bound form of Rho GTPases and Rho-GAPs possess a variety of structural domains not required for their GAP activity, but possibly allowing them to interact with other factors. Thus RhoGAPs might also be effector molecules for Rho and/or these multifunctional proteins may

Table 1. Proteins containing a Rho-GAP domain, and some of their features

Rho-GAP proteins	M.W. (predicted, in kDa)	In vitro GAP specificity for Rho	Rac	Cdc42	Organism	Tissue distribution
p50-RhoGAP	50	+	+	++	Mammalian	Ubiquitous
Bcr	142	−	+	+	"	Brain, hematopoietic cells
Abr	92/97	−	+	+	"	Mainly brain
n-Chimaerin (a1)	38	−	++	+	"	Brain
α2-Chimaerin	50	−	++	+	"	Brain, testes
β1-Chimaerin	34	−	+	−	"	Testes
β2-Chimaerin	54	−	+	−	"	Brain
p85a, p85b	83 and 81	−	−	−	"	Ubiquitous
p190	171	++	+	+	"	Ubiquitous
3BP-1	65	−	+	+	"	Mainly spleen and brain
RLIP76	76	−	+	+	"	Ubiquitous
Myr5	225	++	+	++	"	Ubiquitous
Myosin-IXb	229	?	?	?	"	Mainly myeloid cells
p122-Arp	122	+	−	−	"	?
p190b	172	++	++	+	"	Ubiquitous
p115	115	?	?	?	"	Hematopoietic cells
p58-Mgc	58	+	++	++	"	Mainly testes
Graf	64	+	−	+	Chicken	Ubiquitous
RnRacGap (rotund)	41	?	?	?	*D. melanogaster*	Testes, imaginal discs
Ce-GAP	155	+	+	+	*C. elegans*	?
DdRacGAP	150	/	+	/	*D. discoideum*	/
Bem2	245	+	/	−	*S. cerevisiae*	/
Bem3	125	+	/	++	"	/
Dbm1	112	?	/	?	"	/
LRG1	114	?	/	?	"	/
RGA1	112	?	/	?	"	/

? = not determined; / = does not apply; + and − = tested as positive and negative, respectively.

be required to coordinate signalling downstream of Rho-GTPases through their other molecular partners.

2.1 GAP Activity and Identification of p50-RhoGAP

The existence of GAP activity was shown by McCormick and Trahey after an observation that Ras ability to hydrolyse bound GTP was greatly enhanced in presence of cell extracts [46]. Following purification of the activity and cloning of the cDNA encoding p120-RasGAP [47], a few years later Hall and collaborators identified the first Rho-GAP, as a 29 kDa protein, using cytoplasmic cell extracts and recombinant Rho protein [48, 49]. Partial sequencing of the purified protein from human spleen tissue allowed cDNA cloning and subsequent demonstration that p29 resulted from a proteolytic cleavage of a full-length molecule of 50 kDa, p50-RhoGAP [50]. Further analysis also showed that p50-RhoGAP was identical to a GAP protein purified either from human platelets

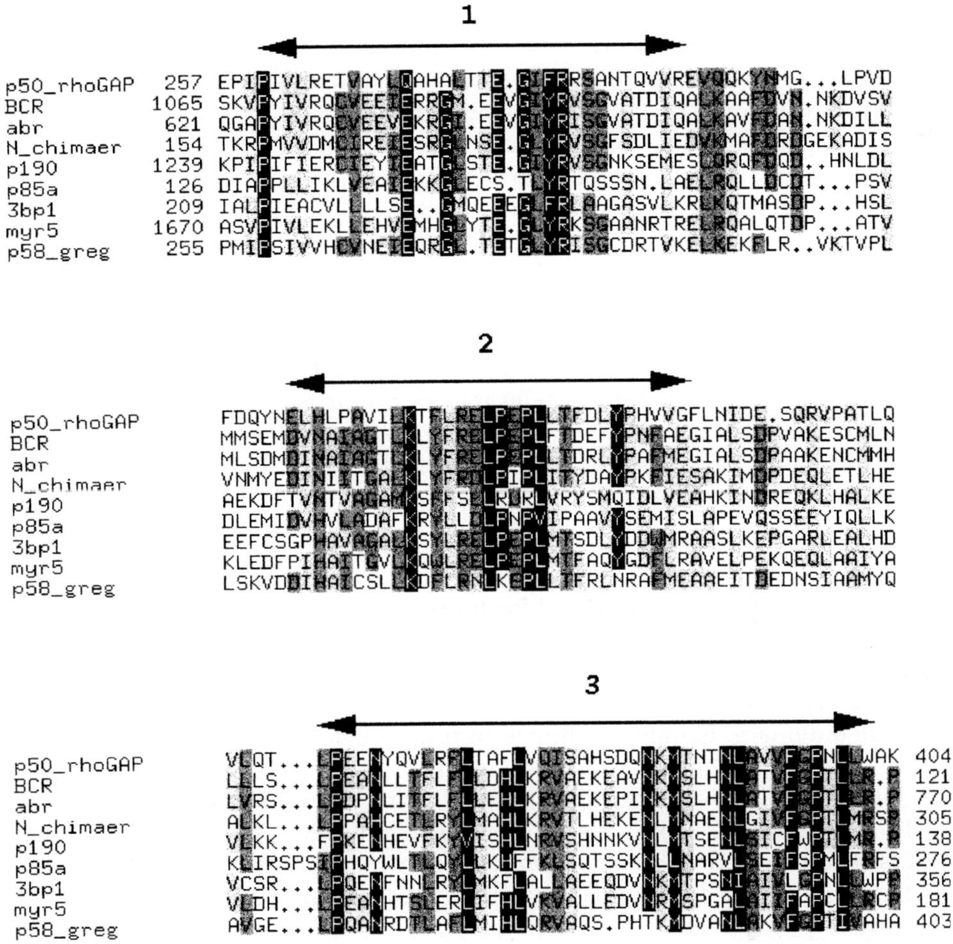

Fig. 2. Alignment of RhoGAP catalytic domains from differents proteins. Conserved positions are in *bold* and the three conserved blocks are delineated by *arrows* p58_greg was renamed p58_Mgc for publication

and acting on Cdc42, or from bovine adrenal gland and acting on Rho [51–53]. The crystal structure of p50-RhoGAP has been described recently; GAP domain was crystallised either alone and its structure resolved at 2 Å, or in complex with GTP-bound Cdc42 GTPase and its structure resolved at 2.7 Å [54, 55]. Crystal analysis showed that the structure of the GAP domain is an unusual arrangement of nine a-helices, the core of which consists of a four-helix bundle. Conserved residues of the RhoGAP domain are mostly located on one face of this bundle, which may constitute an interaction site for GTPases. Cdc42 interacts mainly through its switch I and II regions with a shallow pocket on RhoGAP effectively enriched in conserved residues. The data also suggest that GAP acts on intrinsic GTPase activity both allosterically, by stabilizing the GTPase structure, and directly by contributing to the chemical catalysis of GTP hydrolysis.

Biochemical studies of GAP specificity towards various Ras-like GTPases in vitro have shown that Rho, Rac and Cdc42 were the only targets of p50-RhoGAP. In particular, and despite its name, p50-RhoGAP binds the three Rho GTPases equally well, but has a striking higher GAP activity upon Cdc42 than upon Rac and Rho. It is equally active on Rac and Rho [50, 52, 56]. By contrast, microinjection of cultured fibroblasts with the GAP domain induced specific inhibition of Rho- rather than Rac-mediated effects on the actin cytoskeleton (Cdc42-mediated effects were not tested in the assay) [56]. Sequence analysis of p50-RhoGAP also showed the presence of a potential SH3-binding site consisting of a proline-rich motif located just upstream of the GAP domain, suggesting that interacting proteins may be important for its function on Rho GTPases in vivo. The simple structure of p50-RhoGAP makes it a good candidate as a down-regulator of Rho functions in vivo, but its physiological role remains to be determined.

2.2 Birth of the Rho-GAP Family of Proteins

The sequence of the catalytic domain of p50-RhoGAP led to the observation that several other signalling molecules were sharing a structurally-related domain. Bcr and n-Chimaerin proteins were immediately recognized as bearing a segment with 50% identity to the p50-RhoGAP functional domain, while also sharing its GAP activity for Rho GTPases [57]. In addition to this "sequence comparison" approach, the development of an overlay assay to detect GAP activity from tissue extracts has greatly contributed to the extension of the Rho-GAP family [58]. Sequence comparisons of these various proteins show that they all share a related RhoGAP domain that spans 150 residues of the proteins (Fig. 1) [60]. Consisting of three conserved blocks, with no significant resemblance to the Ras GAP domain, the RhoGAP domain displays binding to Rho GTPases and GAP activity. This conserved region has now been found in numerous and otherwise unrelated multifunctional signalling proteins, most of which possess a GAP activity for Rho GTPases.

2.3 Bcr (Breakpoint Cluster Region) Gene Product

Bcr was originally characterised because of its involvement in a translocation between cht. 9 and 22 generating the Philadelphia chromosome found in myelogenous leukaemia cells and leading to the production of a chimaeric Bcr/Abl oncogene product [59, 60]. Bcr is a 160 kDa phosphoprotein, mainly present in brain and hematopoietic cells, structurally consisting of several distinct functional domains. The N-terminus contains a kinase activity able to phosphorylate serine/threonine in Bcr itself [61]. The central region is homologous to the tandem DH-PH domains found in putative exchange factors for Rho GTPases [62] and indeed, in vitro, it displays a GEF activity upon Rho, Rac and Cdc 42 without affecting other Ras-like proteins [63]. The next identified domain is a C2 region, a potential phosphoinositides receptor. The C-terminus is homologous to the RhoGAP domain and exhibits specific GAP activity on Rac and Cdc42 in vitro [56, 57, 63]. Identical conclusions were obtained after microinjecting fibroblasts with Bcr GAP domain, which inhibited Rac-induced actin cytoskeletal modifications without affecting Rho-induced actin cytoskeletal modifications [66]. Bcr appears therefore as a puzzling complex signalling protein, having at least two opposite regulatory effects on Rho GTPases: on the one hand, using DH domain, it stimulates their activation, on the other hand, using GAP domain, it stimulates their inactivation. To what extent the in vivo activity of these two domains might be differently regulated is unknown.

Although alteration of the *bcr* gene in leukaemia cells with Philadelphia-positive chromosome produces a deregulated chimaeric Bcr protein, studies of these cells have not yet revealed a role for Bcr on Rho GTPases [64, 65].

On the other hand, *bcr*-null mutant mice develop severe tissue injury after an experimentally-induced inflammation and their neutrophils show a pronounced increase in reactive oxygen metabolite production upon activation [66]. This observation is consistent with a negative regulatory role for bcr, through its GAP domain on Rac-induced superoxide production in a cell-free system [67]. Bcr appears therefore to act as a physiological negative-modular or NADPH oxidase activity by down-regulating Rac GTPase. These results point to a specialised role for Bcr in protecting the organism against premature and excessive activation of NADPH oxidase in leukocytes, perhaps relevant to its involvement in leukaemia of this type of cells.

2.4 Abr (Active Bcr-Related) Gene Product

Abr gene was initially isolated by cross-hybridisation with *Bcr* probes [65] and later shown to encode a 92 kDa protein highly enriched in brain. Abr is structurally related to Bcr, with 68% identity, except that it does not present the N-terminal kinase domain [69, 70]. Abr contains both a Dbl-homology domain and a RhoGAP-homology domain, which indeed have been demonstrated to respectively exhibit in vitro GEF activity upon Rho, Rac and Cdc42, and GAP activity upon Rac and Cdc42 [63]. Abr, with its dual activity on GTPases, appears as a potential multiregulator for Rho signalling pathways, but its actual role relative to Rho GTPases is still unknown.

2.5 The Chimaerin Family of Proteins (α1, α2, β1, and β2)

Characterisation of n(α1)-Chimaerin cDNA showed that it encodes a neuronal-specific protein bearing a C-terminal RhoGAP domain and an N-terminal cysteine-rich motif similar to the regulatory domain of protein kinase C [71, 72]. The GAP domain has been shown, in vitro, to act upon Rac, weakly upon Cdc42, and not to act on other Ras-like GTPases [57, 73]. Similarly, microinjections of the GAP domain into fibroblasts inhibited Rac-induced actin structures in a GAP activity-dependent fashion, but did not affect Rho- or Cdc42-induced actin structures [74]. The cysteine-rich motif binds zinc, constitutes a phospholipid/phorbol ester receptor and allows positive and negative regulation of α1-Chimaerin GAP activity through binding with various phospholipids [75, 76]. This modulation of GAP activity by lipids remains to be studied in vivo, but suggests that Chimaerin may link lipid second messengers to Rac-controlled cellular events. Overlay assays allowed detection of more GAPs: one of them was b1-Chimaerin with a GAP activity specific for Rac [77]. Isolation of *β1-Chimaerin* cDNA showed that this new testis-specific gene encoded a 34 kDa protein that can only be detected in late stages of spermatogenesis. β1-Chimaerin has 77% sequence identity with α1-Chimaerin and also bears a cysteine-rich domain followed by a RhoGAP domain.

Further studies done by Lim and collaborators led to the discovery of α2 and β2 isoforms of Chimaerins. α2 and β2 are generated by alternative splicing of *α1-* and *β1-Chimaerin* gene transcripts, respectively, and possess an additional N-terminus SH2 domain [78, 79]. This SH2 domain allows them to interact with various tyrosine-phosphorylated proteins in cells and may modulate their GAP activity for Rac. *α2* and *β2* have a distinct tissue distribution: *α2* is detected in brain regions and in spermatocytes while *β2* is specifically found in cerebellar cells. The Chimaerin family therefore represents a unique class of lipid-regulated GAP proteins acting on Rac GTPase, each with its own pattern of distribution.

Physiological functions of Chimaerins have been studied by microinjection of fibroblasts or neuroblastoma cells with various forms of α1-Chimaerin protein. Whereas the GAP domain alone inhibited Rac-activated lamellipodia formation, full-length α1-Chimaerin stimulated formation of lamellipodia and fillopodia [74]. These α1-Chimaerin-stimulated events required the non-GAP regions, including the N-terminus which is necessary to localise the protein with actin structures, and were dependent on Rac and Cdc42 activity. Moreover, these effects were still observed with a Chimaerin mutant lacking GAP catalytic activity but were lost with a mutant lacking Rac- or Cdc42-binding activity. These data strongly suggest that α1-Chimaerin is less of a down-regulator of Rac than an effector of Rac and Cdc42.

2.6
p85α and p85β Subunits of PI3-Kinase

Phosphatidylinositol 3′-kinase enzyme is made of a p110 catalytic subunit and of a p85α or p85b regulatory subunit [80, 81]. p85, which allows PI3 kinase recruitment at cytoplasmic membranes through interaction of its SH2 domain with activated tyrosine-kinase receptors, also has a RhoGAP and an SH3 domain at its N-terminus. p85α GAP-homology region was the first RhoGAP domain to be crystallised [82]; its structure at 2.0 Å resolution is highly related to the GAP domain of p50-RhoGAP mentioned above.

A remarkable feature is that p85 GAP domains do not exhibit GAP activity for any of the GTPases tested, whereas it is able to specifically bind Cdc42 and Rac. Furthermore, the interaction is restricted to the GTP-bound forms of the GTPases, requires their effector domain and appears to trigger the stimulation of PI3 kinase activity by a few folds [83–85]. These data suggest that p85, through interaction of its GAP-like region with Rac or Cdc42, mediates the stimulation of PI3 kinase activity by these GTPases, which might contribute to their regulation of the cytokeleton.

2.7
p190 (p120-RasGAP-Associated Protein) and p190b

p190, which has been shown to possess GAP activity for Rho GTPases, was first isolated as a tyrosine-phosphorylated protein formins a complex with p120-RasGAP and associating with membrane receptors in mitogenically-stimulated cells [86]. p190 may therefore constitute a link for cells to couple regulation of Ras and Rho dependent-signalling. Cloning of *p190* revealed a weird structure: it has an N-terminus domain encoding a new GTPase-like protein able to bind GTP, a central part homologous to a transcriptional repressor which containins an SH3-binding site, and a C-terminus RhoGAP domain [87, 88]. Its GAP activity in vitro works upon Rho but also at a lower level up Rac and Cdc42. Similar conclusions were drawn from microinjecting experiments showing that its purified GAP domain preferentially inhibited Rho-induced actin stress fibres [24]. p190 function may be to modulate the Rho-dependent actin cytoskeleton [89–91], but this potential role has yet to be demonstrated.

A *p190b* cDNA has been cloned more recently and encodes a protein with 51% identity to p190. It bears an overall structure similar to p190 and exhibits GAP activity in vitro for Rac, Rho and Cdc42 [92]. A noticeable fact is that p190b and Rho are recruited at the plasma membrane upon cell adhesion or integrin cross-linking. This observation indicates that p190b may be a link between integrins and the actin cytoskeleton, but its exact molecular effects on Rho GTPases remain to be examined.

2.8
Other Mammalian Rho-GAPs: 3BP-1, RLIP76, Myr5, Myosin-IXb, p122-Arp, p115 and p58-Mgc

3BP-1 was initially isolated through its ability to bind Abl SH3 domain [93]. In addition to its SH3-binding site, consisting of a proline-rich motif, 3BP-1 contains a RhoGAP-homology domain acting in vitro upon Rac and Cdc42, to a lower extent upon Rho G but not on other Ras-like GTPases [94]. GAP specificity of 3BP-1 was confirmed in intact cells by microinjection of its GAP domain: it inhibits Rac- but not Rho-induced actin structures. The interacting partner(s) of its proline-rich motif and the physiological role of 3BP-1 relative to Rac and Cdc42 functions are still to be determined.

RLIP76, also known as RIP1 or RalBP1, has been identified as a Ral GTPase effector molecule of 76 kDa, bearing a RhoGAP homology domain in addition to its Ral-binding domain [95–97]. Its GAP domain stimulates in vitro GTP-hydrolysing activity of Cdc42 and Rac but not of other GTPases. Since RalGDS, an exchange factor for Ral, seems to be a Ras effector molecule, Ral and RLIP might constitute a connection between Ras and Rac/Cdc42 signalling pathways, downstream of Ras pathways. The exact molecular role of RLIP in Rac/Cdc42-dependent functions in vivo is so far unknown.

Myr5 and Myosin-IXb have been independently isolated in screens for novel myosins and define the ninth class of myosin [98, 99]. These two unconventional myosins are 82% identical, and share distinctive sequences compared with other myosins. Their "head" contains a unique N-terminal extension and a large insertion in the putative actin-binding site, while their "tail" contains a Chimaerin-like structure including a cysteine-rich domain directly followed by a RhoGAP domain. Like all myosins, Myr5 is able to bind actin filaments in an ATP-dependent way. Its GAP domain displays in vitro Gap activity mainly on Rho and Cdc42. Myr5 and Myosin-IXb appear therefore as direct links between Rho GTPases which modulate actin dynamics and the acto-myosin network. Whether the myosin head function is only to allow the down-regulator GAP activity to be localised in the cytoskeleton network, or whether it mediates downstream effects of Rho GTPases on the cytoskeleton is an important goal for future work.

p122-ARP (Adaptor for Rho and Phospholipase C-δ), initially identified through its ability to bind PLC-d1, was later characterized as a dual function molecule, showing phospholipase C-stimulating activity and a RhoGAP domain active in vitro specifically on Rho [100]. Phosphatidylinositol 4,5-biphosphate (PIP2), which is the substrate of PLC-d1, also seems involved in actin cytoskeleton regulation through direct interaction with a number of actin-binding factors. This indicates that PIP2 cleavage by PLC-d1 might be important in cytoskeleton dynamics. p122-ARP therefore represents a potential downstream molecular intermediate of Rho in the regulation of the actin cytoskeleton, but whether this putative effector function is additional to its GAP activity still needs to be experimentally evaluated.

p115 protein was found to be encoded by a human X chromosome-linked gene, *C1*, predominantly expressed in hematopoietic cells [101]. p115, which contains regions of homology to SH3 and RhoGAP domains, is mainly localised in a narrow cytoplasmic area just below the cell membranes. Ectopic expression of the full-length protein in fibroblasts induced inhibition of actin stress fibres. This observation is consistent with a role of p115 in the downregulation, through its GAP activity, of Rho-dependent actin structures in a spatially-defined subcellular location in hematopoietic cells.

p58-Mgc (Male Germ Cell) was recently isolated in a search for Rac targets and then identified as a human protein of 58 kDa specifically expressed in tests, notably in primary spermatocytes [102]. p58-Mgc, whose closest homologue is the *Drosophila* protein RotundRacGAP, bears a Chimaerin-like structure consisting of a cysteine-rich motif directly followed by a RhoGAP domain. In vitro assays demonstrated that its GAP domain exhibits GAP activity only upon Rac and Cdc42. Sequence homologies and pattern of expression suggest that p58-Mgc, like RotundRacGAP, is involved in maturation of male germ cells, but its exact molecular role relative to Rac and Cdc42 signalling requires further analysis.

2.9
Non-Mammalian Rho-GAPs: Graf, RotundRacGAP, CeGAP, DdRacGAP, Bem2, Bem3, Dbm1, LRG1 and RGA1

Several proteins containing a RhoGAP domain have been identified in non-mammalian organisms, and some of them have been shown to be involved in Rho-dependent functions. However, if all proteins so far examined exhibit a GAP domain active on Rho GTPases, we still tend to ignore whether their exact molecular role is to downregulate or to mediate Rho-dependent signalling in vivo.

Graf (GTPase Regulator Associated with Fak) has been isolated from chicken embryo cells as a focal adhesion kinase (Fak)-binding factor [103]. Graf contains a central RhoGAP domain active on Rho and Cdc42 in vitro and a C-terminus SH3 domain which binds a proline-rich motif of Fak. Graf colocalizes with actin-based structures in cells and therefore represents a direct link between Rho GTPases and the actin cytoskeleton. However, the exact molecular role of full-length Graf towards Rho and Cdc42-controlled activities in vivo remains to be elucidated.

RotundRacGAP (RnRacGAP) is a *Drosophila* gene product encoded at the *rotund* locus. This protein, encoded by a gene mainly expressed in primary spermatocytes, bears a Chimaerin-like structure, including a cysteine-rich region directly followed by a RhoGAP domain [104]. Genetic analysis of the *rotund* locus established that *rotund* phenotype includes male infertility. The physiological role of RnRacGAP is apparently linked to maturation of male germ cells as shown by its ability to restore male fertility of *rotund* flies [105].

Ectopic expression of RnRacGAP in flies induced a disruption of actin-based structures and is therefore consistent with a possible downregulator action of RnRacGAP on Rho GTPases in vivo [106].

CeGAP (from *C. Elegans*) contains sequences homologous to PH and RhoGAP domains [107]. Surprisingly, the RhoGAP region exhibited activity on Rho GTPases but also on Ras and on Rab proteins. CeGAP therefore appears involved in signalling of various Ras-like proteins, but this potential integrative role requires further demonstration, particularly in vivo.

DdRacGAP, the first RhoGAP-related protein from *Dictyostelium discoideum* has been cloned by sequence homology [108]. It contains a proline-rich motif, an SH3 region, a tandem of PH-DH domains typical of exchange factors for Rho proteins, and, in addition, a GAP region active on Rho GTPases from *Dictyostelium*. The physiological role of this multifunctional protein relative to Rho GTPases is unknown.

Bem2, Bem3 and Dbml proteins from *S. Cerevisiae* contain a region homologous to the RhoGAP domains. These three factors, identified through genetic screens for proteins involved in polarized cell growth, are notably implicated in bud site selection and in bud formation [109–112]. This process known to implicate Cdc42 and then Rho1p GTPases involves modifications of the cortical cytoskeleton and, in particular, assembly of actin structures at the bud site. In vitro GAP assays have shown that Bem2 works specifically on Rho1p and that Bem3 acts on Cdc42 and weakly on Rho1p [110, 113, 114]. The specificity of Dbm1 has not yet been determined. Genetic data suggest that these RhoGAPs regulate cytoskeletal structures, but whether they act by mediating or down-regulating Rho GTPases effects in vivo is not clear.

LRG1 (LIM and RhoGAP 1 homologous gene product) and RGA1 (Rho GTPase Activating 1 gene product) are proteins from *S. cerevisiae*, containing motifs similar to the RhoGAP domain and cysteine-rich region LIM [115, 116]. LRG1, encoded by a non-essential gene, may play a role during mating, but its function relative to Rho GTPases is so far unknown. RGA1 has been isolated in a genetic screen for mutants inducing a constitutive activation of the pheromone pathway. The pheromone pathway is potentially activated by Cdc42 through stimulation of Ste20 protein kinase, and RGA1 might act as a downregulator of Cdc42 function in vivo.

3 Conclusions

3.1 Too Many Rho GTPase-Regulating Factors?

Rho-GAP domains are found in several multifunctional proteins that are expressed in all types of cells. If they constitute a network of Rho GTPase-downregulating factors, one can wonder how such a dense network would not lead to constitutive inhibition of Rho signalling pathways. A precise regulation of each of the Rho GTPases in a time- and space-dependent fashion is required, and different strategies are used.

One involves GDIs: these molecules are able to prevent the action of Rho-GAPs on Rho GTPases by making "protecting" complexes with GTP-bound forms of the GTPases in vitro [25,118]. In vivo, GDI could then avoid the constitutive inhibition of Rho functions in the presence of Rho-GAPs.

Regulation of Rho-GAPs many also be a consequence of their tissue distribution. All Rho-GAPs are not ubiquitously present, and several of them, including Bcr, Chimaerins, 3BP1, p115, p58-Mgc or Rotund, are only detected in a few cell-types and in some cases just at a given stage during cell differentiation [69, 94, 101–104, 120]. These observations suggest that these proteins are probably involved in a particular function of Rho proteins, specific to the expressing cells.

Specificity is another answer. Rho-GAPs exhibit some preferences for particular Rho GTPases in vitro. Moreover, as suggested by microinjection of GAP domains in living cells [56], GAP specificity in vivo is probably more restricted than in vitro and would allow a given Rho-GAP to regulate only a particular subset of Rho GTPases. Also, substrates for Rho-GAPs in vivo depend on local concentrations of Rho GTPases (and of the GAPs) and are not only determined by affinities observed in vitro.

This points to the spatial distribution of the interacting partners: GDI might here again keep Rho out of reach of GAPs by sequestering them in a subcellular place where there are no GAPs. Reciprocally, GAP localization within cells might be dynamically modulated by molecular partners interacting with the various non-GAP domains of Rho-GAPs. Such restriction in the subcellular environment could promote or abolish its effect on a particular Rho GTPase. Several Rho-GAPs, including Chimaerins, p190, Myosin-IXb, p115, p190b or Graf have indeed been detected in particular subcellular locations [74, 91, 92, 101, 103]. Rho-GAPs interacting partners may also directly modify the enzymatic activity by affecting the protein structure: Chimaerin's GAP activity is regulated through the binding of various phospholipids, using its cysteine-rich region [120]. Apart from this case, direct regulation of Rho-GAPs by interacting partners are unknown. Characterisation of the functional molecular partners of Rho-GAPs is undoubtedly a major goal for further investigations and would probably help to determine the role of these multifunctional proteins.

3.2
A Final Word

The large number of Rho-GDIs and Rho-GAPs may provide a way for cells to regulate every Rho protein in each of their various cellular functions and/or localisations. A non exclusive possibility is that the numerous multifunctional GAP-containing proteins act as effector molecules and participate directly in the diversity of Rho signalling pathways. This possibility, initially described for p120-RasGAP and Ras [121–124], has also been discussed for p85 and a1-Chimaerin [74, 83, 84].

Finally, GAP-containing proteins would not only downregulate Rho proteins; they might be Rho effectors that mediate signalling through their non-GAP domains, acting as sophisticated effector molecules by regulating negative feedback. The possible discrimination in the GAP domain of α1-Chimaerin and p85 between binding to the GTPase and GAP activity [73, 74, 83] suggests that a role for the RhoGAP domain may be no mediate protein interactions, and this is also consistent with Rho-GAPs acting as effector molecules.

Potential effector functions of Rho-GAPs may be directly linked to activation of an enzymatic activity involved in downstream signalling of GTPases, as could occur with p122-Arp and p85 activating PLC and PI3 kinase, respectively [85, 100]. Rho-GAPs may also act by serving as receptors, recruiting activated Rho GTPases in a particular cell environment to allow a given molecular effect. This putative role could be true for α1-Chimaerin, Myosin-IXb, p190b or Graf recruiting Rho GTPases in the vicinity of the actin cytoskeleton network [74, 92, 99, 103].

Rho-GAPs functioning as down-regulators of Rho proteins and Rho-GAPs functioning as mediators of Rho functions with or without a negative regulating activity on Rho GTPases most probably co-exist in cells. Further studies on each full-length molecule as well as its interacting partners in the context of living cells or animals should therefore give precious information for deciphering each Rho-GAP contribution to the finely tuned networks where Rho GTPases are key players.

Added in Proof

1. Since preparation of this text, a new RhoGAP has been identified that is involved in a X-linked mental retardation (Billuard P. et al., Nature, 392: 923–926).
2. Since preparation of this text, Philippe Chavrier's group has shown that although LyGDI is able to interact in vitro with several proteins Rho, none of these Rho proteins is its in vivo partner. This result illustrates another discrepancy between in vitro specificity (usually loose) and in vivo specificity (usually narrow or ill defined) of regulators of Rho proteins (J.-P. Grovel et al., FEBS Lett., 1998, 422: 269–273).

References

1 Whitehead IP, Campbell S, Rossman KL, Der CJ (1997) Dbl family proteins. Biochem Biophys Acta 1332:F1–F23

2 Boguski MS, McCormick F (1993) Proteins regulating Ras and its relatives. Nature 366:643–654

3 Fukumoto Y, Kaibuchi K, Hori H, Fujioka H, Araki S, Ueda T, Kikuchi A, Takai Y (1990) Molecular cloning and characterization of a novel type of regulatory protein (GDI) for the rho proteins, ras p21-like small GTP-binding proteins. Oncogene 5:1321–1328

4 Ueda T, Kikuchi A, Ohga N, Yamamoto J, Takai Y (1990) Purification and characterization from bovine brain cytosol of a novel regulatory protein inhibiting the dissociation of GDP from and the subsequent binding of GTP to rhoB p20, a ras p21-like GTP-binding protein. J Biol Chem 265:9373–9380

5 Lelias JM, Adra CN, Wulf GM, Guillemot JC, Khagad M, Caput D, Lim B (1993) cDNA cloning of a human mRNA preferentially expressed in hematopoietic cells and with homology to a GDP-dissociation inhibitor for the rho GTP-binding proteins. Proc Natl Acad Sci USA 90:1479–1483

6 Scherle P, Behrens T, Staudt LM (1993) Ly-GDI, a dissociation inhibitor of the RhoA GTP-binding protein, is expressed preferentially in lymphocytes. Proc Natl Acad Sci USA 90:7568–7572

7 Adra CN, Ko J, Leonard D, Wirth LJ, Cerione RA, Lim B (1993) Identification of a novel protein with GDP dissociation inhibitor activity for the ras-like proteins CDC42Hs and Rac1. Genes Chromosomes Cancer 8:253–261

8 Zalcman G, Closson V, Camonis J, Honoré N, Rousseau-Merck M-F, Tavitian A, Olofsson B (1996) Rho GDI-3, a new GDP dissociation inhibitor. Identification of a non cytosolic GDI protein interacting with the small GTP-binding proteins RhoB and RhoG. J Biol Chem 271:30366–30374

9 Adra CN, Manor D, Ko JL, Zhu S, Horiuchi T, Van Aelst L, Cerione RA, Lim B (1997) RhoGDIg: a GDP-dissociation inhibitor for Rho proteins with preferential expression in brain and pancreas. Proc Natl Acad Sci USA 94:4279–4284

10 Masuda T, Tanaka K, Nonaka H, Yamochi W, Maeda A, Takai Y (1994) Molecular cloning and characterization of yeast rho GDP dissociation inhibitor. J Biol Chem 269:19713–19718

11 Leffers H, Nielsen MS, Andersen AH, Honoré B, Madsen P, Vandkerckhove J, Celis JE (1993) Identification of two human rho GDP dissociation inhibitor proteins whose overexpression leads to disruption of actin cytoskeleton. Exp Cell Res 209:165–174

12 Gosser YQ, Nomanbhoy TK, Aghazadeh B, Manor D, Combs C, Cerione RA, Rosen MK (1997) C-terminal binding domain of Rho GDP-dissociation inhibitor directs N-terminal inhibitory peptide to GTPases. J Biol Chem 87:814–819

13 Wittinghofer A, Herrmann C (1995) Ras-effector interactions, the problem of specificity. FEBS Lett 369:52–56

14 Ando S, Kaibuchi K, Sazaki T, Hiraoka K, Nishiyama T, Mizuno T, Asada M, Nunoi H, Matsuda I, Matsuura Y, Polakis P, McCormick F, Takai Y (1992) Post-translational processing of rac P21s is important both for their interaction with the GDP/GTP exchange proteins and for their activation of NADPH oxidase. J Biol Chem 267:25709–25713

15 Leonard D, Hart MJ, Platko JV, Eva A, Henzel W, Evans T, Cerrione RA (1992) The identification and characterization of a GDP-dissociation inhibitor (GDI) for the CDC42Hs protein. J Biol Chem 267:22860–22868

16 Regazzi R, Kikuchi A, Takai Y, Wollheim C (1992) The small GTP-binding proteins in the cytosol of insulin-secreting cells are complexed to GDP dissociation inhibitor proteins. J Biol Chem 267:17512–17519

17 Chuang TS, Xu X, Knaus UG, Hart MJ, Bokoch GM (1993) GDP dissociation inhibitor prevents intrinsic and GTPase activating protein-stimulated GTP hydrolysis by the Rac GTP-binding protein. J Biol Chem 268:775–778
18 Sasaki T, Kato M, Takai Y (1993) Consequences of weak interaction of rho GDI with the GTP-bound forms of rho p21 and rac p21. J Biol Chem 268:23959–23963
19 Bourmeyster N, Boquet P, Vignais P (1994) Role of bound GDP in the stability of the RhoA-Rho GDI complex purified from neutrophil cytosol. Biochem Biophys Res Commun 205:174–179
20 Bourmeyster N, Stasia M-J, Garin J, Gagnon J, Boquet P, Vignais P (1992) Copurification of rho protein and the Rho-GDP dissociation inhibitor from bovine neutrophil cytosol. Effect of phosphoinositides on Rho ADP-ribosylation by the C3 exoenzyme of *Clostridium botulinum*. Biochemistry 31:12863–12869
21 Bourmeyster N, Vignais PV (1996) Phosphorylation of RhoGDI stabilizes the RhoA-RhoGDI complex in neutrophil cytosol. Biochem Biophys Res Commun 218:54–60
22 Abo A, Pick E, Hall A, Totty N, Teahan CG, Segal AW (1991) Activation of the NADPH oxidase involves the small GTP-binding protein p21-Rac1. Nature 353:668–670
23 Isomura M, Kikuchi A, Ohga N, Takai Y (1991) Regulation of binding of rhoB p20 to membranes by its specific regulatory protein, GDP dissociation inhibitor. Oncogene 6:119–124
24 Lang P, Gesbert F, Delespine-carmagnat M, Stancou R, Pouchelet M, Bertoglio J (1996) Protein kinase A phosphorylation of RhoA mediates the morphological and functional effects of cyclic AMP in cytotoxic lymphocytes. EMBO J 15:510–519
25 Hancock JF, Hall A (1993) A novel role for Rho-GDI as an inhibitor of GAP proteins. EMBO J 12:1915–1921
26 Hori Y, Kikuchi A, Isomura M, Katayama M, Miura Y, Fujioka H, Kaibuchi K, Takai Y (1991) Post-translational modifications of the C-terminal region of the rho protein are important for its interaction with membranes and the stimulatory and inhibitory GDP/GTP exchange proteins. Oncogene 6:515–522
27 Kikuchi A, Kuroda S, Sasaki T, Kotani K, Hirata K, Katayama M, Takai Y (1992) Functional interactions of stimulatory and inhibitory GDP/GTP exchange proteins and their common substrate small GTP-binding protein. J Biol Chem 267:14611–14615
28 Yaku H, Sasaki T, Takai Y (1994) The Dbl oncogene product as a GDP/GTP exchange protein for the Rho family: its properties in comparison with those of Smg GDS. Biochem Biophys Res Commun 198:811–816
29 Hart MJ, Maru Y, Leonard D, Witte ON, Evans T, Cerione RA (1992) A GDP dissociation inhibitor that serves as a GTPase inhibitor for the Ras-like protein CDC42Hs. Science 258:812–815
30 Keep NH, Barnes M, Barsukov I, Badii R, Lian L-Y, Segal AW, Moody PCE, Roberts GCK (1997) A modulator of rho family G proteins, rhoGDI, binds these G proteins via an immunoglobulin-like domain and a flexible N-terminal arm. Structure 5:623–633
31 Wu W-J, Leonard DA, Cerione RA, Manor D (1997) Interaction between Cdc42Hs and RhoGDI is mediated through the Rho insert region. J Biol Chem 272:26153–26158
32 Takaishi K, Kikuchi A, Kuroda S, Kotani K, Sasaki T, Takai Y (1993) Involvement of rho p21 and its inhibitory GDP/GTP exchange protein (rho GDI) in cell motility. Mol Cell Biol 13:72–79
33 Takaishi K, Sasaki T, Kato M, Yamochi W, Kuroda S, Nakamura T, Takeichi M, Takai Y (1994) Involvement of Rho p21 small GTP-binding protein and its regulator in HGF-induced cell motility. Oncogene 9:273–279
34 Kishi K, Sasaki T, Kuroda S, Itoh T, Takai Y (1993) Regulation of cytoplasmic division of *Xenopus* embryo by rho p21 and its inhibitory GDP/GTP exchange protein (rho GDI). J Cell Biol 120:1187–1195
35 Nishiyama T, Sasaki T, Takaishi K, Kato M, Yaku H, Araki K, Matsuura Y, Takai Y (1994) rac p21 is involved in insulin-induced membrane ruffling and rho p21 is involved in hepatocyte growth factor- and 12-Otetradecanoylphorbol-13-acetate (TPA)-induced membrane ruffling in KB cells. Mol Cell Biol 14:2447–2456

36 Takaishi K, Sasaki T, Kameyama T, Tsukita S, Takai Y (1995) Translocation of activated Rho from the cytoplasm to membrane ruffling area, cell-cell adhesion sites and cleavage furrows. Oncogene 11:39–48
37 Mariot P, O'Sullivann J, Brown AM, Tatham PER (1996) Rho guanine nucleotide dissociation inhibitor protein (RhoGDI) inhibits exocytosis in mast cells. EMBO J 15:6476–6482
38 Boivin D, Beliveau R (1995) Subcellular distrubution and membrane association of Rho-related small GTP-binding protiens in kidney cortex. Am J Physiol (Renal Fluid Electrolyte Physiol.) 269:F180–F189
39 Hirao M, Sato N, Kondo T, Yonemura S, Monden M, Sasaki T, Takai Y, Tsukita S, Tsukita S (1996) Regulation mechanism of ERM (Ezrin/Radixin/Moesin) protein/plasma membrane association: possible involvement of phosphatidylinositol turnover and Rho-dependent signaling pathway. J Cell Biol 135:37–51
40 Takahashi K, Sasaki T, Mammoto A, Takaishi K, Kameyama T, Tsukita S, Takai Y (1997) Direct interaction of the Rho GDP dissociation inhibitor with ezrin/radixin/moesin initiates the activation of the Rho small G protein. J Biol Chem 272:23371–23375
41 Mackay DJG, Esch F, Furthmayr H, Hall A (1997) Rho- and Rac-dependent assembly of focal adhesion complexes and actin filaments in permeabilized fibroblasts: an essential role for ezrin/radixin/moesin proteins. J Cell Biol 138:927–938
42 Platko JV, Leonard DA, Adra CN, Shaw RJ, Cerione RA, Lim B (1995) A single residue can modify target-binding affinity and activity of the functionnal domain of the rho-subfamily GDP dissociation inhibitors. Proc Natl Acad Sci USA 92:2974–2978
43 Danley DE, Chuang T-H, Bokoch GM (1996) Defective Rho GTPase regulation by IL 1b-converting enzyme-mediated cleavage of D4 GDP dissociation inhibitor. J Immunol 157:500–503
44 Na S, Chuang T-S, Cunningham A, Turi TG, Hanke JH, Bokoch GM, Danley DE (1996) D4-DGI, a substrate of CPP32, is proteolyzed during Fas-induced apoptosis. J Biol Chem 271:11209–11213
45 Guillemot J-C, Kruskal BA, Adra CN, Zhu S, Ko J-L, Burch P, Nocka K, Seetoo K, Simons E, Lim B (1996) Targeted disruption of guanosine diphosphate-dissociation inhibitor for rho-related proteins, GDID4: normal hematopoietic differentiation but subtle defect in superoxide production by macrophages derived from in vitro embryonal stem cell differentiation. Blood 88:2722–2731
46 Trahey M, McCormick F (1987) A cytoplasmic protein stimulates normal N-ras p21 GTPase, but does not affect oncogenic mutants. Science 238:542–545
47 Trahey M, Wong G, Halenbeck R, Rubinfeld B, Martin GA, Ladner M, Long CM, Crosier WJ, Watt K, Koths K, McCormick F (1988) Molecular cloning of two types of GAP complementary DNA from human placenta. Science 242:1697–1700
48 Garrett MD, Self AJ, van Oers C, Hall A (1989) Identification of distinct cytoplasmic targets for ras/R-ras and rho regulatory proteins. J Biol Chem 264:10–13
49 Garrett MD, Major GN, Totty N, Hall A (1991) Purification and N-terminal sequence of the p21rho GTPase-activating protein, rho GAP. Biochem J 276:833–836
50 Lancaster CA, Taylor-Harris PM, Self AJ, Brill S, van Erp HE, Hall A (1994) Characterization of rhoGAP. A GTPase-activating protein for rho-related small GTPases. J Biol Chem 269:1137–1142
51 Hart MJ, Shinjo K, Hall A, Evans T, Cerione RA (1991) Identification of the human platelet GTPase activating protein for the CDC42Hs protein. J Biol Chem 266:20840–20848
52 Barfod ET, Zheng Y, Kuang WJ, Hart MJ, Evans T, Cerione RA, Ashkenazi A (1993) Cloning and expression of a human CDC42 GTPase-activating protein reveals a functional SH3-binding domain. J Biol Chem 268:26059–26062
53 Morii N, Kumagai N, Nur EKMS, Narumiya S, Maruta H (1993) rho GAP of 28 kDa (GAP2), but not of 190 kDa (p190), requires Asp65 and Asp67 of rho GTPase for its activation. J Biol Chem 268:27160–27163

54 Barrett T, Xiao B, Dodson EJ, Dodson G, Ludbrook SB, Nurmahomed K, Gamblin SJ, Musacchio A, Smerdon SJ, Eccleston JF (1997) The structure of the GTPase-activating domain from p50rhoGAP. Nature 385:458–461
55 Rittinger K, Walker PA, Eccleston JF, Nurmahomed K, Owen D, Laue E, Gamblin SJ, Smerdon SJ (1997) Crystal structure of a small G protein in complex with the GTPase-activating protein rhoGAP. Nature 388:693–697
56 Ridley AJ, Self AJ, Kasmi F, Paterson HF, Hall A, Marshall CJ, Ellis C (1993) rho family GTPase activating proteins p190, bcr and rhoGAP show distinct specificities in vitro and in vivo. EMBO J 12:5151–5160
57 Diekmann D, Brill S, Garrett MD, Totty N, Hsuan J, Monfries C, Hall C, Lim L, Hall A (1991) Bcr encodes a GTPase-activating protein for p21rac. Nature 351:400–402
58 Manser E, Leung T, Monfries C, Teo M, Hall C, Lim L (1992) Diversity and versatility of GTPase activating proteins for the p21rho subfamily of ras G proteins detected by a novel overlay assay. J Biol Chem 267:16025–16028
59 Heisterkamp N, Stam K, Groffen J, de Klein A, Grosveld G (1985) Structural organization of the bcr gene and its role in the Ph' translocation. Nature 315:758–761
60 Hariharan IK, Adams JM (1987) cDNA sequence for human bcr, the gene that translocates to the abl oncogene in chronic myeloid leukaemia. EMBO J 6:115–119
61 Maru Y, Witte ON (1991) The BCR gene encodes a novel serine/threonine kinase activity within a single exon. Cell 67:459–468
62 Cerione RA, Zheng Y (1996) The Dbl family of oncogenes. Curr Opin Cell Biol 8:216–222
63 Chuang TH, Xu X, Kaartinen V, Heisterkamp N, Groffen J, Bokoch GM (1995) Abr and Bcr are multifunctional regulators of the Rho GTP-binding protein family. Proc Natl Acad Sci USA 92:10282–10286
64 Melo JV, Gordon DE, Cross NC, Goldman JM (1993) The ABL-BCR fusion gene is expressed in chronic myeloid leukemia. Blood 81:158–165
65 Diamond J, Goldman JM, Melo JV (1995) BCR-ABL, ABL-BCR, BCR, and ABL genes are all expressed in individual granulocyte-macrophage colony-forming unit colonies derived from blood of patients with chronic myeloid leukemia. Blood 85:2171–2175
66 Voncken JW, van Schaick H, Kaartinen V, Deemer K, Coates T, Landing B, Pattengale P, Dorseuil O, Bokoch GM, Groffen J, Heisterkamp N (1995) Increased neutrophil respiratory burst in bcr-null mutants. Cell 80:719–728
67 Heyworth PG, Knaus UG, Settleman J, Curnutte JT, Bokoch GM (1993) Regulation of NADPH oxidase activity by Rac GTPase activating protein(s). Mol Biol Cell 4:1217–1223
68 Heisterkamp N, Morris C, Groffen J (1989) ABR, an active BCR-related gene. Nucleic Acids Res 17:8821–8831
69 Heisterkamp N, Kaartinen V, van Soest S, Bokoch GM, Groffen J (1993) Human ABR encodes a protein with GAPrac activity and homology to the DBL nucleotide exchange factor domain. J Biol Chem 268:16903–16906
70 Tan EC, Leung T, Manser E, Lim L (1993) The human active breakpoint cluster region-related gene encodes a brain protein with homology to guanine nucleotide exchange proteins and GTPase-activating proteins. J Biol Chem 268:27291–27298
71 Hall C, Monfries C, Smith P, Lim HH, Kozma R, Ahmed S, Vanniasingham V, Leung T, Lim L (1990) Novel human brain cDNA encoding a 34 000 Mr protein n-chimaerin, related to both the regulatory domain of protein kinase C and BCR, the product of the breakpoint cluster region gene. J Mol Biol 211:11–16
72 Lim HH, Michael GJ, Smith P, Lim L, Hall C (1992) Developmental regulation and neuronal expression of the mRNA of rat n-chimaerin, a p21rac GAP: cDNA sequence. Bhichem J 287:415–422
73 Ahmed S, Lee J, Wen LP, Zhao Z, Ho J, Best A, Kozma R, Lim L (1994) Breakpoint cluster region gene product-related domain of n-chimaerin. Discrimination between Rac-binding and GTPase-activating residues by mutational analysis. J Biol Chem 269:17642–17648

74 Kozma R, Ahmed S, Best A, Lim L (1996) The GTPase-activating protein n-chimaerin cooperates with Rac1 and Cdc42Hs to induce the formation of lamellipodia and filopodia. Mol Cell Biol 16:5069–5080
75 Ahmed S, Kozma R, Lee J, Monfries C, Harden N, Lim L (1991) The cysteine-rich domain of human proteins, neuronal chimaerin, protein kinase C and diacylglycerol kinase binds zinc. Evidence for the involvement of a zinc-dependent structure in phorbol ester binding. Biochem J 280:233–241
76 Ahmed S, Lee L, Kozma R, Best A, Monfries C, Lim L (1993) A novel functional target for tumor-promoting phorbol esters and lysophosphatidic acid. The p21rac-GTPase activating protein n-chimaerin. J Biol Chem 268:10709–10712
77 Leung T, How BE, Manser E, Lim L (1993) Germ cell beta-chimaerin, a new GTPase-activating protein for p21rac, is specifically expressed during the acrosomal assembly stage in rat testis. J Biol Chem 268:3813–3816
78 Hall C, Sin WC, Teo M, Michael GJ, Smith P, Dong JM, Lim HH, Manser E, Spurr NK, Jones TA et al. (1993) Alpha 2-chimerin, an SH2-containing GTPase-activating protein for the ras-related protein p21rac derived by alternate splicing of the human n-chimerin gene, is selectively expressed in brain regions and testes. Mol Cell Biol 13:4986–4998
79 Leung T, How BE, Manser E, Lim L (1994) Cerebellar beta 2-chimaerin, a GTPase-activating protein for p21 ras-related rac is specifically expressed in granule cells and has a unique N-terminal SH2 domain. J Biol Chem 269:12888–12892
80 Otsu M, Hiles I, Gout I, Fry MJ, Ruiz-Larrea F, Panayotou G, Thompson A, Dhand R, Hsuan J, Totty N, Smith AD, Morgan SJ, Courtneidge SA, Parker P, Waterfield MD (1991) Characterization of two 85 kDa proteins that associate with receptor tyrosine kinases, middle-T/pp60c-src complexes, and PI3-kinase. Cell 65:91–104
81 Skolnik EY, Margolis B, Mohammadi M, Lowenstein E, Fischer R, Drepps A, Ullrich A, Schlessinger J (1991) Cloning of PI3 kinase-associated p85 utilizing a novel method for expression/cloning of target proteins for receptor tyrosine kinases. Cell 65:83–90
82 Musacchio A, Cantley LC, Harrison SC (1996) Crystal structure of the breakpoint cluster region-homology domain from phosphoinositide 3-kinase p85 alpha subunit. Proc Natl Acad Sci USA 93:14373–14378
83 Zheng Y, Bagrodia S, Cerione RA (1994) Activation of phosphoinositide 3-kinase activity by Cdc42Hs binding to p85. J Biol Chem 269:18727–18730
84 Tolias KF, Cantley LC, Carpenter CL (1995) Rho family GTPases bind to phosphoinositide kinases. J Biol Chem 270:17656–17659
85 Bokoch GM, Vlahos CJ, Wang Y, Knaus UG, Traynor-Kaplan AE (1996) Rac GTPase interacts specifically with phosphatidylinositol 3-kinase. Biochem J 315:775–779
86 Settleman J, Albright CF, Foster LC, Weinberg RA (1992) Association between GTPase activators for Rho and Ras families. Nature 359:153–154
87 Settleman J, Narasimhan V, Foster LC, Weinberg RA (1992) Molecular cloning of cDNAs encoding the GAP-associated protein p190: implications for a signaling pathway from ras to the nucleus. Cell 69:539–549
88 Foster R, Hu KQ, Shaywitz DA, Settleman J (1994) p190 RhoGAP, the major RasGAP-associated protein, binds GTP directly. Mol Cell Biol 14:7173–7181
89 McGlade J, Brunkhorst B, Anderson D, Mbamalu G, Settleman J, Dedhar S, Rozakis-Adcock M, Chen LB, Pawson T (1993) The N-terminal region of GAP regulates cytoskeletal structure and cell adhesion. EMBO J 12:3073–3081
90 Cheng JC, Frackelton AR Jr, Bearer EL, Kumar PS, Kannan B, Santos-Moore A, Rifai A, Settleman J, Clark JW (1995) Changes in tyrosine-phosphorylated p190 and its association with p120 type I and p100 type II rasGAPs during myelomonocytic differentiation of human leukemic cells. Cell Growth Differ 6:139–148
91 Chang JH, Gill S, Settleman J, Parsons SJ (1995) c-Src regulates the simultaneous rearrangement of actin cytoskeleton, p190RhoGAP, and p120RasGAP following epidermal growth factor stimulation. J Cell Biol 130:355–368

92 Burbelo PD, Miyamoto S, Utani A, Brill S, Yamada KM, Hall A, Yamada Y (1995) p190-B, a new member of the Rho GAP family, and Rho are induced to cluster after integrin cross-linking. J Biol Chem 270:30919–30926
93 Cicchetti P, Mayer BJ, Thiel G, Baltimore D (1992) Identification of a protein that binds to the SH3 region of Abl and is similar to Bcr and GAP-rho. Science 257:803–806
94 Cicchetti P, Ridley AJ, Zheng Y, Cerione RA, Baltimore D (1995) 3BP-1, an SH3 domain binding protein, has GAP activity for Rac and inhibits growth factor-induced membrane ruffling in fibroblasts. EMBO J 14:3127–3135
95 Jullien-Flores V, Dorseuil O, Romero F, Letourneur F, Saragosti S, Berger R, Tavitian A, Gacon G, Camonis JH (1995) Bridging Ral GTPase to Rho pathways. RLIP76, a Ral effector with CDC42/Rac GTPase-activating protein activity. J Biol Chem 270:22473–22477
96 Park SH, Weinberg RA (1995) A putative effector of Ral has homology to Rho/Rac GTPase activating proteins. Oncogene 11:2349–2355
97 Cantor SB, Urano T, Feig LA (1995) Identification and characterization of Ral-binding protein 1, a potential downstream target of Ral GTPases. Mol Cell Biol 15:4578–4584
98 Reinhard J, Scheel AA, Diekmann D, Hall A, Ruppert C, Bahler M (1995) A novel type of myosin implicated in signalling by rho family GTPases. EMBO J 14:697–704
99 Wirth JA, Jensen KA, Post PL, Bement WM, Mooseker MS (1996) Human myosin-IXb, an unconventional myosin with a chimerin-like rho/rac GTPase-activating protein domain in its tail. J Cell Sci 109:653–661
100 Homma Y, Emori Y (1995) A dual functional signal mediator showing RhoGAP and phospholipase C-delta stimulating activities. EMBO J 14:286–291
101 Tribioli C, Droetto S, Bione S, Cesareni G, Torrisi MR, Lotti LV, Lanfrancone L, Toniolo D, Pelicci P (1996) An X chromosome-linked gene encoding a protein with characteristics of a rhoGAP predominantly expressed in hematopoietic cells. Proc Natl Acad Sci USA 93:695–699
102 Touré A, Dorseuil O, Morin L, Timmons P, Jégou B, Reibel L, Gacon G (1998) p58-MgcRacGap, a new human GTPase activating protein for Rac and Cdc42 similar to *Drosophila* rotundRacGAP gene product is expressed in male germ cells. J Biol Chem 273:6019–6023
103 Hildebrand JD, Taylor JM, Parsons JT (1996) An SH3 domain-containing GTPase-activating protein for Rho and Cdc42 associates with focal adhesion kinase. Mol Cell Biol 16:3169–3178
104 Agnel M, Roder L, Vola C, Griffin-Shea R (1992) A *Drosophila* rotund transcript expressed during spermatogenesis and imaginal disc morphogenesis encodes a protein which is similar to human Rac GTPase-activating (racGAP) proteins. Mol Cell Biol 12:5111–5122
105 Hoemann CD, Bergeret E, Guichard A, Griffin-Shea R (1996) Alternative splicing of the *Drosophila melanogaster* rotundRacGAP gene. Gene 168:135–141
106 Guichard A, Bergeret E, Griffin-Shea R (1997) Overexpression of RnRacGAP in *Drosophila melanogaster* deregulates cytoskeletal organisation in cellularising embryos and induces discrete imaginal phenotypes. Mech Dev 61:49–62
107 Chen W, Blanc J, Lim L (1994) Characterization of a promiscuous GTPase-activating protein that has a Bcr-related domain from *Caenorhabditis elegans*. J Biol Chem 269:820–823
108 Ludbrook SB, Eccleston JF, Strom M (1997) Cloning and characterization of a rhoGAP homolog from *Dictyostelium discoideum*. J Biol Chem 272:15682–15686
109 Bender A, Pringle JR (1991) Use of a screen for synthetic lethal and multicopy suppressee mutants to identify two new genes involved in morphogenesis in *Saccharomyces cerevisiae*. Mol Cell Biol 11:1295–1305
110 Zheng Y, Cerione R, Bender A (1994) Control of the yeast bud-site assembly GTPase Cdc42. Catalysis of guanine nucleotide exchange by Cdc24 and stimulation of GTPase activity by Bem3. J Biol Chem 269:2369–2372
111 Wang T, Bretscher A (1995) The rho-GAP encoded by BEM2 regulates cytoskeletal structure in budding yeast. Mol Biol Cell 6:1011–1024

112 Chen GC, Zheng L, Chan CS (1996) The LIM domain-containing Dbm1 GTPase-activating protein is required for normal cellular morphogenesis in *Saccharomyces cerevisiae*. Mol Cell Biol 16:1376–1390
113 Peterson J, Zheng Y, Bender L, Myers A, Cerione R, Bender A (1994) Interactions between the bud emergence proteins Bem1p and Bem2p and Rho-type GTPases in yeast. J Cell Biol 127:1395–1406
114 Zheng Y, Hart MJ, Shinjo K, Evans T, Bender A, Cerione RA (1993) Biochemical comparisons of the *Saccharomyces cerevisiae* Bem2 and Bem3 proteins. Delineation of a limit Cdc42 GTPase-activating protein domain. J Biol Chem 268:24629–24634
115 Muller L, Xu G, Wells R, Hollenberg CP, Piepersberg W (1994) LRG1 is expressed during sporulation in *Saccharomyces cerevisiae* and contains motifs similar to LIM and rho/racGAP domains. Nucleic Acids Res 22:3151–3154
116 Stevenson BJ, Ferguson B, De Virgilio C, Bi E, Pringle JR, Ammerer G, Sprague GF Jr (1995) Mutation of RGA1, which encodes a putative GTPase-activating protein for the polarity-establishment protein Cdc42p, activates the pheromone-reponse pathway in the yeast *Saccharomyces cerevisiae*. Genes Dev 9:2949–2963
117 Hart MJ, Maru Y, Leonard D, Witte ON, Evans T, Cerione RA (1992) A GDP dissociation inhibitor that serves as a GTPase inhibitor for the Ras-like protein CDC42Hs. Science 258:812–815
118 Chuang TH, Xu X, Knaus UG, Hart MJ, Bokoch GM (1993) GDP dissociation inhibitor prevents intrinsic and GTPase activating protein-stimulated GTP hydrolysis by the Rac GTP-binding protein. J Biol Chem 268:775–778
119 Ahmed S, Kozma R, Hall C, Lim L (1995) GTPase-activating protein activity of n(alpha 1)-Chimaerin and effect of lipids. Methods Enzymol 256:114–125
120 Duchesne M, Schweighoffer F, Parker F, Clerc F, Frobert Y, Thang MN, Tocque B (1993) Identification of the SH3 domain of GAP as an essential sequence for Ras-GAP-mediated signaling. Science 259:525–528
121 Martin GA, Yatani A, Clark R, Conroy L, Polakis P, Brown AM, McCormick F (1992) GAP domains responsible for ras p21-dependent inhibition of muscarcinic atrial K+ channel currents. Science 255:192–194
122 Schweighoffer F, Barlat I, Chevallier-Multon MC, Tocque B (1992) Implication of GAP in Ras-dependent transactivation of a polyoma enhancer sequence. Science 256:825–827
123 Tocque B, Delumeau I, Parker F, Maurier F, Multon MC, Schweighoffer F (1997) Ras-GTPase activating protein (GAP): a putative effector for Ras. Cell Signal 9:153–158

Roles of PAK Family Kinases

Edward Manser[1] and Louis Lim[1,2]

1 Introduction

The first identified protein kinase exhibiting direct activation by a Ras-related GTPase was a brain *p*21(Cdc42/Rac) *a*ctivated *k*inase designated p65[PAK] (Manser et al. 1994). The kinase(s) were originally identified in [γ^{32}P]GTP-Cdc42 and [γ^{32}P]GTP-Rac1 overlays of SDS-polyacrylamide gel fractionated brain proteins (Manser et al. 1993). The presence of similar sized target proteins in all tissues suggested that these serine/threonine protein kinases were ubiquitous. In binding the active GTP-forms of Cdc42 and Rac1, the α- and γ-PAKs also decrease intrinsic GTP hydrolysis and block action of GTPase activating proteins (GAPs). This therefore favours signalling to PAK. However, βPAK may be released after activation allowing for amplification of the signal or initiation of other p21 actions (Manser et al. 1994, 1995).

At the time of its discovery, the similarity of mammalian PAK to budding yeast Ste20p, which was known to activate the pheromone-responsive MAP kinase cascade (Leberer et al. 1992; Ramer and Davis 1993) suggested on the one hand that PAK might regulate MAP kinase cascades in mammalian cells, and on the other that Ste20p might be directly regulatable by Cdc42p. Conclusions based on experimental data demonstrating a Cdc42p requirement in pheromone signalling (Simon et al. 1995; Zhao et al. 1995) are tempered by recent detailed studies showing Ste20p can function in both Cdc42-dependent and independent pathways (Peter et al. 1996; Leberer et al. 1997a).

PAK/Ste20p family proteins are indeed closely related in their kinase domains, but in the last few years a growing number of mammalian non-PAK kinases have been uncovered (for example germinal center kinases and Mst kinases) with similarity in the catalytic domain (Sells and Chernoff 1997). Many of these proteins act as regulators of c-Jun amino-terminal kinase pathways.

This review attempts to integrate available information regarding PAK biochemistry, with consideration of the roles of various identified PAK domains, and how these relate to the known functions of this kinase family.

[1] Glaxo-IMCB Group, Institute of Molecular & Cell Biology, 30 Medical Drive, Singapore 117609, Singapore
[2] Institute of Neurology, 1 Wakefield St., London WC1N 1PJ, UK

Progress in Molecular and Subcellular Biology, Vol. 22
Ph. Jeanteur (Ed.)

2 Distribution of PAK Family Kinases

PAK family kinases are distinguished from related kinases by the *PAK N*-terminal (PAN) motif shown in Fig. 1, with their catalytic domain invariably located at the C-terminus. Although a variety of other domains have been identified (see Sect. 8), none of these are universally present. In Fig. 1 the underlined sequences represent residues involved in p21 binding and first identified in PAK by virtue of similarity to the Cdc42-binding domain of ACK (Manser et al. 1993, 1994). A number of other related p21 effectors containing the consensus ISxP(x_{4-6})HxxH have been shown to exhibit Cdc42/Rac interaction and binding through this CRIB motif (Burbelo et al. 1995). Mammalian Cdc42 can function in yeast (Munemitsu et al. 1990); therefore the similarity in target sequences of PAK-family kinases is not surprising.

Three mammalian PAK isoforms have been characterized to date. On the basis of their apparent molecular weights these are designated as αPAK (PAK1, 68 kDa) βPAK (PAK3, 65 kDa) and γPAK (PAK2, 62 kDa). Each isoform varies

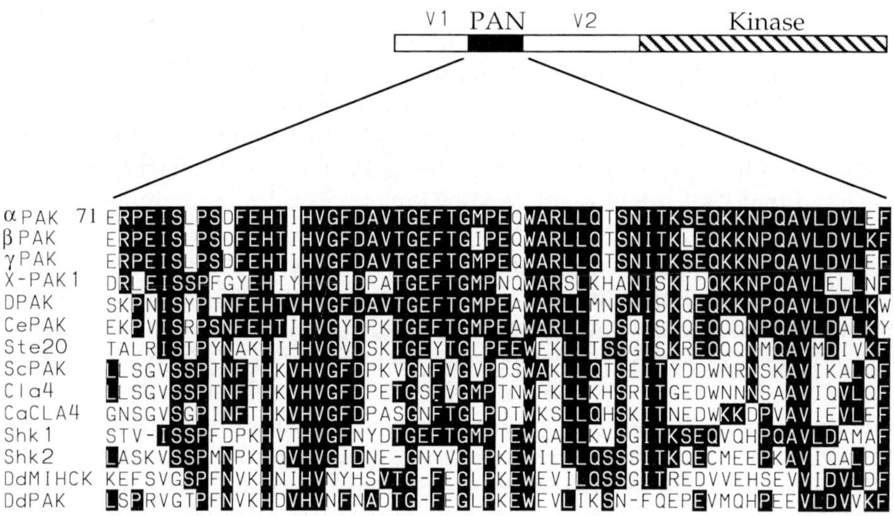

Fig. 1. Alignment of the amino acid sequences of the *PAK N*-terminal (*PAN*) region among known members of the PAK family. A variable region *V1* lies N-terminal to this which, in the Cla4p-related kinases, contains a pleckstrin homology (PH) domain. The PAN is linked to the kinase domain by a second variable (*V2*) region which in mammalian PAK contains a characteristic polyacidic sequence (of unknown function). The sequences shown for mammalian α-, β-, and γ-PAKs are identical among known species. The other PAK-family kinases shown are derived from the following organisms: *Xenopus* X-PAK1; *Drosophila* DPAK; *C. elegans* CePAK; *S. cerevisiae* Ste20p, Cla4p and ScPAK; *C. albicans* CaCla4p; *S. pombe* Shk1 and Shk2; *Dictyostelium* DdMIHCK and DdPAK

only at a few amino acid positions among species: the smallest γPAK is ubiquitously expressed (Martin et al. 1995; Teo et al. 1995). Mammalian PAK kinases exhibit low basal activity in vivo although they can be purified from bacteria in an activated form (Manser et al. 1997). The *Xenopus* X-PAK1 (Faure et al. 1997) shows surprising sequence divergence from mammalian PAKs in the PAN region (Fig. 1), compared with the more closely related invertebrate *C. elegans* and *Drosophila* PAKs.

From the genome sequencing of *Saccharomyces cerevisiae* and cataloging of all protein kinases in this organism (Hunter and Plowman 1997) only three PAK/Ste20p kinases appear to be present. Two of the *S. cerevisiae* PAKs, CLA4p and the related ScPAK contain an N-terminally located pleckstrin homology (PH) domain (Cvrckova et al. 1995): homologues with such a PH domain have not been found in multicellular organisms. Among *Dictyostelium* PAKs there is notable divergence in residues comprising the p21-binding domain, perhaps reflecting the diversity of Cdc42/Rac proteins (at least 7) in this organism (Bush et al. 1993).

3
Regulation of PAK Activity

PAK derived from mammalian sources can be activated 20- to 100-fold in vitro by the GTP-bound forms of Cdc42 or Rac1 but not RhoA (Manser et al. 1994, 1995; Martin et al. 1995). Similarly, DdMIHCK is potently activated by mammalian Rac1 (Lee et al. 1996). This direct and potent activation by GTP-p21 of a kinase in vitro contrasts with the Rho targets ROK and PKN (Leung et al. 1996; Amano et al. 1996) or the Ras target kinase Raf (McCormick 1994). How does binding of the p21 regulate PAK activity? The key initiating event for PAK activation is autophosphorylation. Even in the situation where PAK is activated by proteolysis, there remains a requirement for intra- and intermolecular autophosphorylation to drive full activity (Dennis and Masaracchia 1993). Upon αPAK activation by GTP-Cdc42, six serines in the regulatory N-terminal region, and a single threonine residue in the catalytic domain are phosphorylated (Manser et al. 1997). Using the catalytic fragment of MIHCK, Szczepanowski et al. (1997) have recently demonstrated unique phosphorylation of Ser-627 (corresponding to Thr-422 mapped in αPAK). Alignment of residues between kinase subdomains VII and VIII of representative PAKs from each species (Fig. 2) shows the relevent amino acid to be either threonine or serine. Curiously the alignment shows that *Xenopus* X-PAK1 (Faure et al. 1997) lacks a residue normally flanking this site (corresponding to αPAK Met-423). The two conserved basic residues at positions -2 and -3 to the phosphorylated threonine (starred) may participate in site recognition. In crystal structures this region, known as the 'regulatory loop', often controls access of substrate to the active site (Taylor et al. 1993). Thus, one might predict that Thr-422 (or equivalent sites in other PAKs) is an important player for PAK activation,

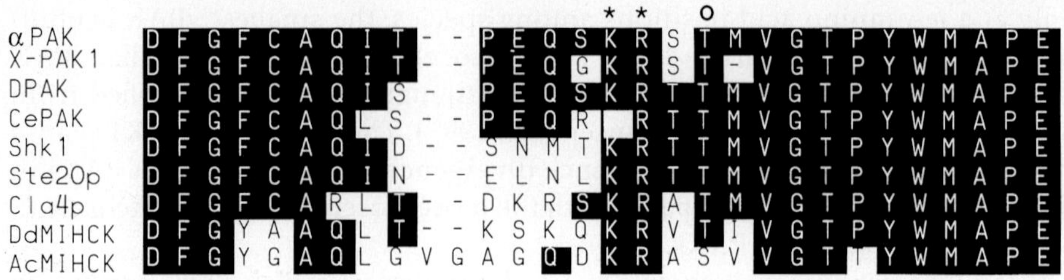

Fig. 2. Sequences comprising the regulatory loop of PAK family kinases. The autophosphorylation site identified in αPAK (Manser et al. 1997) is marked (*o*) and lies at the same position as the phosphorylated serine 627 in *Acanthamoeba* myosin I heavy chain kinase (MIHCK; Szczepanowska et al. 1997). The two conserved basic residues (*starred*) may be required for recognition by the active site. This region plays a key role in regulating protein kinases. (Johnson et al. 1996)

while the N-terminal phosphorylation sites may play supporting or other regulatory roles.

Interestingly, inspection of sequences in the C-terminal half of the PAN region (Fig. 1) indicates a similar extent of cross-species conservation as for the N-terminal region. The function of the C-terminal of the PAN has recently been established as negatively regulating PAK kinase activity (Zhao et al. 1998). This is consistent with two reports that constitutively active PAK mutants can be generated by amino acid substitutions in this region (Bagrodia et al. 1995a; Brown et al. 1996). Thus it is suggested that conformational changes upon p21 binding are transmitted to the flanking sequences to disinhibit the inactive catalytic domain. Removal of these N-terminal inhibitory sequences by proteolysis in vitro (Dennis and Masaracchia, 1993) or by caspase-mediated proteolysis in vivo leads to activation of the kinase (Rudel and Bokoch 1997). *Acanthamoeba* MIHCK is known to be activated by autophosphorylation upon exposure to acidic phospholipids (Brzeska et al. 1996).

4
Functions of Fungal Ste20p/PAK-Like Kinases

Identification of a Cdc42-binding domain in Ste20p (Manser et al. 1994) led to the concept that this and related kinases in budding yeast lie downstream of *CDC42*, which was previously characterized as a gene required for polarized budding (Adams et al. 1990). The relationship between Ste20p and Cdc42p was subsequently tested: Cdc42p does indeed bind to Ste20p, and $Cdc42^{ts}$ strains are entirely defective in the Ste20p-dependent transcriptional pheromone responses mediated by the transcription factor Ste12p (Zhao et al. 1995). GTP-Cdc42 binding was shown to activate Ste20p in vitro (Simon et al. 1995). The significance of the latter result has been questioned by observations that Ste20p is constitutively active when purified from yeast, and that its Cdc42-binding domain is dispensable for pheromone signalling (Peter et al. 1996; Leberer et al. 1997a). Surprisingly, cells containing mutant Ste20p (without ability to bind Cdc42p) were able to induce polarized shmoo formation. Cdc42p must therefore fulfil this role via an uncharacterized pathway downstream of the pheromone receptor (illustrated in Fig. 3). Two-hybrid analysis indicates that the heterotrimeric G-protein β subunit encoded by *STE4* interacts with the Cdc42 guanine nucleotide exchange factor (GEF) Cdc24p (Zhao et al. 1995). Ste20p is also coupled to the pheromone-responsive pathway through Ste5p and Bem1p (the latter protein contains two SH3 domains that contribute to this function; Leeuw et al. 1995), which are themselves directly linked to Ste4p (Fig. 3). This coupling is, however, insufficient to allow signalling to Ste12p; $Cdc24^{ts}$ strains are completely unresponsive to pheromone or Ste4p overexpression (Zhao et al. 1995).

Ste20p plays a morphological role in the switch to pseudohyphal and invasive growth through elements of the same MAP kinase cascade involving Ste11p and Ste7p (Liu et al. 1993). In this case, Ste20p clearly lies downstream of Cdc42p (shown schematically in Fig. 3) in regulating filamentous growth through the same MAP kinase cascade (Mosch et al. 1996). This occurs both in diploid cells, which adopt a filamentous pseudohyphal growth pattern, and in haploid cells, which switch their budding pattern and form filaments that can invade agar plates (Roberts and Fink 1994). The dominant active $RAS2^{V19}$ stimulates filamentous growth (Mosch et al. 1996) which is normally regulated by the nitrogen status of the growth media. In this signalling pathway the Cdc42-binding domain of Ste20p is absolutely essential (Leberer et al. 1997a; Peter et al. 1997).

Ste20p cooperates with the related Cla4p during normal growth: doubly deficient $STE20^{-}/CLA4^{-}$ strains are not viable due to severe defects in bud morphogenesis and cytokinesis (Cvrckova et al. 1995) and rescue with Ste20p requires the protein to interact with Cdc42p (Peters et al. 1996). Although a kinase closely related to Cla4p exists in the *S. cervisiae* genome, it does not apparently share redundant functions with Cla4p – perhaps because it is not constitutively expressed. The tight link between PAK-like kinases and fungal morphological switching is well illustrated in the pathogenic fungus *Candida albicans*. In this organism formation of mycelia is associated with virulence and can be blocked by deletion of the *CaCLA4* gene (Leberer et al. 1997b). In a mouse model for systemic candidiasis, *CaCLA4* deleted strains had suppressed colonization of the kidneys and could survive pathogenic challenge with the mutant.

In fission yeast, pak1 (shk1) deletion or expression of a dominant inactive form of pak1 causes abnormal cell shape and mislocalized actin staining (Marcus et al. 1995; Ottilie et al. 1995). Pak1 has input to the pathway regulating the mating MAP kinase cascade via the MEKK byr2.

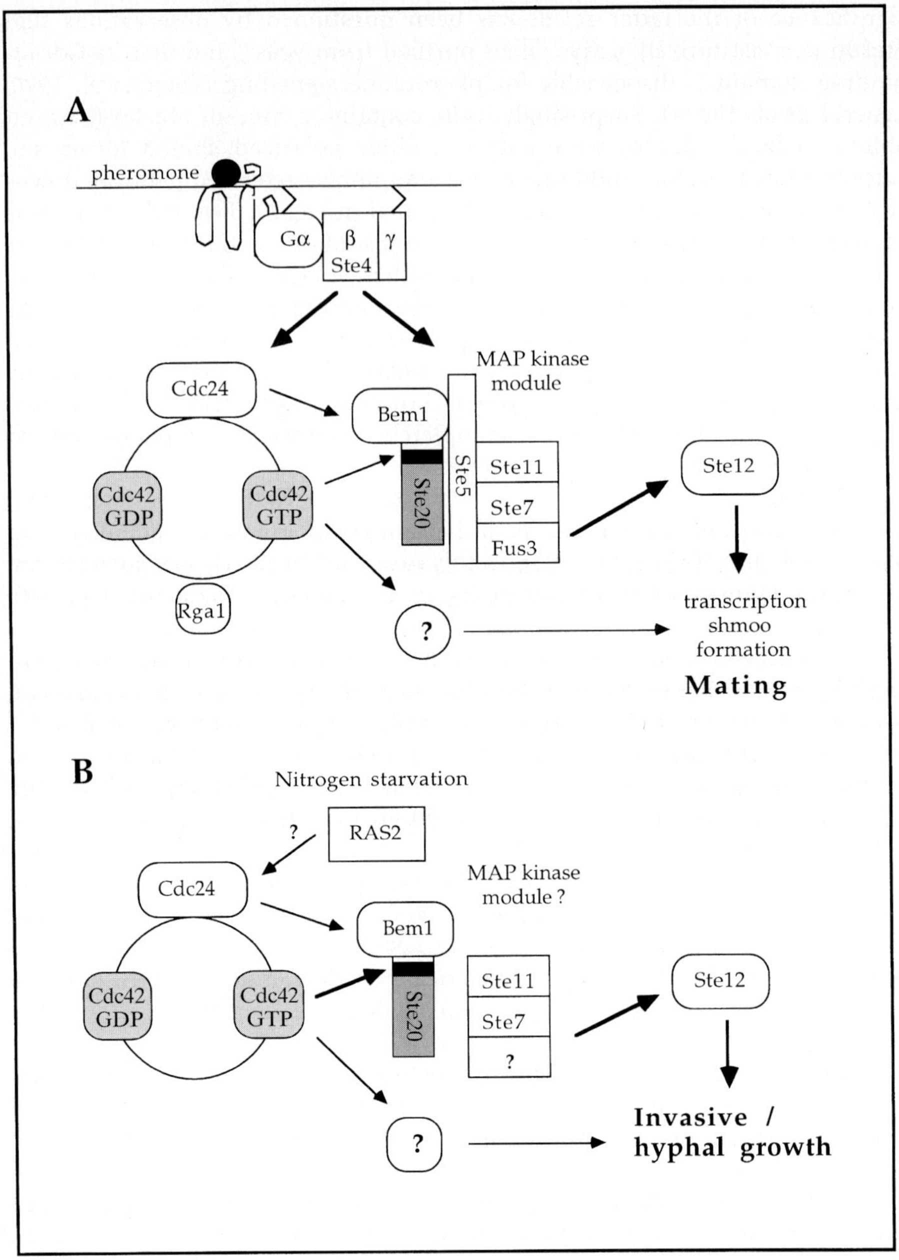
A
pheromone
Gα
β
Ste4
γ
MAP kinase
module
Cdc24
Bem1
Ste5
Ste11
Ste7
Fus3
Ste20
Cdc42
GDP
Cdc42
GTP
Ste12
Rga1
?
transcription
shmoo
formation
Mating
B
Nitrogen starvation
?
RAS2
Cdc24
MAP kinase
module ?
Bem1
Ste11
Ste7
?
Ste20
Cdc42
GDP
Cdc42
GTP
Ste12
?
Invasive /
hyphal growth

5
PAKs and Phosphorylation Pathways

The oxidative burst of phagocytic leukocytes (neutrophils) is mediated by a multi-component NADPH oxidase regulated by Rac (Abo et al. 1991). Both $p67^{phox}$ and PAK are targets for Cdc42 and Rac1 in neutrophils (Prigmore et al. 1995), where PAK may mediate stimulus-activated phosphorylation of $p47^{phox}$, an event required for association of cytosolic components with the membrane (Knaus et al. 1995). Peptides containing residue Ser-328 of $p47^{phox}$ represent good substrates for PAK, this being the site of in vivo phosphorylation upon neutrophil activation (Benna et al. 1994). It would be of interest to determine effects of PAK inhibition on the oxidative burst response of these cells.

An enzyme corresponding to PAK was biochemically identified some time ago as a kinase activity which phosphorylates the C-terminus of the S6 ribosomal subunit (Dennis and Masaracchia 1993). Like the p70 S6 kinase (which apparently also lies downstream of Cdc42 and Rac1; Chou and Blenis 1996) and p90 rsk, purified PAK can phosphorylate a synthetic peptide corresponding to this regulatory region of the S6 protein. The significance of this observation has not yet been addressed. In vitro, PAK also acts on two commonly used basic protein substrates, namely histone H4 and myelin basic protein.

The myosin I heavy chain kinases (MIHCK) from *Dictyostelium* and *Acanthamoeba* represent PAK isoforms in these species (Brzeska et al. 1996; Lee et al. 1996). Myosin 1 isozymes in these organisms are regulated through phosphorylation by MIHCK of a single site in the 'head' which activates maximal actin-dependent MgATPase activity (Lee and Cote 1995). However, sequence alignments of these with mammalian myosin 1 isoforms shows the phosphorylated site to be occupied by glutamate or aspartate residues, suggesting there is no requirement for similar PAK phosphorylation in higher organisms (Bement and Mooseker 1995). Thus, although PAK kinases from man to yeast can phosphorylate *Dictyostelium* and *Acanthamoeba* myosin 1 in

◄

Fig. 3a,b. Signalling pathways involving the budding yeast Cdc42 and the PAK-family kinase Ste20p. The *larger arrows* indicate interactions that have been shown to be absolutely required for signalling. **A** Pheromone responses occur via activation of a heterotrimeric G-protein complex which signals through the Ste4p β-subunit. Ste4p can bind to Cdc24 (Zhao et al. 1995) and two scaffolding proteins Bem1 and Ste5 (Leeuw et al. 1995). The MAP kinase module outputs through Fus3p to Ste12p to regulate transcription. Cdc24 acts to catalyze guanine nucleotide exchange on Cdc42: this complex localizes and is required for formation of the mating projection or shmoo through the uncharacterized pathway (*?*). Ste20p may participate in the process of cell-cell adhesion during conjugation (Leberer et al. 1997a). Cdc42 is negatively regulated by the GAP Rga1p. **B** The switch to filamentous growth in diploid cells in response to nitrogen starvation is regulated by elements of the pheromone response pathway including Cdc42 and Ste20p. However, in this case, Ste5p and Fus3 do not participate in signalling (Mosch et al. 1996). The Ras2p protein may bind and activate Cdc24p, thus initiating the cascade that requires the interaction of Cdc42 and Ste20p (Leberer et al. 1997a; Peter et al. 1997). Transcriptional activation of a different set of genes utilizes the same transcription factor Ste12p

vitro (Wu et al. 1996), the significance of this to mammalian systems is doubtful. In terms of site recognition, the phosphorylation site in a peptide corresponding to the *Acanthamoeba* myosin 1C regulatory site (acted on by both MIHCK and active PAK1) is flanked by an important tyrosine at position +2 and a basic residue at −2 (Brzeska et al. 1997).

The observation that Ste20p in *S. cerevisiae* lies genetically at the top of the pheromone-inducible MAP kinase cascade and phosphorylates the MEKK Ste11p (Wu et al. 1995), has suggested that PAKs play an analogous role in multicellular organisms. Further, Ste20p can induce ERK1 activation in *Xenopus* oocytes, and also activates Jun N-terminal kinase (JNK) as does PAK1 (Polverino et al. 1995). A role for PAK in such signalling is particularly attractive because both Cdc42 and Rac1 can activate JNK (Minden et al. 1995). Experiments in mammalian cells have shown that active PAK can modestly stimulate both JNK and p38 MAP kinase but not the ERK1 pathways (Zhang et al. 1995). Hyperosmotic shock in ventricular myocytes (which activates p38) can activate αPAK (Clerk and Sugden 1997). Recently, Fanger et al. (1997) have shown that MEKK1 and MEKK4 proteins directly bind Cdc42 and Rac1, but dominant inhibitory MEKKs do not block PAK activation of JNK. This suggests that mammalian PAKs do not function like Ste20p through MEKKs (i.e. Ste11p).

6 PAK and Cellular Transformation

The possibility that PAK participates in transformation events has been raised by the observation that Rho p21s cooperate with Ras, and play an essential role in transformation (Qiu et al. 1995; Roux et al. 1997). Activated Cdc42 and Rac also drive cell cycle progression (Olson et al. 1995). In addition to the observation that many RhoGEFs are oncogenic (as reviewed by Cerione and Zheng, 1996) it has recently been shown that fibroblasts expressing constitutively active Cdc42 are anchorage-independent (Qiu et al. 1997) and that a Cdc42(F28L) mutant is transforming (Lin et al. 1997). However, experiments with Rac1 mutants designed to interact selectively with various identified effector proteins have strongly suggested that PAK is not required for these events (Joneson et al. 1996; Lamarche et al. 1996; Westwick et al. 1997). In particular, Rac1 containing mutations Y40C or Y40H fail to bind PAK but show normal cooperativity with Raf-CAAX in foci formation assays. Conversely, the Rac(F37A) mutant which binds and activates PAK normally, cannot promote transformation.

By contrast, the N-terminal region of PAK is capable of blocking the Ras-Rac pathway leading to transformation (Osada et al. 1997). Also, a catalytically inactive version of PAK has been shown to inhibit Ras-driven transformation in Rat-1 cells, though curiously not in NIH3T3 cells (Tang et al. 1997). This inhibition was not through sequestration of active Cdc42/Rac1, suggesting that different (cell-type specific?) signalling molecules are "titrated out" by excess inactive PAK. One candidate protein would be the adaptor Nck, which is itself transforming (Lehman et al. 1990), and could cooperate in Ras transformation. Several groups have documented that the N-terminal 20 residues of mammalian PAKs contain a proline-rich motif which binds Nck in vitro and in vivo (Bagrodia et al. 1995b; Bokoch et al. 1996). It has been suggested that growth factor stimulation, leading to recruitment of Nck, brings PAK to the plasma membrane (Galisteo et al. 1996). Membrane targeting leads to modest activation of PAK (Manser et al. 1997).

Recently, a new class of PAK substrates have been isolated which bind tightly to a PAK proline-rich sequence located between the p21-binding and catalytic domains. These proteins contain, in addition to an N-terminal SH3 domain, a Dbl homology (DH) and pleckstrin homology (PH) domain, and are termed PAK-interacting exchange factors (PIXs; Manser et al. 1998). Within the DH domain, PIX is most closely related to the mammalian Tiam1 protein which acts as an activator of Rac1 (Michiels et al. 1995). Therefore, overexpression of kinase-inactive PAK might function by blocking endogenous PIX recruitment and Rac activation in these cells.

7
Morphological Roles for PAK Downstream of Cdc42 and Rac

The ability of Cdc42 and Rac1 to drive distinct morphological changes involving a coordinated alteration in the actin cytoskeleton, focal adhesion complexes and the plasma membrane (Ridley et al. 1992, Kozma et al. 1995; Nobes and Hall 1995) suggested that the ubiquitous effector PAK would be involved in some common aspect of these rearrangements. Mammalian PAKs are cytoplasmic kinases in resting cells which become recruited to focal complexes (FCs) and membrane ruffles in response to growth factors or activated Cdc42 (Dharmawardhane et al. 1997; Manser et al. 1997). PAK also colocalizes with polymerized actin at the edge of lamellipodia (Dharmawardhane et al. 1997). Cdc42-driven FCs are phenotypically different from RhoA-dependent structures, and enriched for PAK (see Fig. 4). In the *Drosophila* embryo, DPAK is concentrated at the leading edge of epithelial cells undergoing dorsal closure where it co-localizes with phosphotyrosine-rich focal complex-like structures (Harden et al. 1996). The *C. elegans* CePAK colocalizes with its partners CeRac1 and Cdc42Ce at hypodermal cell boundaries in the worm embryo (Chen et al. 1996). These observations have suggested that PAK has a primary role at focal complexes and cell-cell junctions, a site enriched for a number of Rho signalling molecules, including GTPase activating proteins (GAPs) and guanine nucleotide exchange proteins (GEFs; Hildebrand et al. 1996; Hobert et al. 1996).

Microinjection of plasmids encoding constitutively active forms of αPAK in epithelial and fibroblastic cells leads to a dramatic loss of focal adhesions and

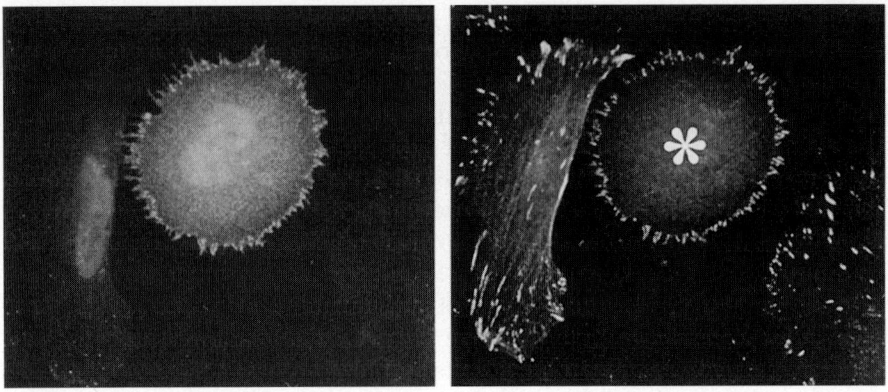

Fig. 4. Cdc42 causes a rearrangement of focal complexes which are enriched for PAK. HeLa cells were cotransfected with expression plasmids encoding $Cdc42^{G12V}$ and αPAK. After 16h, slides were fixed and processed for indirect immunoflourescence as previously described (Manser et al. 1997). Anti-paxillin staining shows the organization of focal adhesions in untransfected cells and in a typical $Cdc42^{G12V}$ expressing cell (*star*) which has become rounded. The anti-PAK staining is enriched in these focal adhesions: αPAK by itself is cytoplasmic

actin stress fibres with concomitant cell contraction (Manser et al. 1997). Opposite effects are seen in fibroblasts injected with wildtype or catalytically inactive PAK (Sells et al. 1997). The loss of stress fibres in cells expressing activated Cdc42 is well documented (Kozma et al. 1995, 1997; Dutarte et al. 1996). By specifically inhibiting PAK, it is possible to demonstrate that $Cdc42^{G12V}$ and $Rac1^{G12V}$ indeed utilize PAK to break down actin stress fibres (Zhao et al. 1998). Additionally, PAK activity is required for other morphological effects of $Cdc42^{G12V}$ expression, namely cell rounding and production of filopodia. Since PAK promotes turnover of focal complexes, what protein is responsible for generating the new, peripherally located Cdc42-type FCs (illustrated in Fig. 4), to which PAK is localized? The recent identification of myotonin-related Cdc42-binding kinases (MRCKs), similar to ROK in the catalytic domain, but with a PAK-like p21-binding domain, may provide an answer (Leung et al. 1998). ROK kinases interact with GTP-RhoA to drive focal adhesion formation through activity of the kinase domain (Leung et al. 1996), and these RhoA functions can be blocked by expressing a catalytically inactive ROK. A dominant inactive version of MRCK blocks formation of focal complexes downstream of Cdc42 but not of RhoA. Thus, in the cellular context, GTP-Cdc42 can recruit two kinases: PAK, which promotes destruction of existing focal adhesions, and MRCK to form new focal complexes. These new focal complexes are themselves enriched for PAK, which is then likely to provide continuous dynamic turnover of the structures. The properties of the tumor invasiveness gene Tiam1, which acts as a RacGEF with metastatic potential (Michiels et al. 1995), could relate to the ability of activated PAK to disrupt normal adherence to the extracellular matrix.

How does PAK localize to focal complexes under the influence of Cdc42 or Rac? PAK does not show significant homology to any characterized focal complex proteins, for example paxillin, vinculin, actinin, and FAK. Deletion analysis of αPAK indicated important sequences for PAK localization were present in residues 1–60 and between 150–250 (Manser et al. 1997). The latter requirement appears to represent binding to the PAK-interacting exchange factor (PIX78) present in cultured cells (Manser et al. 1998). A mutant PAK which is unable to bind PIX is poorly localized to focal complexes generated by $Cdc42^{G12V}$. PIX itself is present at low levels in RhoA-dependent focal adhesions and enriched in Cdc42-dependent peripheral focal complexes, as is its partner PAK. Our present understanding of PAK function suggests the kinase is recruited as a complex with this Rac guanine nucleotide exchange factor (RacGEF), either through direct binding to Cdc42 or through other partners such as Nck. This in turn is likely to cause Rac1 activation and therefore activation of PAK itself, providing a feedback loop.

8 Activities Associated with the Various Domains of PAK

As outlined above, our current understanding is that PAK family kinases function not merely through phosphorylation of target proteins but also through important interactions outside the 'catalytic' domain. Key protein associations occur through conserved proline-rich SH3 binding sites (depicted in Fig. 5). Possible roles for these interactions as well as the other domains of these kinases are given below.

8.1 Proline-Rich Domains

There are three conserved proline-rich (PR) domains comparing sequences of αPAK, DPAK and CePAK while X-PAK1 only has PR1 and PR3 (Fig. 5). The N-terminal site binds to the second SH3 domain of the adaptor proto-oncogene Nck in vitro (Bagrodia et al. 1995b) and in vivo (Galisteo et al. 1996). The *Drosophila* Nck homologue is encoded by the dreadlocks gene (*dock*), which is required for photoreceptor axonal guidance (Garrity et al. 1996): an interaction with DPAK is therefore suspected. Since Drac1 inhibition also has specific effects on *Drosophila* axonal outgrowth (Luo et al. 1994) it is possible that this reflects a loss of DPAK/dock signalling.

The structures induced in 3T3 fibroblasts microinjected with PAK1 protein or PAK1 expression plasmids are apparently independent of both p21 binding activity and kinase activity. Sells et al. (1997) actually observed potentiated formation of polarized filopodia and membrane ruffles (and underlying focal complexes) with p21-binding deficient PAK1(H83,86L) or catalytically inactive PAK1(K299R), compared with wildtype kinase. These rearrangements, which

A

	PR1	PR2	PR3
Rn	DKPPAPPMRNTS	KPLPPNPEL	PPPVIAPRPEHTKSVYTR
Xl	DKPPPPPMRINS		VAPAFPPRPEHTKSIYTR
Dm	DKPPAPPVRLTS	RPLPKEPDL	PPPPVASRPERTKSIYTR
Ce	EKPPAPPIRFSS	KPLPKEPEL	VPPAIPDRPARTLSIYTK

B

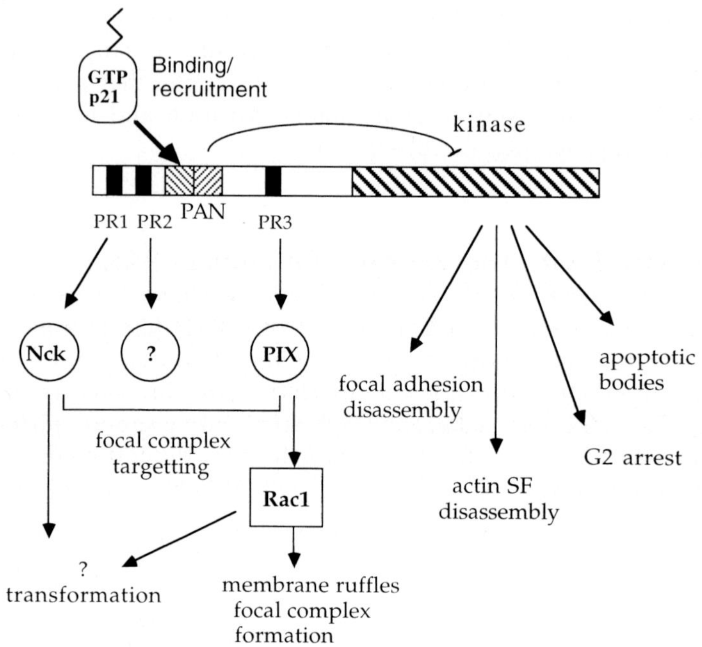

Fig. 5. A schematic diagram of the different roles of vertebrate PAKs. **A** The conserved proline-rich regions which provide binding sites for various SH3 containing proteins (positions marked in **B**). Among PAKs from multicellular organisms there is conservation of these sites including certain amino acids flanking the prolines. **B** The location of the proline-rich (PR) motifs shown in **A** are indicated by *black boxes*. The kinase domain acts to promote a number of activities but the targets in this case are not known. The N-terminal region is complexed to at least two proteins, Nck and PIX which may function to bring PAK to the membrane (Lu et al. 1997). Alternatively, recruitment of PAK by GTP-Cdc42/Rac may bring these proteins onto the membrane

are very reminiscent of Rac1 effects, were somewhat different from those obtained with a PAK mutated in PR1 (Sells et al. 1997): however it seems likely that they also relate to binding of the Rac1 GEF PIX to PR3 (see below).

The PAK PR2 motif is a typical SH3 binding site with consensus RPxPxxP, but the ligand for this has yet to be identified. The PR3 is unique because it does not contain a conventional PxxP motif and yet binds with high affinity to the SH3 domain of PIX (Manser et al. 1998). Thus, cytosolic PAK forms a constitutive complex with a Rac1 GEF and may mediate Cdc42 → Rac signalling upon recruitment to the membrane. A *Drosophila* homologue of PIX has recently been cloned (Werner and Manseau 1997); its interaction with DPAK has not yet been tested.

SH3 domains are also implicated in Ste20p signalling (see Fig. 3): the Bem1p adaptor protein contains two SH3 domains with the second SH3 being required for binding to Ste20p (Leeuw et al. 1995). The target sequence in the kinase has not yet been identified.

8.2 The p21-Binding Domain

This region binds GTP forms of Cdc42 or Rac1 and therefore can be used to specifically inhibit Cdc42/Rac1 signalling, due to its high affinity for these p21s (Osada et al. 1997). Currently, we know of two mechanisms by which PAK activity can be elevated: binding of GTP-p21 to the N-terminal CRIB (as described in Sect. 3) or proteolytic removal of this region. It is not clear whether kinase activation by targeting PAK to the membrane is mediated by p21 binding. Similarly, the mechanism of PAK activation by introduction of the HIV Nef-protein (Sawai et al. 1996) is under investigation: the basis for an association of PAK with Nef immunoprecipitates is unknown (Nunn and Marsh 1996). This could represent an alternate mechanism of PAK activation (for example by phosphorylation of PAK by the p62 Nef-associated kinase).

Two other kinases with related functional CRIB motifs are mixed lineage kinase 3 (Burbelo et al. 1995) and MRCK (see Sect. 7), but direct activation of these kinases by p21 has not been demonstrated. It seems likely that the p21 binding in these cases is primarily a means of localizing the kinases to specific signalling complexes. Results from studies of Ste20p (which is constitutively active – see Sect. 4) indicates that during pseudo-hyphal growth or budding, Cdc42 provides an essential localizing signal for the kinase (Leberer et al. 1997a; Peter et al. 1997). The p21 binding domain therefore has two distinct roles, kinase activation and kinase localization. It is also possible under physiological conditions that mammalian PAKs become associated with p21, yet remain in a relatively inactive state due to specific inhibition of the kinase.

8.3 The Auto-Regulatory Domain

Unlike pseudosubstrate sequences, which can function as relatively short peptides (<20 residue) to compete for substrate binding within the catalytic cleft (Johnson et al. 1996), the PAK inhibitory domain is larger and functions primarily to prevent kinase autophosphorylation (Zhao et al. 1998). An interaction of this inhibitor region with the catalytic domain is likely. Structural studies may be required to provide a clearer understanding of how these sequences transmit changes in the neighbouring p21-binding domain to regulate kinase activity.

8.4
The Kinase Domain

In *Xenopus* oocytes, mammalian PAK has been shown to arrest cleavage at mitotic metaphase (Rooney et al. 1996), though the molecular basis for this activity is unknown. Microinjection of the kinase domain of X-PAK1 (lacking the N-terminal regulatory region) also arrests *Xenopus* oocytes in the G2 phase of the cell-cycle (Faure et al. 1997). Curiously, a catalytically inactive version of this construct was capable of triggering apoptosis. These events were apparently independent of effects due to the N-terminal non-catalytic domain. X-PAK1 itself can be cleaved by caspases in oocyte extracts to release an active catalytic fragment: taken together these results suggest X-PAK1 prevents apoptosis (Faure et al. 1997). This contrasts with the observation that caspase-mediated cleavage of mammalian PAK2, following onset of apoptosis, is required for proper cell condensation and formation of apoptotic bodies (Rudel and Bokoch 1997). Such actions resemble the effects of expressing constitutively active PAK in breaking down actin stress fibres and causing cell condensation (Manser et al. 1997). Recent identification of gelsolin as a morphological effector during apoptosis (Kothakota et al. 1997) points to a key role for the actin cytoskeleton in regulating cell shape changes required for cell rounding and blebbing.

9
Conclusions

PAK kinases are an ancient family of proteins which might be considered the prototype Cdc42 target (Rac not being found in yeast). Although mammalian PAKs are potently activated by Cdc42 and Rac1, Ste20p is apparently constitutively active when immunoprecipitated from yeast cells. Sequence comparison (Fig. 1) appears to indicate that Ste20p does contain a functional negatively regulatory sequence flanking the p21 binding domain, suggesting a different mechanism of activation can occur (independent of Cdc42 function). In multicellular organisms it is likely that PAKs are linked to phosphotyrosine-rich sites through interaction with SH3 domain containing proteins. PAK itself has not been reported to be phosphorylated on tyrosine.

We are now at a stage where further understanding of PAK function requires knowledge of the in vivo substrates of these kinases, and the identity of in vivo PAK-interacting proteins. While it is known that PAKs can be activated via heterotrimeric G-protein signalling (Knaus et al. 1995; Teo et al. 1995) these upstream pathways are also uncharacterized. The discovery that one class of PAK substrate is a tightly complexed RacGEF (PIX) opens up new areas for investigation. Clearly, the identity of other PAK-associated proteins and substrates will allow us to unravel further the complexities of PAK function.

Acknowledgements. This work is supported by the Glaxo Singapore Research Fund.

References

Abo A, Pick E, Hall A, Totty N, Teahan CG, Segal AW (1991) Activation of the NADPH oxidase involves the small GTP-binding protein p21Rac1. Nature 353:668–670

Adams AEM, Johnson DI, Longnecker RM, Sloat BF, Pringle JR (1990) CDC42 and CDC43, two additional genes involved in budding and the establishment of cell polarity in the yeast *Saccharomyces cerevisiae.* J Cell Biol 111:131–142

Amano M, Ito M, Kimura K, Fukata Y, Chihara K, Matsui T, Hamajima Y, Okawa K, Iwamatsu A, Kaibuchi K (1996) Identification of a putative target for Rho as the serine-threonine kinase protein kinase N. Science 271:648–650

Bagrodia S, Derijard B, Davis RJ, Cerione RA (1995a) Cdc42 and PAK-mediated signaling leads to Jun kinase and p38 mitogen activated protein kinase activation. J Biol Chem 270:27995–27998

Bagrodia S, Taylor SJ, Creasy CL, Chernoff J, Cerione RA (1995b) Identification of a mouse $p21^{cdc42/rac}$ activated kinase. J Biol Chem 270:22731–22737

Bement WM, Mooseker MS (1995) TEDS rule: a molecular rationale for differential regulation of myosins by phosphorylation of the heavy chain head. Cell Motil Cytoskeleton 31:87–92

Benna JE, Faust LP, Babior BM (1994) The phosphorylation of the respiratory burst oxidase component p47phox during neurtophil activation. J Biol Chem 269:23431–23436

Benner GE, Dennis PB, Masaracchia RA (1995) Activation of an S6/H4 (PAK65) from human placenta by intramolecular and intermolecular autophosphorylation. J Biol Chem 270:21121–21128

Bokoch GM, Wang Y, Bohl BP, Sells MA, Quilliam LA, Knaus UG (1996) Interaction of the Nck adapter protein with p21-activated kinase (PAK1). J Biol Chem 271:25746–25749

Brown JL, Stowers L, Baer M, Trejo JA, Coughlin S, Chant J (1996) Human Ste20 homologue hPAK1 links GTPases to the JNK MAP kinase pathway. Curr Biol 6:598–605

Brzeska H, Szczepanowska J, Hoey J, Korn ED (1996) The catalytic domain of *Acanthamoeba* myosin I heavy chain kinase. J Biol Chem 271:27056–27062

Brzeska H, Knaus UG, Wang Z-Y, Bokoch GM, Korn ED (1997) p21-activated kinase has substrate specificity similar to *Acanthamoeba* myosin I. Proc Natl Acad Sci USA 94:1092–1095

Bush J, Franek K, Cardelli J (1993) Cloning and characterization of seven novel *Dictyostelium discoideum rac*-related genes belonging to the *rho* family of GTPases. Gene 136:61–68

Burbelo PD, Drechsel D, Hall A (1995) A conserved binding motif defines numerous candidate partner proteins for both Cdc42 and Rac GTPases. J Biol Chem 270:29071–29074

Cerione RA, Zheng Y (1996) The Dbl family of oncogenes. Curr Opin Cell Biol 8:216–222

Chen W, Chen S, Yap S-F, Lim L (1996) The *Caenorhabditis elegans* p21-activated kinase (CePAK) colocalizes with CeRac1 and Cdc42Ce at hypodermal cell boundaries during embryo elongation. J Biol Chem 271:26362–26368

Chou MM, Blenis J (1996) The 70 kDa S6 kinase complexes with and is activated by the Rho family G proteins Cdc42 and Rac1. Cell 17:573–583

Clerk A, Sugden PH (1997) Activation of p21-activated protein kinase α (αPAK) by hyperosmotic shock in neonatal ventricular myocytes. FEBS Lett 403:23–25

Cvrckova F, De Virgilio C, Manser E, Pringle JR, Nasmyth K (1995) Ste20-like protein kinases are required for normal localization of cell growth and for cytokinesis in budding yeast. Genes Dev 9:1817–1830

Dennis PB, Masaracchia RA (1993) Activation of an S6 kinase from human placenta by autophosphorylation. J Biol Chem 268:19833–19841

Dharmawardhane S, Sanders LS, Martin SS, Daniels RH, Bokoch GM (1997) Localization of p21-activated kinase 1 (PAK1) to pinocytotic vesicles and cortical actin structures in stimulated cells. J Cell Biol 138:1265–1278

Dutarte H, Davoust J, Gorvel J-P, Chavrier P (1996) Cytokinesis arrest and redistribution of actin-cytoskeleton regulatory components in cells expressing the Rho GTPase Cdc42Hs. J Cell Sci 109:367–377

Fanger GR, Johnson NL, Johnson GL (1997) MEK kinases are regulated by EGF and selectively interact with Rac/Cdc42. EMBO J 16:4961–4972

Faure S, Vigneron S, Doree M, Morin N (1997) A member of the Ste20/PAK family of protein kinases is involved in both arrest of *Xenopus* oocytes at G2/prophase of the first meiotic cell cycle and in prevention of apoptosis. EMBO J 16:5550–5561

Galisteo ML, Chernoff J, Su Y-C, Skolnik EY, Schlessinger J (1996) The adaptor protein Nck links receptor tyrosine kinases with the serine-threonine kinase Pak1. J Biol Chem 271:20997–21000

Garrity PA, Rao Y, Salecker I, McGlade J, Pawson T, Zipursky SL (1996) *Drosophila* photoreceptor axon guidance and targeting requires the dreadlocks SH2/SH3 adaptor protein. Cell 85:639–650

Harden N, Lee J, Loh H-Y, Ong Y-M, Tan I, Leung T, Manser E, Lim L (1996) A *Drosophila* homolog of the Rac- and Cdc42-activated serine/threonine kinase PAK is a potential focal adhesion and focal complex protein that co-localizes with dynamic actin structures. Mol Cell Biol 16:1896–1908

Hildebrand JD, Taylor JM, Parsons JT (1996) An SH3 domain-containing GTPase-activating protein for Rho and Cdc42 associates with focal adhesion kinase. Mol Cell Biol 16:3169–3178

Hobert O, Schilling JW, Beckerle MC, Ullrich A, Jallal B (1996) SH3 domain dependent interaction of the proto-oncogene product Vav with the focal contact protein zyxin. Oncogene 12:1577–1581

Hunter T, Plowman GD (1997) The protein kinases of budding yeast: six score and more. Trends Biochem Sci 22:18–22

Johnson LN, Noble MEM, Owen DJ (1996) Active and inactive protein kinases: structural basis for regulation. Cell 85:149–158

Joneson T, McDonugh M, Bar-Sagi D, Van Aelst L (1996) Rac regulation of actin polymerization and proliferation by a pathway distinct from Jun kinase. Science 274:1374–1376

Knaus UG, Morris S, Dong H-J, Chernoff J, Bokoch GM (1995) Regulation of human leukocyte p21-activated kinases through G-protein-coupled receptors. Science 269:221–223

Kothakota S, Azuma T, Reinhard C, Klippel A, Tang J, Chu K, McGarry TJ, Kirschner MW, Koths K, Kwiatkowski DJ, Williams LT (1997) Caspase-3 generated fragments of gelsolin: effector of morphological change in apoptosis. Science 278:294–298

Kozma R, Ahmed S, Best A, Lim L (1995) The Ras-related protein Cdc42Hs and bradykinin promote formation of peripheral actin microspikes and filopodia in Swiss 3T3 fibroblasts. Mol Cell Biol 15:1942–1952

Kozma R, Sarner S, Ahmed S, Lim L (1997) Rho family GTPases and neuronal growth cone remodelling: relationship between increased complexity induced by Cdc42Hs, Rac1, and acetylcholine and collapse induced by RhoA and lysophosphatidic acid. Mol Cell Biol 17:1201–1211

Lamarche N, Tapon N, Stowers L, Burbelo PD, Aspenstrom P, Bridges T, Chant J, Hall A (1996) Rac and Cdc42 induce actin polymerization and G1 cell cycle progression independently of $p65^{PAK}$ and the JNK/SAPK MAP kinase cascade. Cell 8:519–529

Leberer E, Dignard D, Harcus D, Thomas DY, Whiteway M (1992) The protein kinase homologue Ste20p is required to link the yeast pheromone response G-protein $\beta\gamma$ subunits to downstream signalling components. EMBO J 11:4815–4824

Leberer E, Wu C, Leeuw T, Fourest-Lieuvin A, Segall JE, Thomas DY (1997a) Functional characterization of the Cdc42p binding domain of yeast Ste20p protein kinase. EMBO J 16:83–97

Leberer E, Ziegelbauer Schmidt A, Harcus D, Dignard D, Ash J, Johnson L, Thomas DY (1997b) Virulence and hyphal formation of *Candida albicans* require the Ste20p-like protein kinase CaCla4p. Curr Biol 7:539–546

Lee S-F, Cote GP (1995) Purification and characterization of a *Dictyostelium* protein kinase required for actin activation of the Mg^{2+}ATPase activity of *Dictyostelium* myosin 1D. J Biol Chem 170:11776–11782

Lee S-F, Egelhoff TT, Mahasneh A, Cote GP (1996) Cloning and characterization of a *Dictyostelium* myosin I heavy chain kinase activated by Cdc42 and Rac. J Biol Chem 271:27044–27048

Leeuw T, Fourest-Lieuvin A, Wu C, Chenevert J, Clark K, Whiteway M, Thomas DY, Leberer E (1995) Pheromone response in yeast: association of Bem1p with proteins of the MAP kinase cascade and actin. Science 270:1210–1213

Lehman JM, Riethmuller G, Johnson JP (1990) Nck, a melanoma cDNA encoding a cytoplasmic protein consisting of Src homology units SH2 and SH3. Nucleic Acid Res 18: 1048

Leung T, Chen X-Q, Manser E, Lim L (1996) The p160 RhoA-binding kinase ROKa is a member of a kinase family and is involved in the reorganization of the cytoskeleton. Mol Cell Biol, 16:5313–5327

Leung T, Chen X-Q, Tan I, Manser E, Lim L (1998) Myotonic dystrophy kinase-related Cdc42-binding kinase acts as a Cdc42 effector in promoting cytoskeletal reorganization. Mol Cell Biol 18:130–140

Lin R, Bagrodia S, Cerione R, Manor D (1997) A novel Cdc42Hs mutant induces cellular transformation. Curr Biol 7:794–797

Liu H, Styles C, Fink GR (1993) Elements of the yeast pheromone response pathway required for filamentous growth of diploids. Science 262:1741–1744

Lu W, Katz S, Gupta R, Mayer BJ (1997) Activation of Pak by membrane localization mediated by an SH3 domain from the adaptor protein Nck. Curr Biol 7:85–94

Luo L, Liao Y-J, Jan L-Y, Jan Y-N (1994) Distinct morphogenic functions of similar small GTPases: *Drosophila* Drac1 is involved in axonal outgrowth and myoblast fusion. Genes Dev 8:1787–1802

Manser E, Leung T, Salihuddin H, Tan L, Lim L (1993) A non-receptor tyrosine kinase that inhibits the GTPase activity of $p21^{cdc42}$. Nature 363:364–367

Manser E, Leung T, Salihuddin H, Zhao Z-S, Lim L (1994) A brain serine/ threonine protein kinase activated by Cdc42 and Rac1. Nature 367:40–46

Manser E, Chong C, Zhao Z-S, Leung T, Michael G, Hall C, Lim L (1995) Molecular cloning of a new member of the p21-Cdc42/Rac-activated kinase (PAK) family. J Biol Chem 270:25070–25078

Manser E, Huang H-Y, Loo T-H, Chen X-Q, Dong J-M, Leung T, Lim L (1997) Expression of constitutively active a-PAK reveals effects of the kinase on actin and focal complexes. Mol Cell Biol 17:1129–1143

Manser E, Loo T-H, Koh C-G, Zhao Z-S, Chen X-Q, Tan L, Tan I, Leung T, Lim L (1998) PAK kinases are directly coupled to the PIX family of nucleotide exchange factors. Mol Cell 1:183–192

Marcus S, Polverino A, Chang E, Robbins D, Cobb MH, Wigler MH (1995) Shk1, a homolog of the *Saccharomyces cerevisiae* Ste20 and mammalian p65PAK protein kinases, is a component of a Ras/Cdc42 signaling module in the fission yeast *Schizosaccharomyces pombe.* Proc Natl Acad Sci U S A 92:6180–6184

Martin GA, Bollag G, McCormick FA, Abo A (1995) A novel serine kinase activated by Rac/ Cdc42Hs-dependent autophosphorylation is related to PAK65 and Ste20. EMBO J 14:1970–1978

McCormick F (1994) Raf: the holy grail of Ras biology? Trends Cell Biol 4:347–350

Michiels F, Habets GG, Stam JC, van der Kammen RA, Collard JG (1995) A role for Rac in Tiam1-induced membrane ruffling and invasion. Nature 375:338–340

Minden A, Lin A, Claret F-X, Abo A, Karin M (1995) Selective activation of the JNK signalling cascade and c-Jun transcriptional activity by the small GTPases Rac and Cdc42Hs. Cell 81:1147–1157

Mosch H-U, Roberts RL, Fink GR (1996) Ras2 signals via the Cdc42/Ste20/ mitogen-activated protein kinase module to induce filamentous growth in *Saccaromyces cerevisiae.* Proc Natl Acad Sci U S A 93:5352–5356

Munemitsu S, Innis MA, Clark R, McCormick F, Ullrich A, Polarkis P (1990) Molecular cloning and expression of a G25K cDNA, the human homolog of the yeast cell cycle gene *CDC42.* Mol Cell Biol 10:5977–5982

Nobes CD, Hall A (1995) Rho, Rac, and Cdc42 GTPases regulate the assembly of multi-molecular focal complexes associated with actin stress fibres, lamellipodia, and filopodia. Cell 81:53–62

Nunn MF, Marsh JW (1996) Human immunodeficiency virus type 1 Nef associates with a member of the p21-activated kinase family. J Virol 70:6157–6161

Olson MF, Ashworth A, Hall A (1995) An essential role for Rho, Rac, and Cdc42 GTPases in cell cycle progression through G_1. Science 269:1270–1272

Osada S, Azawa M, Koyama T, Hirai S, Ohno S (1997) A domain containing the Cdc42/Rac interactive binding (CRIB) region of p65(PAK) inhibits transcriptional activation and cell transformation by the Ras-Rac pathway. FEBS Lett 404:227–233

Ottilie S, Miller PJ, Johnson DI, Creasy CL, Sells MA, Bagrodia S, Forsburg SL, Chernoff J (1995) Fission yeast pak1+ encodes a protein kinase that interacts with Cdc42p and is involved in the control of cell polarity and mating. EMBO J 14:5908–5919

Peter M, Neiman AM, Park H-O, van-Lohuizen M, Herskowitz I (1996) Functional analysis of the interaction between the small GTP binding protein Cdc42 and the Ste20 protein kinase in yeast. EMBO J 15:7046–7059

Polverino A, Frost J, Yang P, Hutchison M, Neiman AM, Cobb MH, Marcus S (1995) Activation of mitogen-activated protein kinase cascades by p21-activated protein kinases in cell-free extracts of *Xenopus* oocytes. J Biol Chem 270:26067–26070

Prigmore E, Ahmed S, Best A, Kozma R, Manser E, Segal A, Lim L (1995) A 68 kDa kinase and p67 neutrophil oxidase are targets for Cdc42Hs and Rac1 in neutrophils. J Biol Chem 270:10717–10722

Qui RG, Chen J, Kirn D, McCormick F, Symons M (1995) An essential role for Rho in Ras transformation. Nature 374:457–459

Qiu R-G, Abo A, McCormick F, Symons M (1997) Cdc42 regulates anchorage-independent growth and is necessary for Ras transformation. Mol Cell Biol 17:3449–3458

Ramer SW, Davis RW (1993) A dominant truncation allele identified a gene Ste20, that encodes a putative protein kinase necessary for mating in *Saccharomyces cerevisiae.* Proc Natl Acad Sci U S A 90:452–456

Ridley AJ, Paterson HF, Johnston CL, Diekmann D, Hall A (1992) The small GTP-binding protein Rac regulates growth factor-induced membrane ruffling. Cell 70:401–410

Roberts RL, Fink GR (1994) Elements of a single MAP kinase cascade in *Saccharomyces cerevisiae* mediate two developmental programs in the same cell type: mating and invasive growth. Genes Dev 8:2974–2985

Rooney RD, Tuazon PT, Meek WE, Carrol EJ, Hagen JJ, Gump EL, Monnig CA, Lugo T, Traugh JA (1996) Cleavage arrest of early frog embryos by the G-protein-activated protein kinase PAK1. J Biol Chem 271:21498–21504

Roux P, Gauthier-Rouviere C, Doucet-Brutin S, Fort P (1997) The small GTPases Cdc42Hs, Rac1 and RhoG delineate Raf-independent pathways that cooperate to transform NIH3T3 cells. Curr Biol 7:629–637

Rudel T, Bokoch GM (1997) Membrane and morphological changes in apoptotic cells regulated by caspase-mediated activation of PAK2. Science 276:1571–1574

Sawai ET, Khan IH, Montbriand PM, Peterlin BM, Cheng-Mayer C, Luciw PA (1996) Activation of PAK by HIV and SIV Nef: importance for AIDS in rhesus macaques. Curr Biol 6:1519–1527

Sells MA, Chernoff J (1997) Emerging from the Pak: the p21-activated protein kinase family. Trends Cell Biol 7:162–167

Sells MA, Knaos UG, Bagrodia S, Ambrose DM, Bokoch GM, Chernoff J (1997) Human p21-activated kinase (Pak1) regulates actin organization in mammalian cells. Curr Biol 7:202–210

Simon M-N, de Virgilio C, Souza B, Pringle JR, Abo A, Reed SI (1995) Role for the Rho-family GTPase Cdc42 in yeast mating-pheromone signal pathway. Nature 376:702–705

Szczepanowska J, Zhang X, Herring CJ, Qin J, Korn ED, Brzeska H (1997) Identification by mass spectrometry of the phosphorylated residue responsible for activation of the catalytic domain of myosin I heavy chain kinase, a member of the PAK/Ste20 family. Proc Natl Acad Sci USA 94:8503–8508

Tang Y, Chen Z, Ambrose D, Liu J, Gibbs JB, Chernoff J, Fields J (1997) Kinase-deficient Pak1 mutants inhibit Ras transformation of Rat-1 fibroblasts. Mol Cell Biol 17:4454–4464

Taylor SS, Knighton DR, Sheng J, Sowadski JM, Gibbs CS, Zoller MJ (1993) A template for the protein kinase family. TIBS 18:84–89

Teo M, Manser E, Lim L (1995) Identification and molecular cloning of a $p21^{cdc42/rac1}$ -activated serine/threonine kinase that is rapidly activated by thrombin in platelets. J Biol Chem 270:26690–26697

Wang ZY, Brzeska H, Baines IC, Korn ED (1995) Properties of *Acanthamoeba* myosin I heavy chain kinase bound to phospholipid vesicles. J Biol Chem 270:27969–27976

Werner LA, Manseau LJ (1997) A *Drosophila* gene with predicted rhoGEF, pleckstrin homology and SH3 domains is highly expressed in morphogenic tissue. Gene 187:107–114

Westwick JK, Lambert QT, Clarke GJ, Symons M, Van Aelst L, Pestell RG, Der CJ (1997) Rac regulation of transformation, gene expression and actin organization by multiple PAK-independent pathways. Mol Cell Biol 17:1324–1335

Wu C, Whiteway M, Thomas DY, Leberer E (1995) Molecular characterization of Ste20p, a potential mitogen-activated protein or extracellular signal-regulated kinase (MEK) kinase kinase from *Saccharomyces cerevisiae.* J Biol Chem 270:15984–15992

Wu C, Lee S-F, Furmaniak-Kazmierczak E, Cote GP, Thomas DY, Leberer E (1996) Activation of myosin-I by members of the Ste20p protein kinase family. J Biol Chem 271:31787–31790

Zhang S, Han J, Sells MA, Chenoff J, Knaus UG, Ulevitch RJ, Bokoch GM (1995) Rho family GTPases regulate p38 mitogen-activated protein kinase through the down-stream mediator Pak1. J Biol Chem 270:23934–23936

Zhao Z-S, Leung T, Manser E, Lim, L (1995) Pheromone signalling in *Saccharomyces cerevisiae* requires the small GTP-binding protein Cdc42p and its activator CDC24. Mol Cell Biol 15:5246–5257

Zhao Z-S, Manser E, Chen X-Q, Chong C, Leung T, Lim L (1998) A conserved negative regulatory region in αPAK: inhibition of the PAK kinases reveals their morphological roles downstream of Cdc42 and Rac1. 18:2153–2163

Zheng Y, Bender A, Cerione RA (1995) Interactions among proteins involved in bud-site selection and bud site assembly in *Saccharomyces cerevisiae.* J Biol Chem 270:626–630

Rac and Cdc42 Effectors

Benjamin Boettner and Linda Van Aelst

1
Introduction

Rac and Cdc42, like all members of the Ras superfamily, act as molecular switches, cycling between an inactive GDP-bound state and an active GTP-bound state. Until recently, members of the Rho-subfamily were believed to be primarily involved in the regulation of cytoskeletal organization in response to extracellular growth factors. However, studies from numerous groups over the past few years have revealed that the Rho-GTPases play crucial roles in a wide variety of biological processes such as transcriptional regulation, cell growth control, membrane trafficking including endocytosis, exocytosis and phagocytosis, as well as development (Fig. 1). The signal transduction pathways mediating these biological phenomena are complex and intricately interwoven. A major challenge has been the identification of downstream effector molecules by which Rac and Cdc42 mediate these activities. Numerous targets of Rac and Cdc42 have been identified (Fig. 2) and further characterization of some of them has led to a better understanding of the function Rac and Cdc42 display at the molecular level. The different Rac and Cdc42 specific effectors and our current understanding of their physiological roles are the major focus of this chapter.

2
Effector Proteins of Rac and Cdc42

One of the hallmarks of an effector protein for the Ras superfamily of GTPases is a binding preference for the GTP bound state of these molecular switches. In all of the small GTPases, a highly conserved region encompassing amino acids 30–40 (according to Ras nomenclature) is present and GDP-GTP exchange promoted by specific GEFs causes a significant conformational shift in this stretch that enables the interaction with various effector molecules conveying the signal from the GTPase to molecules downstream in the signaling pathways. Intrinsic or enzymatically enhanced GTPase activity is thought to revert

Cold Spring Harbor Lab., 1 Bungtown Road, Cold Spring Harbor, 11724 New York, USA

Progress in Molecular and Subcellular Biology, Vol. 22
Ph. Jeanteur (Ed.)

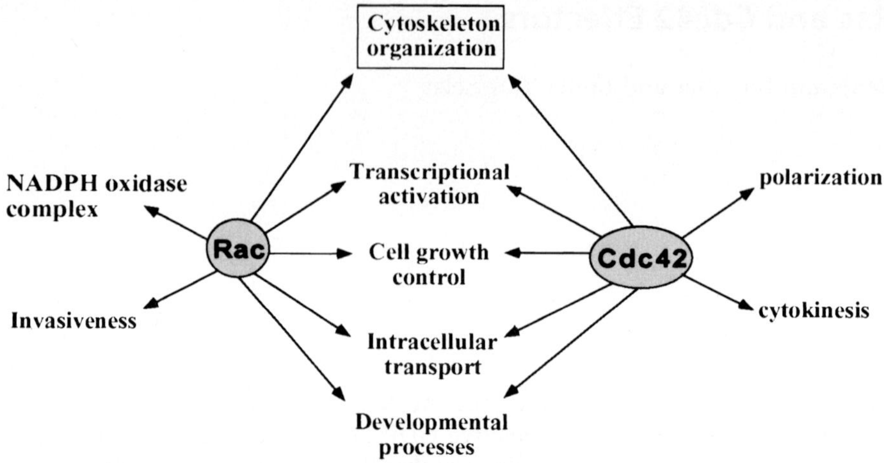

Fig. 1. Rac and Cdc42 play a role in multiple biological processes. In addition to the common functions these GTPases share, Rac has been shown to regulate the activity of the NADPH oxidase complex in phagocytes, and to confer an invasive phenotype to T-cells, whereas Cdc42 has been implicated in polarization and cytokinesis

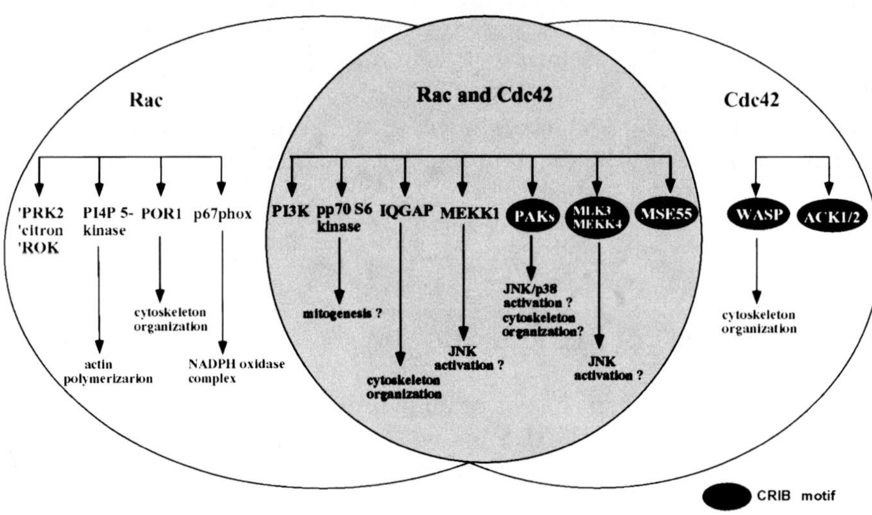

Fig. 2. Mammalian targets of Rac and Cdc42. Rac and Cdc42 share a variety of common targets. The PAKs, MLK3, MEKK4, MSE55, WASP and ACKs share a common Rac/Cdc42 motif (CRIB); (POR1) partner of Rac; (IQGAP) GAP containing Ile-Gln motifs; PRK2, citron and ROK also interact with Rho

the active conformation to the inactive state which consequently results in disruption of the productive signaling complex between the GTPase and a specific effector (Wittinghofer and Nassar 1996). Introduction of mutations in the above mentioned effector loop has provided insights into the relevance of specific GTPase/effector interactions, since those mutants appear to interact selectively with certain effector molecules. The first example presented in the literature was the Ras-paradigm (White et al. 1995; Joneson et al. 1996b), but later on similar experiments have been conducted to unravel Rac and Cdc42 functions (Joneson et al. 1996a; Lamarche et al. 1996; Westwick et al. 1997). It should be mentioned here that for some of the GTPases, a second effector loop has been postulated which in the case of Rac is constituted by residues 143–175 (Diekmann et al. 1995) and whose composition might also affect specific interactions. Here also, a change in conformation seems to be the basis for the generation of a binding competent site in the protein. Noteworthy, however, is the observation that GAP proteins like bcrGAP at least partially require the same regions for interaction, in this case the Rac NH_2-terminal effector loop, but otherwise differ in their requirements for interactive sites on behalf of the GTPase (Diekmann et al. 1995).

2.1 Potential Effectors of Rac and Cdc42 Involved in Cytoskeletal Rearrangements

One of the major functions ascribed to Rac and Cdc42 is their role in the organization of the actin cytoskeleton (for detailed discussion, see Chap. by Ann Ridley). The effects of Rac and Cdc42 on the cytoskeleton have been most extensively studied in fibroblasts. In these cells, it has been demonstrated that activation of Rac by extracellular growth factors such as PDGF, EGF or insulin elicits actin polymerization at the plasma membrane to produce lamellipodia and membrane ruffles, whereas activation of Cdc42 (e.g. by bradykinin) triggered the formation of filopodial protrusions and microspikes at the cell periphery (Ridley et al. 1992; Kozma et al. 1995; Nobes and Hall 1995). Furthermore, as judged by the respective cytoskeletal rearrangements in Swiss 3T3 fibroblasts, Cdc42, Rac and Rho can be placed in a cascade wherein Cdc42 activates Rac, which in turn activates Rho (Nobes and Hall 1995). Evidence has also been provided for the involvement of Rac and Cdc42 in the control of additional actin based structures in other cell types. To give a few examples, Cdc42 has been shown to regulate polarization of T cells toward antigenpresenting cells (Stowers et al. 1995), and in the yeasts, *Saccharomyces cerevisiae* and *Schizosaccharomyces pombe*, Cdc42 has been demonstrated to control cytoskeletal polarization during vegetative division or mating, and cell morphology respectively (Chant 1996; Miller and Johnson 1994). In *Drosophila*, Rac is required for actin assembly at the adherens junctions of the wing disc epithelium, whereas Cdc42 controls elongation and maintenance of the polarized cell shape (Eaton et al. 1995). Furthermore, both Rac and Cdc42 have also been shown to induce the assembly of multimolecular focal complexes at the plasma membrane of fibroblasts (Nobes and Hall 1995).

Although actin polymerization and focal complex formation were thought to be interdependent processes, recent studies by Machesky and Hall demonstrated that the integrin-dependent focal complex assembly is not required for the Rac-induced lamellipodia formation or membrane ruffles, suggesting that different effector pathways are involved in mediating these activities (Machesky and Hall 1997). The cytoskeletal rearrangements and focal complex formation triggered by Rac and Cdc42 appear to have important implications on numerous biological activities such as cell adhesion, cell motility, phagocytosis, cytokinesis and apoptosis (see Ridley, this Vol.). The formation of the specific F-actin structures induced by Cdc42 and Rac involves signaling pathways which regulate actin polymerization and actin cross-linking (Zigmond 1996). It is not inconceivable that coordinated effects of multiple effector molecules, which may be different depending on cell types and physiological conditions are required for the induction of the aforementioned actin structures. Although the molecular mechanisms underlying the responses of activated Rac and Cdc42 still remain elusive, several proteins have been identified as potential mediators of Rac and Cdc42-induced effects on the cytoskeleton (see Fig. 1).

A potential mechanism by which Rac controls actin polymerization in platelets has recently been provided by Hartwig and colleagues. Using permeabilized platelets, they demonstrated that recombinant Rac increases the availability of actin filament barbed ends, which in intact cells normally precedes actin polymerization. Furthermore, they showed that Rac also stimulated phosphatidyl inositol bisphosphate (PIP2) synthesis and that Rac-induced uncapping of actin filaments was inhibited by PIP2-binding peptides derived from gelsolin (Hartwig et al. 1995). These results, together with the findings which showed that PI4P 5-kinase, catalyzing the conversion of PI4P into PIP2, interacted with Rac (Tolias et al. 1995), suggest a model in which activation of Rac triggers the activation of PI4P 5-kinase and thus an increase in PIP2 levels; the generated PIP2 leads to an enhancement of actin polymerization by inducing the uncapping of actin filaments and thereby increasing the number of free barbed ends. Whether such a mechanism could apply also for the induction of Cdc42 induced actin structures or could be functional in other cell types is not clear. In studies using fibroblast cell extracts, Rho but not Rac was found to stimulate PI4P 5-kinase activity (Chong et al. 1994). Recently, Zigmond et al. (1997) demonstrated that addition of GTPγ-loaded Cdc42 to high speed supernatant of lysates of polymorphonuclear leukocytes or *Dictyostelium discoideum* amoebe was effective in inducing actin polymerization. However, in these systems they did not find a correlation between actin polymerization and PIP2 synthesis. It should be noted that in these specific settings, GTPγ-loaded Rac did not stimulate actin polymerization.

Several lines of evidence have been provided which suggest a role for the Wiskott-Aldrich syndrome protein (WASP) as the mediator of Cdc42 effects on the cytoskeleton in T-cells. The Wiskott-Aldrich syndrome is an X-linked disorder, characterized by thrombocytopenia, recurrent infections due to defects in T-cell and B-cell function and eczema (Kirchhausen and Rosen 1996). Notably, T-cells and platelets of affected males exhibit cytoskeletal abnormalities, suggesting that defects in the cytoskeleton may lie at the basis of the syndrome (Kenney et al. 1986; Molina et al. 1992). More recently, several groups demonstrated that WASP interacts directly with Cdc42 in a GTP-dependent manner (Aspenstrom et al. 1996; Kolluri et al. 1996; Symons et al. 1996). Also, weak interaction with Rac could be observed, but none with Rho. Expression of WASP in normal rat kidney (NRK) epithelial cells or Jurkat cells resulted in the induction of ectopic actin polymerization at sites that are enriched in WASP (Symons et al. 1996). This clustering could be inhibited by co-expression with dominant negative Cdc42, Cdc42N17. No inhibition, however, was observed when dominant negative Rac (RacN17) was co-expressed, indicating that although Rac weakly interacts with WASP, the latter is unlikely to be a physiological target of Rac (Symons et al. 1996). Interestingly, Cdc42 has been shown to regulate the localization of actin to sites of cell-cell contact during antigen presentation in T-cells (Stowers et al. 1995). It is therefore tempting to speculate that the cytoskeletal abnormalities observed in T-lymphocytes of WAS patients result from a defect in Cdc42 signaling and that WASP plays an important role in Cdc42 regulated T-cell polarization. Whereas Cdc42 appears to be ubiquitously distributed and has been demonstrated to induce cytoskeletal alterations in a variety of cell types, WASP is exclusively found in hematopoietic cells (Symons et al. 1996). This raises the question of whether other Cdc42 specific target molecules that could perform this function in non-hematopoietic cells may exist. Indeed, additional targets of Cdc42 have been identified (see below). Alternatively, it is possible that isoforms of WASP exist in other cell types. In support of the second possibility, a homologue of WASP, designated N-WASP, which is strongly expressed in brain and weakly in other tissues, has been identified (Miki et al. 1996). WASP and N-WASP share several functional domains:

1. Both proteins harbor a CRIB (Cdc42/Rac interactive binding) motif. This motif was first identified in a family of serine/threonine kinases, referred to as PAKs (p21-activated kinases), which bind both Cdc42 and Rac in a GTP-dependent manner and become activated upon binding, and as discussed below were suggested by some investigators to play a role in transcriptional activation (Bagrodia et al. 1995a,b; Knaus et al. 1995; Manser et al. 1995; Zhang et al. 1995; Brown et al. 1996; for more detail see Chap. by Manser). The CRIB domain appears to be sufficient for binding to the GTPases (Manser et al. 1995). Furthermore, a computer assisted search revealed that this sequence motif is present in >25 proteins from a wide range of species (Burbelo et al. 1995).

2. An amino terminal domain with homology to the pleckstrin homology (PH) domain. Miki et al. (1996) demonstrated the ability of the PH domain contained in N-WASP to bind PIP2.
3. At the carboxy-terminal of the CRIB motif, there is a proline rich region shown to interact with SH3 domains of several signaling molecules which may play an important role in the regulation of WASP function (Rivero et al. 1995; Banin et al. 1996; She et al. 1997).
4. Two WH (WASP homology) motifs at the N- and C-terminal regions of the proteins. These motifs in other proline rich proteins have been shown to play a role in actin cytoskeleton organization (Abo and Symons 1997).

In addition to the above shared domains, N-WASP harbors IQ motifs which are not present in WASP and which, as previously shown for other IQ-motif containing proteins, also interacted with calmodulin (Miki et al. 1996). However, the ability of N-WASP to interact with Cdc42 remains to be demonstrated. Overexpression of N-WASP in COS7 cells also induced cytoskeletal rearrangements, although they were phenotypically different to those observed for WASP; N-WASP elicited a marked accumulation of actin filaments in cortical areas. Furthermore, treatment of serum starved N-WASP expressing cells with EGF resulted in a disappearance of the accumulated actin filaments and formation of microspikes (Miki et al. 1996). Additional evidence for a role of WASP in cytoskeletal organization came from the observation that disruption mutants of the *S. cerevisiae* WASP homologue, BEE1, exhibited a striking change in the organization of actin filaments resulting in defects in budding and cytokinesis (Li 1997). Disruption of BEE1 abolished the formation of cortical actin patches in the bud, and instead actin was assembled into thick cables that were not restricted to the cell cortex. Furthermore, the actin bundles in the buds of bee1 disruption mutants differed in their biochemical composition from those of cortical actin patches (Li 1997). Thus, it appears that WASP proteins play a role in actin remodeling and may mediate some of Cdc42 effects on the cytoskeleton in specific cell types. Whether WASP mediates Cdc42-induced filopodial protrusions in fibroblasts is questionable, since studies by Lamarche et al. (1996) showed that a mutant form of Cdc42 (Cdc42L61,C40) which was unable to bind WASP, was still able to induce filopodia formation.

The same Cdc42 mutant also failed to interact with the above mentioned family of serine/threonine kinases (PAKs), suggesting that they are not likely to be mediators of Cdc42 induced filopodia formation in fibroblasts (Lamarche et al. 1996). Similar mutant analysis studies suggested that PAKs are also not responsible for mediating Rac induced formation of lamellipodia and membrane ruffling in fibroblasts, since Rac mutants (RacV12,H40; RacL61,C40) which failed to bind PAKs were still able to induce membrane ruffles and lamellipodia and, vice versa, Rac mutants (RacV12,L37; RacL61,A37) which were still able to bind PAKs, nevertheless failed to induce membrane ruffles and lamellipodia (Joneson et al. 1996a; Lamarche et al. 1996). It should be noted, however, that expression of activated alleles of PAKs in fibroblasts was observed to induce actin reorganization (Manser et al. 1997; Sells et al. 1997). One possible explanation for this observation is that Cdc42/Rac independent pathways may trigger the activation of PAKs. Alternatively, PAKs might mediate effects on the cytoskeleton induced by Rac and/or Cdc42 which are different from those mentioned above. Studies in yeast and *Drosophila* are supportive for a role of PAKs in cytoskeletal organization, since in both organisms PAK homologues appear to play important roles in morphogenesis (Sells and Chernoff 1997). Furthermore, mutational analysis of Ste20p (*S. cerevisiae* PAK homologue) showed that Cdc42p binding to Ste20p is required for the induction of pseudohyphal growth and the control of polymerized growth during budding. However, in contrast to what was previously suggested, this interaction is not required for the activation of the pheromone-responsive MAPK pathway, suggesting that indeed Cdc42-dependent and independent pathways exist for the activation of Ste20 (Peter et al. 1996; Leberer et al. 1997).

Another group of proteins, IQGAPs, were more recently identified as possible effectors of Cdc42 and Rac, and they possess some features which distinguish them as potential regulators of the cytoskeleton. IQGAP1, originally isolated by a RNA-PCR method (Weissbach et al. 1994), was more recently identified by numerous groups performing affinity chromatography on immobilized Cdc42-GTP and/or Rac-GTP (Hart et al. 1996; Kuroda et al. 1996; McCallum et al. 1996). In addition to IQGAP1, a close homologue (62% identity) referred to as IQGAP2, has been identified (Brill et al. 1996; McCallum et al. 1996). In spite of finding 25% sequence identity between both IQGAPs and almost the entire *S. pombe* sar1 protein, a RasGAP homologue, they failed to bind Ras, nor did they exhibit GAP activity towards Hras. They have been shown, however, to inhibit the intrinsic GTPase activity of Cdc42 and Rac, but do not show GAP activity toward Rac and Cdc42 (Brill et al. 1996; Hart et al. 1996). Both IQGAP1 and 2 also failed to bind Rho. In addition to the carboxy-terminal rasGAP-related domain, IQGAPs have several interesting motifs. The amino-terminal region of both IQGAPs harbors a calponin homology domain and the central region shares similarity with IQ domains of unconventional myosins, and furthermore contains a WW motif and a polyproline binding domain. A calponin homology motif has also been found in the Vav proto-oncoprotein, a Dbl related Rho GTPase exchange factor (Cerione and Zheng 1996), and a similar motif is present in members of the spectrin, filamin and fimbrin families and appear to function as F-actin binding domains (Castresana and Saraste 1995). WW motifs have been found in numerous proteins, including dystrophin, the SH3-binding Yes-associated protein and the Dbl related protein 38D4, and have been shown to mediate binding to proline-rich sequences (Sudol et al. 1995; Chan et al. 1996). Finaly, the IQ motif is a calmodulin binding motif, which was first identified in unconventional myosins, including myr5 (Reinhard et al. 1995). The latter is an actin binding protein which possesses rhoGAP activity. The N-terminal domain of IQGAP, which contains the IQ domains does indeed associate with calmodulin. Joyal et al. (1997) demonstrated that the binding of calmodulin to IQGAP occurs both in the presence and absence of Ca^{2+}. Interestingly, calmodulin is able to modulate the interaction between IQGAP and Cdc42. Preincubation of Ca^{2+}/calmodulin with IQGAP prior to the addition of Cdc42 resulted in the disruption of the IQGAP/Cdc42 interaction. This effect was shown to be Ca^{2+}-dependent since Ca^{2+}-free calmodulin did not prevent the association of IQGAP1 with Cdc42 (Joyal et al. 1997). Several groups observed the colocalization of IQGAP with actin-rich ruffling membranes and lamellipodia but not with stress fibers, which is consistent with the ability of IQGAP to associate with Cdc42 and Rac but not with Rho (Hart et al. 1996; Kuroda et al. 1996; Bashour et al. 1997; Erickson et al. 1997). In addition, IQGAP has been shown to bind F-actin directly. Bashour et al. (1997) were able to demonstrate that purified IQGAP causes cross linking of F-actin into a network that exhibited gel-like properties and that calmodulin has an inhibitory effect on the actin binding activity of IQGAP. Furthermore, it appears that cycling of Cdc42 between its active and inactive states is required for a formation of a stable Cdc42/IQGAP/actin complex (Erickson et al. 1997). Altogether, the above findings support a model in which IQGAP is part of a multimolecular complex that regulates the organization of the actin cytoskeleton. The exact mechanism however still remains elusive. A role for IQGAP in cellular events such as cell-cell adhesion and cytokinesis has been suggested. Both activities are mediated by Rac and/or Cdc42 and involve cytoskeletal rearrangements (see Ridley, this Vol.). Accumulation of IQGAP at cell-cell junctions and co-localization of IQGAP with α-catenin was observed (Kuroda et al. 1996). Evidence for the involvement of IQGAP in cytokinesis came from studies with the *Dictyostelium* IQGAP related proteins (Adachi et al. 1997). Three IQGAP related proteins, DGAP1, DdRasGAP1 and GAPA1 were identified in *Dictyostelium*. These three proteins appear to function in distinct pathways regulating cytokinesis and probably involve different GTPases. Cells overexpressing DGAP1 become multinucleated and divide through traction-mediated cytofission when placed on a substratum. They phenotypically resemble those cells deficient in RacE and myosin II (Faix and Dittrich 1996). Thus, DGAP1 might interact with RacE, although this remains to be proven. Cells lacking GAPA1 grow as giant multinucleated cells both on a substratum and in suspension. Unlike RacE and probably DGAP1, GAPA1 is not required during the early stages of cytokinesis, but participates in signaling events required for cleavage of the midbody (Adachi et al. 1997). It has not been tested yet which GTPase it interacts with. DdRasGAP1 was isolated in a two-hybrid screen designed to identify Ras-GTP interacting proteins. Gene disruption of DdRasGAP caused cytokinesis defects but only in a suspension culture (Lee et al. 1997).

Additional proteins which interact either with Cdc42 or Rac have been identified and shown to play a role in cytoskeletal organization. A protein, called CIP4, with homology to the non-kinase domain of FER was isolated in a yeast two-hybrid screen for proteins that bind to the activated mutant form of Cdc42, Cdc42L61 (Aspenstrom 1997). No interaction with Rac or Rho could be observed. In vitro binding analysis demonstrated that CIP4 binds directly to Cdc42 in a GTP dependent manner. Overexpression of CIP4 in serum stimulated Swiss 3T3 cells caused a reduction of stress fiber formation. The same effect was observed when CIP4 expressing cells were stimulated with PDGF or bradykinin. Interestingly, clustering of CIP4 into a large number of foci was observed when Cdc42 and Rac were co-expressed, or upon stimulation with PDGF or bradykinin. It has been postulated that the Cdc42/CIP4 complex may play a role in signaling pathways leading to development of the limbs, based on the fact that CIP4 shows approximately 50% identity with a mouse protein, FBP17 (Aspenstrom 1997). This protein was isolated from an expression library in a search for molecules that could bind to a proline rich peptide derived from formin, a protein previously shown to play a role in limb formation in mice (Chan et al. 1996).

A 34 kDa protein, POR1 (partner of Rac) was isolated in a two-hybrid screen and was shown to play a role in Rac induced membrane ruffling (Van Aelst et al. 1996). POR1 interacted with Rac, but not with Cdc42 in a GTP dependent manner. Deletion analysis of POR1 demonstrated that the carboxy-terminal region of the protein but not the amino terminal region was necessary for binding to Rac. Both deletion mutants, however, inhibited the induction of membrane ruffles by RacV12, while a synergistic effect of wild type POR1 with RasV12 was observed for the induction of membrane ruffling. Consistent with a role of POR1 in membrane ruffling was the observation that a mutant Rac, RacV12,L37, which failed to bind POR1, also failed to induce membrane ruffling (Joneson et al. 1996a). Interestingly, POR1 has more recently been found to interact with the GTPase, ARF6 (D'Souza-Schorey et al. 1997). ARF6 is the least conserved member of the ARF family of GTPases (Tsuchiya et al. 1991). In addition to its role in regulating peripheral membrane trafficking (Peters et al. 1995; D'Souza-Schorey et al. 1997), ARF6 and its activated mutant, ARF6(Q67L), have been shown to elicit cytoskeletal rearrangements at the cell surface (Radhakrishna et al. 1996; D'Souza-Schorey et al. 1997). Cytoskeletal rearrangements induced by ARF6(Q67L) could be inhibited by co-expression of the above described deletion mutants of POR1, but not with the dominant negative mutant of Rac, Rac(S17N) (D'Souza-Schorey et al. 1997). These findings indicate that ARF6 and Rac function on separate signaling pathways to mediate cytoskeletal reorganization, and suggest a role for POR1 as an important regulatory element in orchestrating cytoskeletal rearrangements at the cell periphery induced by ARF6 and Rac. It is conceivable that, depending on the nature of the extracellular stimuli, POR1 could interact with either ARF6 or Rac or both, to establish highly specified patterns of cytoskeletal rearrangements at the plasma membrane. The molecular mechanism of POR1 function, however, remains to be defined.

Two proteins, ROK and PRK2, originally identified as Rho-interacting proteins, were also found to interact with Rac (see Kaibuchi, this Vol.). Vincent and Settleman showed that PRK2 kinase, a close relative of PKN, interacts with and is activated by Rho in a nucleotide-independent manner, whereas the interaction with (and activation by) Rac is completely GTP dependent. Expression of a kinase deficient mutant form of PRK2 in fibroblasts resulted in the disruption of stress fibers, suggesting a role for PRK2 in cytoskeletal organization (Vincent and Settleman 1997). $p160^{ROCK}$ and its isoform ROKα have been shown to interact in vitro and in the two-hybrid system with Rac (Joneson et al. 1996a,b; Lamarche et al. 1996). The biological relevance of the interaction of Rac with either $p160^{ROCK}$, ROKα or PRK2 remains elusive. Interestingly, a putative effector for Cdc42, called Genghis Khan (Gek), has been identified in *Drosophila* which, like $p160^{ROCK}$ (ROKα), exhibits sequence similarity with the human myotonic dystrophy kinase (DMPK). Studies using loss-of-function mutants of Gek suggest that the function of WT Gek is to negatively regulate actin polymerization, since the most prominent phenotype in egg chambers homozygous for gek mutations is ectopic actin polymerization (Luo et al. 1997). Like $p160^{ROCK}$, ROKα and PRK2, citron binds to Rac and Rho, but not to Cdc42. Citron was isolated in a two-hybrid screen for potential RhoC targets. It harbors a C6H12 zinc finger, a PH domain and a long coiled-coil forming region (Madaule et al. 1995). However, the role of citron in Rac (or Rho) signaling awaits establishment. Somewhat puzzling is the observation that different GTPases share similar effector targets. It is possible, however, that the effector response promoted by each complex is different due to differences in subcellular localization or in substrate specificity.

Recent studies by Peppelenbosh et al. (1995) have provided a potential link between Rac and Rho for the induction of stress fibers. Evidence was provided that Rac is required for the activation of arachidonic acid metabolism by epidermal growth factor (EGF) and insulin. The induction of arachidonic acid release and subsequent leukotriene production results in the formation of stress fibers and this event requires functional Rho proteins (Peppelenbosch et al. 1995). Since cytosolic PLA2 is one of the major enzymes responsible for the production of arachidonic acid (Dennis 1997), it is most likely that cPLA2 functions downstream of Rac in the pathway leading to Rho activation.

Taken together, multiple effectors of Rac and Cdc42 appear to exist for mediating their effects on the cytoskeleton. This may, however, not be that surprising since different physiological requirements and morphologies are certainly based on variances in cytoskeletal organization. Moreover, the Rac and Cdc42 effectors may also be cell type dependent. Further characterization of the above-described effectors and their interacting proteins will be indispensable in obtaining clearer insights into the exact mechanism whereby Rac and Cdc42 affect the cytoskeleton.

2.2
Potential Effectors of Rac and Cdc42 Induced Transcriptional Activation

Recently, Rac and Cdc42 have been shown to be potent activators of the JNK/ SAPK (c-jun N-terminal kinase/stress-activated protein kinase) and p38 MAPKs in several cellular systems (Coso et al. 1995; Minden et al. 1995). The JNK and p38 isoforms are activated in a presumably differential manner in response to proinflammatory cytokines, a variety of cellular stresses and to a minor degree as well to mitogenic stimuli. Targets of the JNK isoenzymes include the transcriptional activators c-Jun, ATF-2 and Elk1 (Pulverer et al. 1991; Derijard et al. 1994; Gille et al. 1995; Gupta et al. 1995), whereas p38 proteins which are homologous to the yeast HOG1, a component of the stress-activated response pathway, also possess other substrates such as MAPKAP kinase-2 (Rouse et al. 1994), which is involved in phosphorylation of the small heat shock proteins Hsp25/HSP27 in response to inflammatory cytokines, physiological stress and growth factors (Stokoe et al. 1992). In addition, Mxi2, a p38 isoform was isolated as a protein which, in vitro, could phosphorylate Max, the Myc heterodimerization partner (Zervos et al. 1995). Another target is the transcription factor CHOP/GADD153, a protein bearing homology to CREB (Wang and Ron 1996). As regards upstream elements in the activating pathways connecting JNK/p38 to the small GTPases, several proteins have been identified as potential mediators. Since JNK and p38 molecules belong to the MAPK subfamily of protein kinases, they could be anticipated to act as a component of a three part MEKK/MEK/MAPK module, analogous to the well-defined cascades in yeast and the Ras induced MAPK module in mammalian cells. In this respect, kinases have been isolated which specifically stimulate the activity of SAPK, but only display a low affinity towards the p42/p44 MAPKs. This group of activators include the homologous murine and human proteins SAPK/ERK kinase-1 (SEK1) and MKK4, respectively (Sanchez et al. 1994; Derijard et al. 1995). Overexpression of a kinase-deficient SEK1 mutant in situ specifically abrogates JNK activation, but has no influence on p42/p44 kinase activity. Furthermore, when coexpressed in mammalian cells SEK1, in contrast to MEK1, interacts with JNK as was shown in co-immunoprecipitaion experiments (Sanchez et al. 1994). There is convincing evidence, however, that SEK1 might not be the only direct JNK activator. Experiments with extracts prepared from osmotically shocked 3Y1 cells revealed additional chromatographically distinguishable activities triggering JNK (Moriguchi et al. 1995). More recently, the identification of MKK7 was reported and shown to directly phosphorylate and activate JNK (Holland et al. 1997). It has not yet been resolved whether SEK1 could also activate p38 in situ. Rather, MKK3 and MKK6 appear to function as p38 specific activators at the MAPKK level (Sanchez et al. 1994; Derijard et al. 1995; Han et al. 1996). Moving further upstream in the activation pathway, MEKK1, a protein which also binds to the Ras GTPase in vitro, has a marked preference for JNK over p42/p44 activation (Xu et al. 1995). Subsequently, homologues of MEKK1 have been cloned which also seem to activate

JNK. However, differences in their primary sequences also suggest variable modes of activation, and analyses supported by confocal fluorescence microscopy of COS cells demonstrated that the MEKK1 and 4 isoforms specifically associated with Rac/Cdc42, wheras MEKK2 and 3 did not (Fanger et al. 1997). MEKK4 possesses a modified CDC42/Rac interacting binding domain (CRIB) and dominant negative versions of MEKK1 and MEKK4 in the same experimental settings abolished Rac/Cdc42 induced JNK activation. This interference, however, seems separable from the above mentioned group of Rac/Cdc42 interacting proteins, the p21-activated kinases (PAKs) which also contain CRIB motifs and have been discussed as candidates linking the GTPases to the downstream modules leading to JNK and p38 (Fanger et al. 1997). Constitutively active PAK1 and PAK3 mutants have been reported to stimulate JNK/p38 activity in transfection experiments and, like MEKK1 and MEKK4, PAK proteins also seem irrelevant for p42/p44 activation (Bagrodia et al. 1995a,b; Ottilie et al. 1995; Zhang et al. 1995; Brown et al. 1996). Furthermore, overexpressed PAK1 is activated via autophosphorylation following IL-1β stimulation, and overexpression of the PAK65 N-terminal regulatory domain inhibits JNK activation by Rac/Cdc42, although this might also be explicable by a tethering effect. Notably, a somewhat contradictory result was obtained by Westwick et al. (1997) who investigated the role of PAK in Rac function. These investigators showed that an effector loop mutant of Rac, namely RacL61,D43, which no longer binds PAK, could still activate the JNK/p38 proteins in a transient transfection assay. Other experiments conducted with Rac/Cdc42 effector loop mutants suggest PAK proteins to be dispensable for Rac and Cdc42 induced actin remodeling and G1 progressive signaling, but do not explicitly exclude them from JNK/p38 activating pathways (Joneson et al. 1996; Lamarche et al. 1996). Teramoto et al. (1996) recently reported that the newly identified mixed lineage kinase 3 (MLK3) can activate JNK without affecting p42/p44 and p38. Moreover, they demonstrated direct association between MLK3 and Rac/Cdc42 (Teramoto et al. 1996). In summary, all these different lines of evidence suggest the existence of more than one effector protein coupling Rac/Cdc42 to JNK and p38 activation and that marked differences in the expression and compartmentalisation patterns of these proteins between cell types might dictate which routes are used under specific circumstances.

Rac1, Cdc42 and RhoA cooperatively activate transcription at the serum response element (SRE) in a variety of promotors, e.g. the c-fos promotor (Hill et al. 1995). Efficient stimulation of the SRE requires formation of a ternary complex between Elk-1/TCF, SRF and the respective DNA sequence. Whereas recruitment of Elk-1/TCF in this context requires input from the mitogen stimulated MAPK cascade, SRF stimulation depends on signaling through Rho GTPases. Interestingly, RhoA, which is stimulated in response to LPA and AlF^-, activates the SRF complex in a Rac/Cdc42 independent fashion and vice versa, implicating at least two different effector pathways emanating from the Rho GTPases (Hill et al. 1995). Hence, it appears that, in this regard, these members of the Rho GTPase subfamily function in independent cascades, as opposed to a Rho/Rac/Cdc42 cascade proposed earlier for the tight control of cytoskeletal rearrangements. The nature of these pathways, however, remains elusive, since neither JNK nor p38 efficiently stimulate transcription at the SRE. First insights came from studies performed by Kim et al. which suggested phospholipase A2 and its substrate arachidonic acid as crucial for transcription at an SRE-luciferase reporter element. Furthermore, lipoxygenase, an enzyme using arachidonic acid as a substrate to generate other compounds also appears a relevant player in that pathway (Kim and Kim 1997). Application of mepacrine, a specific phospholipase A2 inhibitor, or expression of lipocortin1, a phospholipase inhibiting peptide, markedly interfered with this activation mechanism. Concerning signaling components specific for RhoA, it is noteworthy that Chihara et al. reported an elevated transcriptional activity at an SRE-containing promotor employing constitutively activated Rho kinase (ROK), a protein previously implicated in RhoA triggered cytoskeletal events (Chihara et al. 1997).

More recently, Cdc42 and Rac (and RhoA) have been implicated in NF-κB signaling (Sulciner et al. 1996; Perona et al. 1997). Activation of Cdc42 and Rac results in phosphorylation of IκB, the inhibitory subunit of p50/p50 and p50/p65 dimers, its subsequent degradation and the translocation of the transcriptionally active dimers into the nucleus. NF-κB, like JNK, is a target molecule in TNFα signaling and dominant negative versions of RhoA and Cdc42 indeed inhibit NF-κB activation following TNFα stimulation of COS cells in a transfection assay using an HIV-Luciferase reporter construct. Interestingly, dominant negative Rac was inactive in this respect. Sulciner et al. (1996) proposed a model in which Rac upregulates the production of reactive oxygen species (ROS) and that an elevated level of this compound subsequently activates NF-κB. NF-κB is indeed potently activated by ROS. As discussed below, Rac triggers the NADPH oxidase system in phagocytic cells and probably also in B- and T-cells, but NADPH originated ROS production is restricted to the phagosome, a membrane bound subcompartment of the cell. Whether this or a similar system could account for a parallel effect in other cell types is at present unclear.

2.3 Potential Effector Pathways of Rac and Cdc42 Involved in Cell Growth Control

Compelling evidence has been provided which indicates that Rac, Cdc42 and Rho are involved in multiple aspects of cell growth control. A first hint for a potential role of these GTPases in cell growth control was the finding that a large number of exchange factors for the Rho family members are potent oncoproteins (Cerione and Zheng 1996). More recently, all three GTPases have been shown to play a critical role in mitogenesis and transformation. With respect to the former phenomenon, microinjection of activated mutant forms of Rac, Cdc42 and Rho in Swiss 3T3 cells stimulated cell cycle progression through G1 and DNA synthesis, while expression of dominant negative forms of these GTPases blocked stimulation of DNA synthesis in response to growth factors (Olson et al. 1995). It has been found that activated mutant forms of Rac (and Rho) induce tumorigenic transformation of some cell types. For example, activated mutant forms of Rac were sufficient to cause malignant transformation of rodent fibroblasts (Khosravi et al. 1995; Qiu et al. 1995). More recently, a role for Cdc42 in the control of cell transformation has also been provided. Qui et al. showed that fibroblasts expressing constitutively active Cdc42 (Cdc42V12) were anchorage independent and proliferated in nude mice, but in contrast to RacV12-expressing Rat1 fibroblasts, Cdc42 cell lines failed to show enhanced growth in low serum (Qiu et al. 1997). Lin et al. (1997) observed that expression of a constitutively active Cdc42 mutant in NIH-3T3 cells caused pronounced growth inhibition. However, when they used another Cdc42 mutant, Cdc42(F28L), which undergoes GTP-GDP exchange in the absence of a guanine nucleotide exchange factor (GEF), they observed that fibroblasts expressing this mutant exhibited anchorage independent growth and lower dependency on serum for growth, suggesting that complete cycling is required for the Cdc42 growth promoting signal. The differences observed between the two groups may be due to different levels of expression induced by Cdc42V12 and Cdc42(F28L) in the respective cell lines. Furthermore, evidence implicating Rac, Cdc42 (and Rho) as essential components of Ras-induced malignant transformation has been provided and includes the following observations: (1) dominant negative mutants of Rac (RacN17), Cdc42 (Cdc42N17) and Rho (RhoN19) inhibit Ras focus formation; (2) activated Rac and Rho mutants synergize with activated Raf to induce foci (Khosravi et al. 1995; Prendergast et al. 1995; Qiu et al. 1995a,b, 1997; Lebowitz et al. 1997; Qiu 1997); and (3) farnesyltransferase inhibitors suppress Ras transformation by interfering with the activity of Rho (Lebowitz et al. 1995). It appears that the pathways mediated by Rac, Cdc42 and Rho in the control of cell growth and Ras transformation are independent. In addition to its role in mitogenesis and proliferation, Rac has been implicated in the induction of an invasive and metastatic potential of T-lymphoma cells. Initially, a potential role for Rac in invasiveness was suggested by the observation that the tumor invasiveness gene Tiam1 encodes a guanine-nucleotide exchange factor (GEF) for Rac (Michiels et al. 1995). Tiam1 was identified by combining retroviral insertional mutagenesis with efficient in vitro selection for invasive T-lymphoma cell variants (Habets et al. 1994). More recently, activated Rac, when expressed in lymphoma cells, has been shown to confer an invasive phenotype and metastatic potential to these cells. Furthermore, some lines of evidence suggest the involvement of Rac and Cdc42 in the control of cell death (Lebowitz et al. 1995; Gulbins et al. 1996; Brenner et al. 1997; Chuang et al. 1997).

The signaling pathways utilized by Rac and Cdc42 (and also Rho) to control mitogenesis, proliferation and invasiveness still remain poorly understood. Although the evidence is far from complete, it has been postulated that transcriptional activation induced by Rac and Cdc42, and/or morphological changes elicited by Rac and Cdc42, may contribute to their transforming abilities. Studies utilizing effector domain mutants of Rac and Cdc42, which display differential binding to their effector proteins, demonstrated however that JNK activation induced by these GTPases is dispensable for their effects on mitogenesis and transformation. Rac mutants (RacV12,H40 or RacL61,C40) which failed to bind and activate PAK and JNK were still able to stimulate DNA synthesis or to mediate transformation, and vice versa Rac mutants (RacV12,L37 or RacL61,A37), although able to bind and activate PAK and JNK, failed to stimulate DNA synthesis and mediate transformation (Joneson et al. 1996a; Lamarche et al. 1996). Similarly, a Cdc42 mutant (Cdc42L61,C40) which failed to bind and activate PAK and JNK, retained the ability to stimulate DNA synthesis. Notably, the RacL61,C40 and Cdc42L61,C40 mutants were not only shown to be defective in PAK binding but also failed to bind other CRIB containing proteins, including WASP, MSE55 and the *C. elegans* protein FO9F7.5, suggesting that the CRIB containing proteins are not required for Rac and Cdc42 mitogenic and transforming potential (Lamarche et al. 1996). Interestingly, Chuang et al. (1997) reported that Cdc42 and to a lesser extent Rac induced an apoptopic cell death program in Jurkat cells, and that this effect can be inhibited by dominant negative components of the JNK cascade. The involvement of the JNK cascade in the signaling pathway from the Fas receptor to Rac2 leading to apoptosis was also postulated (Gulbins et al. 1996; Brenner et al. 1997). Thus, it appears that in some settings the Rac and/or Cdc42 induced JNK activation may contribute to the apoptotic program. Using a similar genetic approach as described above, Westwick et al. (1997) demonstrated that activation of SRF is dispensable for Rac's growth promoting activity. Sulciner et al. (1996) suggested a model in which the growth promoting activity of Rac may involve the production of ROS and potentially NF-κB activation. They showed that the mitogenic activity of Ras can be blocked by agents that block ROS production and that the superoxide production in response to Ras was independent of the MAPK pathway but involved a Rac mediated signaling pathway. Whether this model proves to be the case requires further investigation, since ROS production and NF-κB activation have also been correlated with apoptosis. A potential link to mitogenic signaling pathways may involve pp70^{S6kinase}, which has been shown to bind to and be activated by Rac and Cdc42 (Chou and Blenis 1996). Although a role for pp70^{S6kinase} in cell cycle progression has been demonstrated (Proud 1996), it remains to be proven whether this kinase mediates Rac and/or Cdc42 effects on mitogenesis. In regard to Rac's role in invasiveness, it is conceivable that Rac's ability to induce activation of PEA3 may contribute to its ability to confer an invasive phenotype (O'Hagan et al. 1996). It is known that the promoters of most metalloproteases contain binding sites for both AP-1 and PEA3, and that increased metalloprotease expression is correlated with malignant expression.

Since a key feature of a transformed cell is an altered organization of the actin cytoskeleton, it has been postulated that the RhoGTPase signaling pathways that mediate cytoskeletal rearrangements contribute to the general phenomenon of transformation. Accumulating data, however, point out that the short-term induction of cytoskeletal rearrangements by Rac and Cdc42 do not contribute to their effects on transformation (Symons 1996). For example, expression of activated Rho induces stress fibers, however, in Ras transformed cells the formation of these structures is inhibited. Deletion mutants of Tiam (GEF for Rac) unable to induce membrane ruffles are still able to transform NIH-3T3 cells. Furthermore, a Rac effector mutant (RacL61,D43) that still exhibited strong lamellipodia formation is impaired in its transforming capabilities of NIH 3T3 (UNC) cells (Westwick et al. 1997). Whether the Rac induced formation of lamellipodia contributes to its ability to confer an invasive potential to T-cells remains to be elucidated.

Further identification and characterization of effector proteins in combination with mutant analysis will be required to unravel the signaling pathways which mediate the effects of Rac and Cdc42 on cell growth control.

2.4
Phox 67, a Rac Target in the Superoxide Formation of Phagocytic Cells

In "professional phagocytes" (macrophages, neutrophils, eosinophils) and apparently also in B- and T-cells, Rac has been shown to regulate the activity of a membrane associated NADPH oxidase complex (Segal and Abo 1993; Bokoch 1995). This multiprotein complex derives electrons from NADPH on its cytosolic face and pumps them into the lumen of phagosomes where they consequently serve to generate superoxide anion (O_2^-) and subsequently toxic oxidants like hydrogen peroxide (H_2O_2), hydroxyl radicals (OH) as well as hypochlorous acid (HOCl). The latter compounds represent the killing agents of the cell in the fight against phagocytosed material. The basic e- transport function of the complex resides in the unusual cytochrome b_{558}, consisting of two subunits, $gp91^{phox}$ and $p22^{phox}$, which, are integrated into the plasma membrane. Proper regulation of the complex, however, requires two additional soluble proteins, $p47^{phox}$ and $p67^{phox}$, which, upon activation, are tethered to the membrane components of the complex. The only known features of these proteins are the SH3 domains present in duplicate in each of them (Leto et al. 1994; de Mendez et al. 1996). Interaction and recruitment of $p47^{phox}$ and $p67^{phox}$ seems to be dependent on the SH3 domains as has been demonstrated by de Mendez et al. (1996) who performed thorough deletion studies on the two proteins. Furthermore, $p47^{phox}$ translocation appeared to coincide with multiple phosphorylation events directed to critical Serine residues of the proteins.

The overall significance of the aforementioned components was further emphasized by studying patients suffering from Chronic Granulomatous Disease (CGD) which is marked by a lack of the "oxidative burst" function in phagocytes. On the molecular level, an X-linked form of the disease was shown to result from the absence of $gp91^{phox}$ and $p22^{phox}$ subunits, whereas an autosomal recessive form is due to a lack of $p47^{phox}$ in phagocytes (Segal et al. 1983; Teahan et al. 1987; Smith and Curnutte 1991). Carriers of such lesions are significantly predisposed to microbial infections, often with a fatal result. Later on, another player in the system, Rac, was demonstrated to exert a regulatory influence on the complex, and the activity of the complex also correlates with the movement of Rac from a cytosolic to a membrane-attached location (Abo et al. 1991, 1994; Quinn et al. 1993). The positive regulatory effect of Rac on the oxidase is restricted to the GTP-bound form of the protein as was demonstrated in a cell-free system and could be mimicked by non-hydrolysable GTP-γS as well as with a V12 mutant lacking intrinsic GTP-ase activity.

In search of a specific effector molecule of Rac in the oxidase complex, Diekmann et al. (1994) revealed $p67^{phox}$ as a Rac interacting protein. This interaction appeared to be restricted to the GTP-bound form of Rac and could be narrowed down to the very N-terminal 199 amino acids of $p67^{phox}$, a region not involving one of its SH3 domains. Effector loop mutants of Rac harboring amino acid substitutions at positions 35 ($T \rightarrow A$), 38 ($D \rightarrow A$) and 40 ($Y \rightarrow K$) abolished O_2-production when introduced into cells, indicating that $p67^{phox}$ binds to the Rac effector loop region. Using chimeric proteins comprised of different Rac and Rho specific portions, the same investigators could map a second $p67^{phox}$ interacting site in Rac located between residues 143 and 175. Interestingly, the same regions were crucial for $p65^{PAK}$ binding (Diekmann et al. 1994) of the GTPase and, adding another level of complexity, Knaus et al. showed that a PAK kinase phosphorylates $p47^{phox}$, an event – as mentioned before – coinciding with $p47^{phox}$ translocation and the overall activity of the complex (Knaus et al. 1995). Furthermore, the recently described insert region in Rac (residues 124–135), although not involved in $p67^{phox}$ binding, has been shown to affect the activity of the NADPH complex, most likely by interacting with another component of the complex (Nisimoto et al. 1997).

Many details of this scenario have been uncovered. Many others remain to be answered: above all, how triggering stimuli as diverse as specific lipid compounds, soluble peptides like complement component C5a and formyl peptides derived from opsonized molecules can all converge on Rac as the key regulator of the complex. Other questions can be asked about the mechanistic details of the system. Which functions exactly fulfill the $p47^{phox}$ and $p67^{phox}$ subunits of the complex? The protein sequence, apart from the SH3 motifs, does not give any hints as to the possible function of these components. Another enigma is the role of the GTP binding protein, Rap1A. This protein is associated with the complex, and when introduced into cells as a dominant positive or negative mutant can up- or downregulate the activity of the oxidase, respectively (Quinn et al. 1989; Maly et al. 1994). Some of the above mentioned stimuli also result in a remodeling of cortical actin and since it is well known that Rac constitutes a key regulator of membrane ruffling as well as other actin related events it is tempting to speculate upon an intimate connection between such different functions as O_2^- synthesis and actin rearrangements in stimulated phagocytically active cells.

3
Concluding Remarks

Over the past years, it has become clear that the Rho GTPases participate in a wide variety of biological activities. Major efforts of numerous groups have been focused on the identification of proteins interacting with specific Rho GTPases, mediating their actions. As a result of this tremendous experimental input, a large number of Rac and Cdc42 interacting proteins has been identified. However, their actual contribution to the above outlined effects and pathways have been resolved to a satisfactory degree only in a few cases. Description of more target molecules and elucidating downstream factors of the one already known will represent a major goal for the future. This, in conjunction with biochemical efforts addressing the molecular and cellular fine tuning present in each of the pathways, will serve to pave a way to a better understanding.

Acknowledgments. We thank M. Symons, C. D'Souza-Shorey and M. McDonough for discussions and helpful comments. L.V.A. is a recipient of an award from the V Foundation and the Sidney Kimmel Foundation for Cancer Research.

References

Abo A, Symons M (1997) Wiskott-Aldrich syndrome protein (WASP). In: Kreis T, Vale R (eds) Guidebook to the cytoskeletal and motor proteins (in press)

Abo A, Pick E, Hall A, Totty N, Teahan CG, Segal AW (1991) Activation of the NADPH oxidase involves the small GTP-binding protein p21rac1. Nature 353:668–670

Abo A, Webb MR, Grogan A, Segal AW (1994) Activation of NADPH oxidase involves the dissociation of p21rac from its inhibitory GDP/GTP exchange protein (rhoGDI) followed by its translocation to the plasma membrane. Biochem J 3:585–591

Adachi H, Takahashi Y, Hasebe T, Shirouzu M, Yokoyama S, Sutoh K (1997) *Dictyostelium* IQGAP-related protein specifically involved in the completion of cytokinesis. J Cell Biol 137:891–898

Aspenstrom P (1997) A Cdc42 target protein with homology to the non-kinase domain of FER has a potential role in regulating the actin cytoskeleton. Curr Biol 7:479–487

Aspenstrom P, Lindberg U, Hall A (1996) Two GTPases Cdc42 and Rac, bind directly to a protein implicated in the immunodeficiency disorder Wiskott-Aldrich syndrome. Curr Biol 6:70–75

Bagrodia S, Derijard B, Davis RJ, Cerione RA (1995a) Cdc42 and PAK-mediated signaling leads to Jun kinase and p38 mitogen-activated protein kinase activation. J Biol Chem 270:27995–27998

Bagrodia S, Taylor SJ, Creasy CL, Chernoff J, Cerione RA (1995b) Identification of a mouse p21Cdc42/Rac activated kinase. J Biol Chem 270:22731–22737

Banin S, Truong O, Katz DR, Waterfield MD, Brickell PM, Gout I (1996) Wiskott-Aldrich syndrome protein (WASp) is a binding partner for c-Src family protein-tyrosine kinases. Curr Biol 6:981–988

Bashour AM, Fullerton AT, Hart MJ, Bloom GS (1997) IQGAP1, a Rac- and Cdc42-binding protein, directly binds and cross-links microfilaments. J Cell Biol 137:1555–1566

Bokoch GM (1995) Regulation of the phagocyte respiratory burst by small GTP-binding proteins. Trends Cell Biol 5:109–113

Brenner B, Koppenhoefer U, Weinstock C, Linderkamp O, Lang F, Gulbins E (1997) Fas- or ceramide-induced apoptosis is mediated by a Rac1-regulated activation of Jun N-terminal kinase/p38 kinases and GADD153. J Biol Chem 272:22173–22181

Brill S, Li S, Lyman CW, Church DM, Wasmuth JJ, Weissbach L, Bernards A, Snijders AJ (1996) The Ras GTPase-activating-protein-related human protein IQGAP2 harbors a potential actin binding domain and interacts with calmodulin and Rho family GTPases. Mol Cell Biol 16:4869–4878

Brown JL, Stowers L, Baer M, Trejo J, Coughlin S, Chant J (1996) Human Ste20 homologue hPAK1 links GTPases to the JNK MAP kinase pathway. Curr Biol 6:598–605

Burbelo PD, Drechsel D, Hall A (1995) A conserved binding motif defines numerous candidate target proteins for both Cdc42 and Rac GTPases. J Biol Chem 270:29071–29074

Castresana J, Saraste M (1995) Does Vav bind to F-actin through a CH domain? FEBS Lett 374:149–151

Cerione RA, Zheng Y (1996) The Dbl family of oncogenes. Curr Opin Cell Biol 8:216–222

Chan DC, Bedford MT, Leder P (1996) Formin binding proteins bear WWP/WW domains that bind proline-rich peptides and functionally resemble SH3 domains. EMBO J 15:1045–1054

Chant J (1996) Generation of cell polarity in yeast. Curr Opin Cell Biol 8:557–565

Chihara K, Amano M, Nakamura N, Yano T, Shibata M, Tokui T, Ichikawa H, Ikebe R, Ikebe M, Kaibuchi K (1997) Cytoskeletal rearrangements and transcriptional activation of c-fos serum response element by rho-kinase. J Biol Chem 272:25121–25127

Chong LD, Traynor KA, Bokoch GM, Schwartz MA (1994) The small GTP-binding protein Rho regulates a phosphatidylinositol 4-phosphate 5-kinase in mammalian cells. Cell 79:507–513

Chou MM, Blenis J (1996) The 70 kDa S6 kinase complexes with and is activated by the Rho family G proteins Cdc42 and Rac1. Cell 85:573–583

Chuang TH, Hahn KM, Lee JD, Danley DE, Bokoch GM (1997) The small GTPase Cdc42 initiates an apoptotic signaling pathway in Jurkat T lymphocytes. Mol Biol Cell 8:1687–1698

Coso OA, Chiariello M, Yu JC, Teramoto H, Crespo P, Xu N, Miki T, Gutkind JS (1995) The small GTP-binding proteins Rac1 and Cdc42 regulate the activity of the JNK/SAPK signaling pathway. Cell 81:1137–1146

D'Souza-Schorey C, Boshans R, McDonough M, Stahl PD, Van Aelst L (1997) A role for POR1 a Rac1 interacting protein in ARF6 mediated cytoskeletal rearrangements. EMBO J 16:5445–5454

De Mendez I, Adams AG, Sokolic RA, Malech HL, Leto TL (1996) Multiple SH3 domain interactions regulate NADPH oxidase assembly in whole cells. EMBO J 15:1211–1220

Dennis EA (1997) The growing phospholipase A2 superfamily of signal transduction enzymes. Trends Biochem Sci 22:1–2

Derijard B, Hibi M, Wu IH, Barrett T, Su B, Deng T, Karin M, Davis RJ (1994) JNK1: a protein kinase stimulated by UV light and Ha-Ras that binds and phosphorylates the c-Jun activation domain. Cell 76:1025–1037

Derijard B, Raingeaud J, Barrett T, Wu IH, Han J, Ulevitch RJ, Davis RJ (1995) Independent human MAP-kinase signal transduction pathways defined by MEK and MKK isoforms. Science 267:682–685

Diekmann D, Abo A, Johnston C, Segal AW, Hall A (1994) Interaction of Rac with p67phox and regulation of phagocytic NADPH oxidase activity. Science 265:531–533

Diekmann D, Nobes CD, Burbelo PD, Abo A, Hall A (1995) Rac GTPase interacts with GAPs and target proteins through multiple effector sites. EMBO J 14:5297–5305

Eaton S, Auvinen P, Luo L, Jan YN, Simons K (1995) CDC42 and Rac1 control different actin-dependent processes in the *Drosophila* wing disc epithelium. J Cell Biol 131:151–164
Erickson JW, Cerione RA, Hart MJ (1997) Identification of an actin cytoskeletal complex that includes IQGAP and the cdc42 GTPase. J Biol Chem 272:24443–24447
Faix J, Dittrich W (1996) DGAP1, a homologue of rasGTPase activating proteins that controls growth, cytokinesis, and development in *Dictyostelium discoideum*. FEBS Lett 394:251–257
Fanger GR, Johnson NL, Johnson GL (1997) MEK kinases are regulated by EGF and selectively interact with Rac/Cdc42. EMBO J 16:4961–4972
Gille H, Strahl T, Shaw PE (1995) Activation of ternary complex factor Elk-1 by stress-activated protein kinases. Curr Biol 5:1191–1200
Gulbins E, Coggeshall KM, Brenner B, Schlottmann K, Linderkamp O, Lang F (1996) Fas-induced apoptosis is mediated by activation of a Ras and Rac protein-regulated signaling pathway. J Biol Chem 271:26389–26394
Gupta S, Campbell D, Derijard B, Davis RJ (1995) Transcription factor ATF2 regulation by the JNK signal transduction pathway. Science 267:389–393
Habets GG, Scholtes EH, Zuydgeest D, van der Kammen RA, Stam JC, Berns A, Collard JG (1994) Identification of an invasion-inducing gene, Tiam-1, that encodes a protein with homology to GDP-GTP exchangers for Rho-like proteins. Cell 77:537–549
Han J, Lee JD, Jiang Y, Li Z, Feng L, Ulevitch RJ (1996) Characterization of the structure and function of a novel MAP kinase kinase (MKK6). J Biol Chem 271:2886–2891
Hart MJ, Callow MG, Souza B, Polakis P (1996) IQGAP1, a calmodulin-binding protein with a rasGAP-related domain, is a potential effector for cdc42Hs. EMBO J 15:2997–3005
Hartwig JH, Bokoch GM, Carpenter CL, Janmey PA, Taylor LA, Toker A, Stossel TP (1995) Thrombin receptor ligation and activated Rac uncap actin filament barbed ends through phosphoinositide synthesis in permeabilized human platelets. Cell 82:643–653
Hill CS, Wynne J, Treisman R (1995) The Rho family GTPases RhoA, Rac1, and CDC42Hs regulate transcriptional activation by SRF. Cell 81:1159–1170
Holland PM, Suzanne M, Campbell JS, Noselli S, Cooper JA (1997) MKK7 is a stress-activated mitogen-activated protein kinase kinase functionally related to hemipterous. J Biol Chem 272:24994–24998
Joneson T, McDonough M, Bar-Sagi D, Van Aelst L (1996a) RAC regulation of actin polymerization and proliferation by a pathway distinct from Jun kinase. Science 274:1374–1376
Joneson T, White MA, Wigler MH, Bar-Sagi D (1996b) Stimulation of membrane ruffling and MAP kinase activation by distinct effectors of RAS. Science 271:810–812
Joyal JL, Annan RS, Ho YD, Huddleston ME, Carr SA, Hart MJ, Sacks DB (1997) Calmodulin modulates the interaction between IQGAP1 and Cdc42. Identification of IQGAP1 by nanoelectrospray tandem mass spectrometry. J Biol Chem 272:15419–15425
Kenney D, Cairns L, Remold ODE, Peterson J, Rosen FS, Parkman R (1986) Morphological abnormalities in the lymphocytes of patients with the Wiskott-Aldrich syndrome. Blood 68:1329–1332
Khosravi FR, Solski PA, Clark GJ, Kinch MS, Der CJ (1995) Activation of Rac1, RhoA, and mitogen-activated protein kinases is required for Ras transformation. Mol Cell Biol 15:6443–6453
Kim BC, Kim JH (1997) Nuclear signalling by rac GTPase: essential role of phospholipase A2. Biochem J 326:333–337
Kirchhausen T, Rosen FS (1996) Disease mechanism: unravelling Wiskott-Aldrich syndrome. Curr Biol 6:676–678
Knaus UG, Morris S, Dong HJ, Chernoff J, Bokoch GM (1995) Regulation of human leukocyte p21-activated kinases through G protein – coupled receptors. Science 269:221–223
Kolluri R, Tolias KF, Carpenter CL, Rosen FS, Kirchhausen T (1996) Direct interaction of the Wiskott-Aldrich syndrome protein with the GTPase Cdc42. Proc Natl Acad Sci USA 93:5615–5618
Kozma R, Ahmed S, Best A, Lim L (1995) The Ras-related protein Cdc42Hs and bradykinin promote formation of peripheral actin microspikes and filopodia in Swiss 3T3 fibroblasts. Mol Cell Biol 15:1942–1952

Kuroda S, Fukata M, Kobayashi K, Nakafuku M, Nomura N, Iwamatsu A, Kaibuchi K (1996) Identification of IQGAP as a putative target for the small GTPases, Cdc42 and Rac1. J Biol Chem 271:23363–23367

Lamarche N, Tapon N, Stowers L, Burbelo PD, Aspenstrom P, Bridges T, Chant J, Hall A (1996) Rac and Cdc42 induce actin polymerization and G1 cell cycle progression independently of p65PAK and the JNK/SAPK MAP kinase cascade. Cell 87:519–529

Leberer E, Wu C, Leeuw T, Fourest LA, Segall JE, Thomas DY (1997) Functional characterization of the Cdc42p binding domain of yeast Ste20p protein kinase. EMBO J 16:83–97

Lebowitz PF, Davide JP, Prendergast GC (1995) Evidence that farnesyltransferase inhibitors suppress Ras transformation by interfering with Rho activity. Mol Cell Biol 15:6613–6622

Lebowitz PF, Du W, Prendergast GC (1997) Prenylation of RhoB is required for its cell transforming function but not its ability to activate serum response element-dependent transcription. J Biol Chem 272:16093–16095

Lee S, Escalante R, Firtel RA (1997) A Ras GAP is essential for cytokinesis and spatial patterning in *Dictyostelium*. Development 124:983–996

Leto TL, Adams AG, de Mendez I (1994) Assembly of the phagocyte NADPH oxidase binding of Src homology 3 domains to proline-rich targets. Proc Natl Acad Sci USA 91:10650–10654

Li R (1997) Bee1, a yeast protein with homology to Wiscott-Aldrich syndrome protein, is critical for the assembly of cortical actin cytoskeleton. J Cell Biol 136:649–658

Lin R, Bagrodia S, Cerione R, Manor D (1997) A novel Cdc42Hs mutant induces cellular transformation. Curr Biol 7:794–797

Luo L, Lee T, Tsai L, Tang G, Jan LY, Jan YN (1997) Genghis Khan (Gek) as a putative effector for *Drosophila* Cdc42 and regulator of actin polymerization. Proc Natl Acad Sci USA 94:12963–12968

Machesky LM, Hall A (1997) Role of actin polymerization and adhesion to extracellular matrix in Rac- and Rho-induced cytoskeletal reorganization. J Cell Biol 138:913–926

Madaule P, Furuyashiki T, Reid T, Ishizaki T, Watanabe G, Morii N, Narumiya S (1995) A novel partner for the GTP-bound forms of rho and rac. Febs Lett 377:243–248

Maly FE, Quilliam LA, Dorseuil O, Der CJ, Bokoch GM (1994) Activated or dominant inhibitory mutants of Rap1A decrease the oxidative burst of Epstein-Barr virus-transformed human B lymphocytes. J Biol Chem 269:18743–18746

Manser E, Chong C, Zhao ZS, Leung T, Michael G, Hall C, Lim L (1995) Molecular cloning of a new member of the p21-Cdc42/Rac-activated kinase (PAK) family. J Biol Chem 270:25070–25078

Manser E, Huang HY, Loo TH, Chen XQ, Dong JM, Leung T, Lim L (1997) Expression of constitutively active alpha-PAK reveals effects of the kinase on actin and focal complexes. Mol Cell Biol 17:1129–1143

McCallum SJ, Wu WJ, Cerione RA (1996) Identification of a putative effector for Cdc42Hs with high sequence similarity to the RasGAP-related protein IQGAP1 and a Cdc42Hs binding partner with similarity to IQGAP2. J Biol Chem 271:21732–21737

Michiels F, Habets GG, Stam JC, van der Kammen RA, Collard JG (1995) A role for Rac in Tiam1-induced membrane ruffling and invasion. Nature 375:338–340

Miki H, Miura K, Takenawa T (1996) N-WASP, a novel actin-depolymerizing protein, regulates the cortical cytoskeletal rearrangement in a PIP2-dependent manner downstream of tyrosine kinases. EMBO J 15:5326–5335

Miller PJ, Johnson DI (1994) Cdc42p GTPase is involved in controlling polarized cell growth in *Schizosaccharomyces pombe*. Mol Cell Biol 14:1075–1083

Minden A, Lin A, Claret FX, Abo A, Karin M (1995) Selective activation of the JNK signaling cascade and c-Jun transcriptional activity by the small GTPases Rac and Cdc42Hs. Cell 81:1147–1157

Molina IJ, Kenney DM, Rosen FS, Remold ODE (1992) T cell lines characterize events in the pathogenesis of the Wiskott-Aldrich syndrome. J Exp Med 176:867–874

Moriguchi T, Kawasaki H, Matsuda S, Gotoh Y, Nishida E (1995) Evidence for multiple activators for stress-activated protein kinase/c-Jun amino-terminal kinases. Existence of novel activators. J Biol Chem 270:12969–12972

Nisimoto Y, Freeman J, Motalebi SA, Hirshberg M, Lambeth JD (1997) Rac binding to p67(phox). Structural basis for interactions of the Rac1 effector region and insert region with components of the respiratory burst oxidase. J Biol Chem 272:18834–18841

Nobes CD, Hall A (1995) Rho, rac, and cdc42 GTPases regulate the assembly of multimolecular focal complexes associated with actin stress fibers, lamellipodia, and filopodia. Cell 81:53–62

O'Hagan RC, Tozer RG, Symons M, McCormick F, Hassell JA (1996) The activity of the Ets transcription factor PEA3 is regulated by two distinct MAPK cascades. Oncogene 13:1323–1333

Olson MF, Ashworth A, Hall A (1995) An essential role for Rho, Rac, and Cdc42 GTPases in cell cycle progression through G1. Science 269:1270–1272

Ottilie S, Miller PJ, Johnson DI, Creasy CL, Sells MA, Bagrodia S, Forsburg SL, Chernoff J (1995) Fission yeast pakl+ encodes a protein kinase that interacts with Cdc42p and is involved in the control of cell polarity and mating. EMBO J 14:5908–5919

Peppelenbosch MP, Qiu RG, de V, Smits AM, Tertoolen LG, de LS, McCormick F, Hall A, Symons MH, Bos JL (1995) Rac mediates growth factor-induced arachidonic acid release. Cell 81:849–856

Perona R, Montaner S, Saniger L, Sanchez PI, Bravo R, Lacal JC (1997) Activation of the nuclear factor-kappaB by Rho, CDC42, and Rac-1 proteins. Genes Dev 11:463–475

Peter M, Neiman AM, Park HO, van LM, Herskowitz I (1996) Functional analysis of the interaction between the small GTP binding protein Cdc42 and the Ste20 protein kinase in yeast. EMBO J 15:7046–7059

Peters PJ, Hsu VW, Ooi CE, Finazzi D, Teal SB, Oorschot V, Donaldson JG, Klausner RD (1995) Overexpression of wild-type and mutant ARF1 and ARF6: distinct perturbations of nonoverlapping membrand compartments. J Cell Biol 128:1003–1017

Prendergast GC, Khosravi FR, Solski PA, Kurzawa H, Lebowitz PF, Der CJ (1995) Critical role of Rho in cell transformation by oncogenic Ras. Oncogene 10:2289–2296

Proud CG (1996) p70 S6 kinase: an enigma with variations. Trends Biochem Sci 21:181–185

Pulverer BJ, Kyriakis JM, Avruch J, Nikolakaki E, Woodgett JR (1991) Phosphorylation of c-jun mediated by MAP kinases. Nature 353:670–674

Qiu, R-G (1997) The role of Rho-like GTPases in cell proliferation and Ras transformation. Thesis

Qiu RG, Chen J, Kirn D, McCormick F, Symons M (1995a) An essential role for Rac in Ras transformation. Nature 374:457–459

Qiu RG, Chen J, McCormick F, Symons M (1995b) A role for Rho in Ras transformation. Proc Natl Acad Sci USA 92:11781–11785

Qiu RG, Abo A, McCormick F, Symons M (1997) Cdc42 regulates anchorage-independent growth and is necessary for Ras transformation. Mol Cell Biol 17:3449–3458

Quinn MT, Evans T, Loetterle LR, Jesaitis AJ, Bokoch GM (1993) Translocation of Rac correlates with NADPH oxidase activation. Evidence for equimolar translocation of oxidase components. J Biol Chem 268:20983–20987

Quinn MT, Parkos CA, Walker L, Orkin SH, Dinauer MC, Jesaitis AJ (1989) Association of a Ras-related protein with cytochrome b of human neutrophils. Nature 342:198–200

Radhakrishna H, Klausner RD, Donaldson JG (1996) Aluminum fluoride stimulates surface protrusions in cells overexpressing the ARF6 GTPase. J Cell Biol 134:935–946

Reinhard J, Scheel AA, Diekmann D, Hall A, Ruppert C, Bahler M (1995) A novel type of myosin implicated in signalling by rho family GTPases. EMBO J 14:697–704

Ridley AJ, Paterson HF, Johnston CL, Diekmann D, Hall A (1992) The small GTP-binding protein rac regulates growth factor-induced membrane ruffling. Cell 70:401–410

Rivero LO, Marcilla A, Sameshima JH, Robbins KC (1995) Wiskott-Aldrich syndrome protein physically associates with Nck through Src homology 3 domains. Mol Cell Biol 15:5725–5731

Rouse J, Cohen P, Trigon S, Morange M, Alonso LA, Zamanillo D, Hunt T, Nebreda AR (1994) A novel kinase cascade triggered by stress and heat shock that stimulates MAPKAP kinase-2 and phosphorylation of the small heat shock proteins. Cell 78:1027–1037

Sanchez I, Hughes RT, Mayer BJ, Yee K, Woodgett JR, Avruch J, Kyriakis JM, Zon LI (1994) Role of SAPK/ERK kinase-1 in the stress-activated pathway regulating transcription factor c-Jun. Nature 372:794–798

Segal AW, Abo A (1993) The biochemical basis of the NADPH oxidase of phagocytes. Trends Biochem Sci 18:43–47

Segal AW, Cross AR, Garcia RC, Borregaard N, Valerius NH, Soothill JF, Jones OT (1983) Absence of cytochrome β-245 in chronic granulomatous disease. A multicenter European evaluation of its incidence and relevance. N Engl J Med 308:245–251

Sells MA, Chernoff J (1997) Emerging from the Pak: the p21-activated protein kinase family. Trends Cell Biol 7:162–167

Sells M, Knaus U, Bagrodia S, Ambrose DM, Bokoch GM, Chernoff J (1997) Human p21-activated kinase (Pak1) regulates actin organization in mammalian cells. Curr Biol 7:202–210

She H, Rockow S, Tang J, Nishimura R, Skolnik EY, Chen M, Margolis B, Li W (1997) Wiskott-Aldrich syndrome protein is associated with the adapter protein Grb2 and the epidermal growth factor receptor in living cells. Mol Biol Cell 8:1709–1721

Smith RM, Curnutte JT (1991) Molecular basis of chronic granulomatous disease. Blood 77:673–686

Stokoe D, Engel K, Campbell DG, Cohen P, Gaestel M (1992) Identification of MAPKAP kinase 2 as a major enzyme responsible for the phosphorylation of the small mammalian heat shock proteins. FEBS Lett 313:307–313

Stowers L, Yelon D, Berg L, Chant J (1995) Regulation of the polarization of T-cells toward antigen presenting cells by the Ras-related GTPase Cdc42. Proc Natl Acad Sci USA 92:5027–5031

Sudol M, Chen HI, Bougeret C, Einbond A, Bork P (1995) Characterization of a novel protein-binding module – the WW domain. FEBS Lett 369:67–71

Sulciner DJ, Irani K, Yu ZX, Ferrans VJ, Goldschmidt CP, Finkel T (1996) rac1 regulates a cytokine-stimulated, redox-dependent pathway necessary for NF-kappaB activation. Mol Cell Biol 16:7115–7121

Symons M (1996) Rho family GTPases: the cytoskeleton and beyond. Trends Biochem Sci 21:178–181

Symons M, Derry JM, Karlak B, Jiang S, Lemahieu V, Mccormick F, Francke U, Abo A (1996) Wiskott-Aldrich syndrome protein, a novel effector for the GTPase CDC42Hs, is implicated in actin polymerization. Cell 84:723–734

Teahan C, Rowe P, Parker P, Totty N, Segal AW (1987) The X-linked chronic granulomatous disease gene codes for the beta- chain of cytochrome b-245. Nature 327:720–721

Teramoto H, Coso OA, Miyata H, Igishi T, Miki T, Gutkind JS (1996) Signaling from the small GTP-binding proteins Rac1 and Cdc42 to the c- Jun N-terminal kinase/stress-activated protein kinase pathway. A role for mixed lineage kinase 3/protein-tyrosine kinase 1, a novel member of the mixed lineage kinase family. J Biol Chem 271:27225–27228

Tolias KF, Cantley LC, Carpenter CL (1995) Rho family GTPases bind to phosphoinositide kinases. J Biol Chem 270:17656–17659

Tsuchiya M, Price SR, Tsai S-C, Moss J, Vaughan M (1991) Molecular identification of ADP-ribosylation factor mRNAs and their expression in mammalian cells. J Biol Chem 266:2772–2777

Van Aelst L, Joneson T, Bar-Sagi D (1996) Identification of a novel Rac1-interacting protein involved in membrane ruffling. EMBO J 15:3778–37786

Vincent S, Settleman J (1997) The PRK2 kinase is a potential effector target of both Rho and Rac GTPases and regulates actin cytoskeletal organization. Mol Cell Biol 17:2247–2256

Wang XZ, Ron D (1996) Stress-induced phosphorylation and activation of the transcription factor CHOP (GADD153) by p38 MAP Kinase. Science 272:1347–1349

Weissbach L, Settleman J, Kalady MF, Snijders AJ, Murthy AE, Yan YX, Bernards A (1994) Identification of a human rasGAP-related protein containing calmodulin-binding motifs. J Biol Chem 269:20517–20521

Westwick JK, Lambert QT, Clark GJ, Symons M, Van Aelst L, Pestell RG, Der CJ (1997) Rac regulation of transformation, gene expression, and actin organization by multiple, PAK-independent pathways. Mol Cell Biol 17:1324–1335

White MA, Nicolette C, Minden A, Polverino A, Van Aelst L, Karin M, Wigler MH (1995) Multiple Ras functions can contribute to mammalian cell transformation. Cell 80:533–541

Wittinghofer A, Nassar N (1996) How Ras-related proteins talk to their effectors. Trends Biochem Sci 21:488–491

Xu S, Robbins D, Frost J, Dang A, Lange CC, Cobb MH (1995) MEKK1 phosphorylates MEK1 and MEK2 but does not cause activation of mitogen-activated protein kinase. Proc Natl Acad Sci USA 92:6808–6812

Yang W, Cerione RA (1997) Cloning and characterization of a novel Cdc42-associated tyrosine kinase, ACK-2, from bovine brain. J Biol Chem 275:24819–24824

Zervos AS, Faccio L, Gatto JP, Kyriakis JM, Brent R (1995) Mxi2, a mitogen-activated protein kinase that recognizes and phosphorylates Max protein. Proc Natl Acad Sci USA 92:10531–10534

Zhang S, Han J, Sells MA, Chernoff J, Knaus UG, Ulevitch RJ, Bokoch GM (1995) Rho family GTPases regulate p38 mitogen-activated protein kinase through the downstream mediator Pak1. J Biol Chem 270:23934–23936

Zigmond SH (1996) Signal transduction and actin filament organization. Curr Opin Cell Biol 8:66–73

Zigmond SH, Joyce M, Borleis J, Bokoch GM, Devreotes PN (1997) Regulation of actin polymerization in cell-free systems by GTPgammaS and Cdc42. J Cell Biol 138:363–374

Small GTPases of the Rho Family and Cell Transformation

Philippe Fort

1 Introduction

The Rho GTPases form a distinct subgroup of the Ras superfamily of low molecular weight GTP binding proteins. These proteins are implicated in signal transduction leading to changes in membrane structures and cytoskeletal reorganisation associated with changes in cell shape. Like other Ras-related proteins, Rho GTPases are thought to adopt either an active GTP-bound conformational state or an inactive GDP-bound state. Although cycling between these states is controlled by several regulatory proteins, mutations in Rho proteins can favour a specific status: an asparagine substitution in Rho at a position homologous to Ras threonine 17 leads to a drop in its affinity for GTP. This mutated protein acts as an inhibitor by sequestering positive regulatory factors, thereby preventing activation of the endogenous Rho GTPase. Conversely, substitutions of residues similar to those found in oncogenic Ras proteins (e.g. G12V or Q61L) leads to constitutively active Rho proteins, due to a reduced GTP hydrolysis. Once loaded with GTP, the GTPase gains the ability to bind cognate effector downstream targets, which converts the input signal into a specific set of activations.

Over the past few years, it has been shown that Rho GTPases play a role in the organisation of the actin cytoskeleton, and also have critical functions in the control of cell proliferation. In the present review, I address the implication of the Rho family in cell transformation and apoptosis, as well as in the regulatory cross-talk between pathways controlled by Rho and Ras proteins.

1.1 The Rho Family

Rho proteins are key elements in the regulation of numerous functions such as the assembly of the cytoskeleton (Ridley 1994), cell motility (Aepfelbacher et al. 1994), smooth muscle contraction (Hirata et al. 1992), metastasis (Habets 1994), apoptosis (Jimenez et al. 1995), as well as various aspects of cell polarity.

Institut de Génétique Moléculaire de Montpellier, UMR, CNRS 5535, 1919, route de Mende, F-34293 Montpellier, cedex 05, France

Progress in Molecular and Subcellular Biology, Vol. 22
Ph. Jeanteur (Ed.)

The Rho family is made of two branches, one comprising RhoA (Madaule and Axel 1985), RhoB and RhoC (Chardin et al. 1988), RhoD (Murphy et al. 1996), RhoE (Foster et al. 1996) and RhoL (Murphy and Montel 1996), the other comprising Rac1 and Rac2 (Didsbury et al. 1989), Rac3 (Courjal et al. 1997; Haataja et al. 1997), RhoG (Vincent et al. 1992), Cdc42Hs (Shinjo et al. 1990), TC10 (Drivas et al. 1990) and TTF (Dallery et al. 1995). RhoA, RhoB and RhoC control the formation of focal adhesions and actin stress fibres in fibroblasts (Flescher et al. 1993; Ridley and Hall 1992; Ridley et al. 1992), while RhoD causes rearrangements of the actin cytoskeleton and controls early endosome motility and distribution (Murphy et al. 1996). Rac proteins are required for growth factor-induced membrane ruffling and lamellipodia formation in fibroblasts (Ridley 1994; Ridley et al. 1992). Additionally, in neurons, their activity is required for axonal outgrowth (Luo et al. 1994) while in phagocytic cells, they play a role in the activation of NADPH oxidase (Abo et al. 1991, 1994; Knaus et al. 1991, 1992). Cdc42 is involved in the establishment of cell polarity in yeast (Adams et al. 1990; Johnson and Pringle 1990) and the mammalian homologue Cdc42Hs is required in the polarization of helper T cells toward antigen-presenting cells (Stowers et al. 1995). Cdc42Hs also regulates the formation of filopodia in growth stimulated fibroblastic cells (Nobes and Hall 1995). In addition to their role in cell morphology, Rho, Rac and Cdc42Hs have been shown to promote cell cycle progression through G_1, trigger DNA synthesis (Olson et al. 1995) and regulate the activation of the ubiquitous transcription factor SRF (serum response factor; Hill et al. 1995). However, the effect induced by Cdc42Hs is still controversial, as other reports indicate that its expression causes cytokinesis arrest (Dutartre et al. 1996), and inhibits serum-stimulated cell cycle progression at G_1/S through a mechanism requiring the MAP kinase p38/RK (Molnar et al. 1997). Rac, Cdc42Hs and RhoG also stimulate the c-Jun kinase JNK/SAPK, an enzyme essential in the signalling pathways from IL-1 or TNFα receptors (Coso et al. 1995; Roux et al. 1997).

1.2
Regulators of the Rho Family and their Oncogenic Properties

The transition between GTP-bound and GDP-bound conformational states of the Rho GTPases is controlled by a wide array of regulatory proteins: guanine nucleotide exchange factors (GEFs), GTPase-activating proteins (GAPs) and GDP-dissociation inhibitory factors (GDIs). So far, only three GDI proteins have been characterized, namely RhoGDI, D4/LyGDI and γ-GDI/RhoGDI3. RhoGDI is ubiquitously expressed in mammalian tissues and organs (Fukumoto et al. 1990), while D4/LyGDI expression is restricted to hematopoietic tissues (Adra et al. 1993; Scherle et al. 1993). γ-GDI/RhoGDI3 is preferentially expressed in the brain and pancreas (Adra et al. 1997), and has a binding specificity for RhoB and RhoG (Zalcman et al. 1996).

A high number of GAPs acting on the Rho family have been isolated. They include p190 (Settleman et al. 1992), p190-B (Burbelo et al. 1995), RhoGAP (Garrett et al. 1991), abr (Heisterkamp et al. 1993), bcr (Heisterkamp et al. 1983), Cdc42Hs-GAP (Hart et al. 1991b; Barfod et al. 1993), 3BP-1 (Cicchetti et al. 1995), p85 (Cooper and Kashishian 1993), n-chimaerin (Hall et al. 1990), p115 (Tribioli et al. 1996), RIP1/RLIP76 (Jullien-Flores et al. 1995; Park and Weinberg 1995), myr-5 (Reinhard et al. 1995), the myosin IXb (Wirth et al. 1996), graf (Hildebrand et al. 1996), and IQGAP1 (Brill et al. 1996). These proteins exhibit a wide range of tissue specificity and act differentially on the members of the Rho family (Manser et al. 1992; Ridley et al. 1993).

The first characterized exchange factor was Dbl, an oncoprotein isolated from diffuse B cell lymphomas (Hart et al. 1991a, 1994; Ron et al. 1991). Dbl was shown to share a conserved 250 amino acid domain (DH domain) with Cdc24, an exchange factor for yeast Cdc42 (Zheng et al. 1994). Since the initial characterization of Dbl and Cdc24, additional members of the Dbl family have been characterized, including oncoproteins such as Bcr (Heisterkamp et al. 1983), Vav (Katzav et al. 1989), Ost (Horii et al. 1994), Tiam-1 (Habets et al. 1994), ect2 (Miki et al. 1993), Tim (Chan et al. 1994), fgd1 (Pasteris et al. 1994), abr (Heisterkamp et al. 1993), dbs (Whitehead et al. 1995a), lbc (Toksoz and Williams 1994), lfc (Whitehead et al. 1995b) and lsc (Whitehead et al. 1996). Like GAPs, GEFs exhibit a wide diversity in their tissue distribution and specificity towards Rho members.

Thus, only proteins capable of activating one or several Rho proteins have oncogenic properties (Cerione and Zheng 1996). A direct link between activation of Rho GTPases and cell transformation has been demonstrated by the finding that the GEF activity of the dbl exchange factor is required for cell transformation (Hart et al. 1994). Since GEFs may activate one or several Rho GTPases, it is hypothesized that one or several pathways controlled by Rho proteins are involved in cell transformation.

2 Rho Proteins and Cell Transformation

2.1 Intrinsic Transforming Properties of Rho Proteins

During the last 10 years, several studies have investigated the transforming potential of the Rho family members. It was first observed that although expression of activated RhoA (V14-RhoA) in NIH3T3 cells did not cause focus formation in monolayers or growth in soft agar, it led to a reduced dependence on serum for growth, a higher saturation density and a tumourigenic potential in nude mice (Avraham and Weinberg 1989). Since then, many Rho proteins have been examined for their focus forming activity in NIH3T3 cells. Interestingly, like V14-RhoA, wild-type RhoA was also shown to confer a reduced anchorage- and serum-dependent growth (Perona et al. 1993). V14-RhoA and wild-type RhoA transfected cells were able to induce tumours in nude mice, albeit at a much lower efficiency for the wild-type RhoA. RhoA-induced tumours consist of well-differentiated fibrosarcomas, which exhibit intersecting bundles of spindle cells. Similar tumour types were obtained in nude mice transplanted with V14-RhoB expressing cells (Prendergast et al. 1995).

Analysis of Racl (Qiu et al. 1995a) and RhoG (Roux et al. 1997) showed that expression of either protein in NIH3T3 cells led to a partial loss of cell contact inhibition and a reduced dependence on serum. In addition, overexpression of constitutively active Racl (V12-Racl) induced invasiveness of BW5147 T-lymphoma cells in a culture assay system, as did the Rac-specific exchange factor Tiaml (Michiels et al. 1995; van Leeuwen et al. 1995). V12-Racl-expressing NIH3T3 cells were found to be polymorphic, appearing either as large flat cells with extensive membrane ruffling, similar to dbl-transformed cells (Ron et al. 1991), or spindle shaped and refractile, as observed for V14-RhoA-transformed cells. These cells were also characterized by a high proportion of multinucleated cells (Qiu et al. 1997). Subsequent transplantation in nude mice induced tumours at the same efficiency as RhoA (Qiu et al. 1995a, b).

Although expression of activated Cdc42Hs protein (V12-Cdc42Hs) in Rat1 or NIH3T3 cells led to a dramatic loss of anchorage dependence, it had no effect on serum-dependent growth and on contact inhibition (Roux et al. 1997; Qiu et al. 1997). Interestingly, V12-Cdc42Hs-expressing Rat1 cells readily produced tumours in nude mice (Qiu et al. 1997) but did not form foci (Roux et al. 1997). These cells showed a high membrane activity, were generally multinucleated, had a rounded phenotype, and a reduced adhesion to the substratum (Qiu et al. 1997).

It thus appears that members of the Rho family have similar transforming potentials: decreased anchorage-dependent cell growth (Cdc42Hs and RhoA), reduction of cell contact inhibition and serum dependence (Rac1 and RhoG), and tumour formation in nude mice (Cdc42Hs, Rac1 and RhoA). These results led to the question of whether these proteins activate distinct pathways leading to similar phenotypic changes or whether their shared properties are mediated by a single pathway. Such a situation was described for the control of the actin cytoskeleton, where activated Cdc42Hs activates Rac1, which in turn can modulate the activity of Rho (Ridley et al. 1992; Nobes and Hall 1995; Tapon and Hall 1997).

2.2 Cross-Talk Between Rho Controlled Pathways

In contrast to the low focus forming activity of cells transfected with GTPases, cells expressing exchange factors display a high focus forming activity. As exchange factors can activate several distinct GTPases, this suggests a probable cooperation between Rho-controlled pathways. For instance, the Ost exchange factor potentially acts on RhoA and Cdc42Hs, and also has the ability to bind the GTP-bound form of Rac1 (Horii et al. 1994). Thus, the activated Ost protein might express its full oncogenic properties through the activation of three distinct Rho-dependent pathways.

Analysis of coordinated cross-talk has been worked out by coexpressing various combinations of constitutively active and dominant negative Rho protein variants. Whereas constitutively active individual Rho GTPases showed very weak focus forming activities, their simultaneous expression results in much higher focus formation (Roux et al. 1997): coexpression of V12-Cdc42Hs and V12-Rac1, or V12-Cdc42Hs and V12-RhoG produced a high focus forming activity, in the range of that observed for Ost (Horii et al. 1994). Coexpression of V12-RhoG and V12-Rac1 elicited a focus forming activity an order of magnitude lower (Roux et al. 1997). Interestingly, coexpression of N17-Rac1, a dominant negative Rac, caused an 80% inhibition in the focus forming activity of cells expressing V12-Cdc42Hs and V12-RhoG, while expression of N17-RhoG and N17-Cdc42Hs only marginally reduced the focus formation elicited by V12-Rac1/V12-Cdc42Hs and V12-Rac1/V12-RhoG, respectively. This suggests that Rac acts downstream of RhoG in a pathway independent from Cdc42Hs. Along the same lines, coexpression of V12-Cdc42Hs and N17-Rac1 strongly inhibited the generation of multinucleated cells as well as PDGF-induced lamellipodia (Qiu et al. 1997). In contrast, anchorage-independent growth in soft agar was not significantly inhibited. This suggests that Cdc42Hs might control at least two distinct pathways: one pathway responsible for lamellipodia and cleavage furrow formation where Cdc42Hs is upstream of Rac, and a second Rac-independent pathway, which controls anchorage-independent cell growth.

Thus, although Cdc42Hs was shown to act upstream of Rac1 in pathways signalling cytoskeletal reorganization of various cell types (Allen et al. 1997; Machesky and Hall 1996; Nobes and Hall 1995; Tapon and Hall 1997), we can conclude that Rac1 and Cdc42Hs delineate at least two independent pathways that cooperate in cell transformation (Fig. 1). Cdc42Hs controls the anchorage-dependent cell growth, while Rac1 controls cell contact inhibition. RhoG appears to be involved in some of the same pathways as Rac1, probably acting upstream of Rac. However, RhoG is also involved in different pathways, as it cooperates independently with Rac1 and Cdc42Hs in focus formation.

The observation that Rho-dependent pathways leading to cell transformation follow a scheme that does not superimpose with the regulatory cascade described for the control of actin cytoskeleton (see end of Sect. 2.1) suggests that the expression of V12-Cdc42Hs only partially activates the endogenous Rac1 protein, e.g. by acting only on Rac1 localized in specific subcellular domains or by leading to an overall lower amount of GTP-bound Rac1.

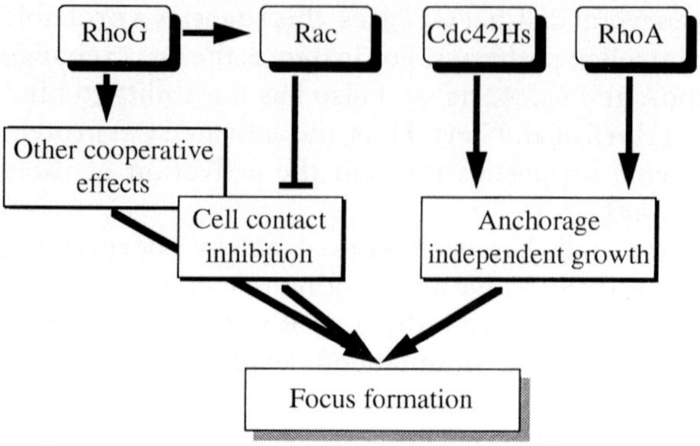

Fig. 1. Model for cooperativity of Rho GTPases in focus formation. Cdc42Hs and RhoA promote anchorage-independent growth, while RhoG and Rac reduce cell contact inhibition. The simultaneous activation of both pathways leads to a high focus forming activity. RhoG is thought to act mainly upstream of Rac, although additional cooperative effects are suspected

3 Cross-Talk Between Ras and Rho-Dependent Pathways in Cell Transformation

3.1 Cooperation Between Ras and Rho Pathways

Several lines of evidence led to the hypothesis that Rho proteins are activated in Ras-controlled pathways leading to cell transformation:

1. Microinjection of constitutively active or wild-type Ras proteins has long been known to elicit extensive membrane ruffles and phase-fluid pinocytosis in resting cells (Bar-Sagi and Feramisco 1986). This feature has been shown to require an active endogenous Rac1 protein (Ridley et al. 1992).
2. In cells stimulated by growth factors, the Rho-GAP p190 protein (Settleman et al. 1992) is rapidly tyrosine-phosphorylated and clusters with the Ras-GAPp120/p62^{Dok} complex (Pronk et al. 1993).
3. More recently, a direct functional connection between Ras signalling and Rho protein activity was reported in the yeast *S. pombe*, in which signals that controls normal morphology and mating are conveyed from Ras1 to Cdc42Sp (Chang et al. 1994).

The direct implication of Rho-dependent pathways in Ras-mediated transformation was further investigated by coexpressing oncogenic Ras proteins with dominant inhibitory and activated versions of Rho proteins. Ras-induced focus formation was found to be inhibited upon expression of all tested dominant negative mutants of Rho family members, i.e. Rac1, RhoA, RhoB, Cdc42 and RhoG (Khosravi-Far et al. 1995; Prendergast et al. 1995; Qiu et al. 1995a,b, 1997; Roux et al. 1997). More accurate analyses revealed differences in the effects of these inhibitory proteins on Ras-induced transformation (Qiu et al. 1997). N17-Cdc42 was shown to revert the transformed morphology of Ras-expressing cells, whereas N17-Rac1 did not. Conversely, expression of N17-Rac1 strongly inhibited low-serum growth of Ras-transformed cells, whereas N17-Cdc42Hs had a limited effect. Both proteins strongly inhibited cell growth in soft agar.

The overall inhibited effect of N17 variants was not due to a toxic effect, as their expression at similar levels induced minimal changes in NIH3T3 cells: N17-Rac1 and N17-RhoG expression led to a reduction in cell saturation density but did not modify cell growth in 10% foetal calf serum (Qiu et al. 1995a; Roux et al. 1997), while no changes in morphology or cell growth were reported in N17-Cdc42Hs-expressing cells (Qiu et al. 1997; Roux et al. 1997). Expression of N19-RhoA did not reduce cell cloning efficiency, although it led to a reduction in stress fibre formation (Qiu et al. 1995b). In addition to inhibitory mutants, overexpression of p190-RhoGAP [which has a marked preferential activity for Rho (A,B,C) in vitro (Ridley et al. 1993)] was recently shown to suppress Ha-Ras-mediated cell transformation (Wang et al. 1997). This demonstrates that the amount of intracellular GTP-bound Rho is critical for Ras transformation. Thus, although the possibility remains that some mutants inhibit non specifically distinct Rho proteins, these data suggest that multiple pathways controlled by Rho proteins are necessary for full Ras-dependent transformation.

The implication of Rho-dependent pathways in cell transformation was further strengthened by the phenotypic changes observed in cells coexpressing activated Ras and Rho proteins. L63-RhoA or I115-Rac1 expression led to a twofold increase in the number of L61-Ras-induced foci (Khosravi-Far et al. 1995). Similarly, a three fold to four fold increase in activated Ras focus forming activity was observed upon coexpression with V12-Rac1, and to a lower extent, with V12-RhoG and V120-Cdc42Hs (Roux et al. 1997). In addition, the morphology of the resulting foci was dramatically altered. Whereas cells expressing L61-Ras were highly refractile and spindle-shaped, coexpression with L63-RhoA or I115-Rac1 produced foci with rounded refractile and poorly adherent cells.

Thus, these data are consistent with the hypothesis that distinct Rho-dependent pathways are important in Ras-mediated transformation. Even though Rho GTPases do not induce focus formation on their own, they synergize with Ras, suggesting that Rho proteins modulate the susceptibility of NIH3T3 cells to Ras-mediated transformation (Fig. 2).

3.2
Cooperation Between Raf and Rho Pathways

Ras-dependent signal transduction has been extensively studied over the past 10 years. Upon activation by ligand-stimulated tyrosine kinase receptors, activated Ras complexes with several effector proteins (reviewed in Marshall

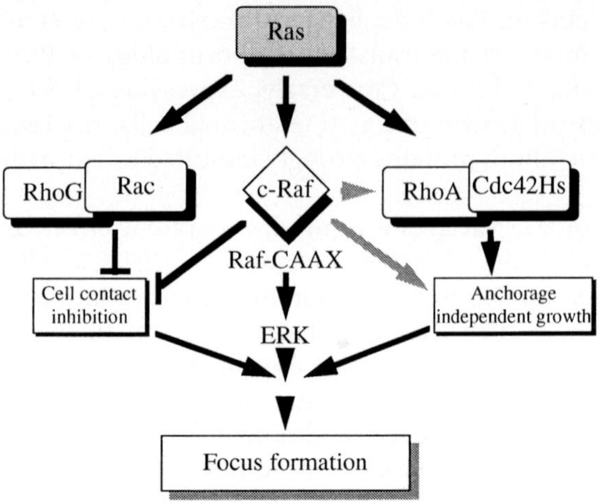

Fig. 2. Role of the Rho family in Ras- and Raf-mediated transformation. The activity of all Rho members is required for Ras transformation. Depending on the cell system used, RhoA and Cdc42Hs activities are dispensable or required for Raf transformation (*grey arrows*). All activated Rho GTPases cooperate with Ras and Raf. The mechanisms by which constitutively activated Raf by passes Rho protein pathways are still unknown

1996), including Raf-1, phosphatidyl-inositol-3-OH kinase (PI3K) and MEK kinase 1. Once phosphorylated at the plasma membrane, the activated Raf-1 kinase phosphorylates the MAPK kinases MEK1 and MEK2, which in turn activate p42 and p44 MAPK (extracellular signal-regulated kinases, ERKs; reviewed in Marais and Marshall 1996). Phosphorylated ERKs then translocate into the nucleus, where they activate nuclear transcription factors, eventually leading to gene activation and mitogenesis. As Rho proteins involved in Ras-mediated transformation might affect several pathways downstream of Ras, it was crucial to determine the relationships between Rho-dependent pathways and the Raf-1/ERK pathway. Such an analysis was worked out by coexpressing dominant negative or constitutively active versions of Rho (A, B), RhoG, Rac1 and Cdc42Hs proteins with activated forms of Raf-1. Coexpression of the dominant negative N19-RhoB in Rat1 fibroblasts did not reduce foci formation elicited by the v-Raf oncogene (Prendergast et al. 1995). In NIH3T3 cells, the focus forming activity of Raf-CAAX, which constitutively activates the MAP-kinase pathway (Leevers et al. 1994), was not reduced upon expression of N17-Rac1 (Qiu et al. 1995a). Identical results were obtained upon coexpression of N17-Rac1, N17-RhoG and N17-Cdc42Hs proteins with the v-Raf oncogene (Roux et al. 1997). These results therefore suggest that Rho-dependent pathways involved in Ras transformation are distinct from the Raf-1/ERK pathway. This was further confirmed by assaying for cooperation of both types of pathways in transformation. Expression of each activated Rho GTPase led to a two to five fold increase in v-Raf or Raf-CAAX focus forming activity (Khosravi-

Far et al. 1995; Prendergast et al. 1995; Qiu 1995a,b, 1997; Roux et al. 1997). However, conflicting results were obtained with RhoA and Cdc42Hs, where dominant negative proteins inhibited Raf-CAAX-induced focus formation (Qiu et al. 1995b, 1997). Although not investigated, a reason for this discrepancy might lie in different behaviour between v-Raf and Raf-CAAX.

The distinct nature of Rho-dependent and ERK pathways was strengthened by the observation that individual activated Rho GTPases failed to activate ERK (Coso et al. 1995; Hill et al. 1995; Khosravi-Far et al. 1995; Roux et al. 1997). Furthermore, ERK activation could not be detected upon coexpression of V12-Rac1, V12-RhoG and V12-Cdc42Hs, whereas their associated expression elicited strong focus formation, up to 20% of that achieved with V12-Ras (Roux et al. 1997). These data are consistent with studies using two activated Ras variants mutated in their effector sites (V12-G37-Ras and V12-C40-Ras, (Khosravi-Far et al. 1996)). Both variants are defective in Raf-1 binding and subsequent ERK activation, but still trigger cellular transformation morphologically indistinguishable from that induced by constitutively activated Rho proteins. In addition, they both activate JNK, as do Rho family members, and their coexpression results in a synergistic cooperation of their transforming activities.

In conclusion, the recent literature indicates that Ras transformation is mediated by at least two distinct Raf/ERK–independent pathways, which might be controlled by Rho proteins (Fig. 2). It now remains to determine which pathways might be involved in cell transformation. As previously mentioned, Rho proteins have been implicated in the control of numerous processes, and in particular, in the activation of MAP kinases distinct from ERK and in the reorganization of the actin cytoskeleton. The relevance of these biological processes in cell transformation is discussed in the next section.

4 Signalling Pathways Downstream of Rho Proteins Involved in Cell Transformation

4.1 Rho-Dependent Activation of the JNK/SAPK Pathway

In addition to ERK, two novel classes of mammalian enzymes closely related to MAPK cascades have been identified. One class includes RK/p38 kinase (Han et al. 1994), which shares similarity with the yeast *S. cerevisiae* HOG1 kinase, involved in protection from hyperosmotic solutions (reviewed in Cano and Mahadevan 1995; Herskowitz 1995; Kyriakis and Avruch 1996). The second class is made up of a family of closely related kinases activated by cellular stress, either named stress-activated protein kinases (SAPKs; Kyriakis et al. 1994) or c-jun N-terminal kinases (JNKs; Derijard et al. 1994). The JNK/SAPK pathway is activated by a wide range of stimuli, such as protein synthesis inhibitors, inflammatory cytokines, changes in osmolarity, ultraviolet irradiation or heat shock. Once activated, JNK/SAPK proteins phosphorylate the transactivating domain of c-Jun, thereby modulating AP-1 activity and gene expression. JNK/SAPK proteins are also activated in growth-stimulated cells in a Ras-dependent but Raf-1-independent manner (Minden et al. 1994). Additionally, recent reports have shown that JNK/SAPK is associated with cell transformation (Raitano et al. 1995; Xu et al. 1996; Galley et al. 1997; Rodrigues et al. 1997). Characterization of the kinases acting upstream of JNK/SAPK led to the identification of MEKK1 and PAK, homologues of the yeast STE11 and STE20 proteins, respectively (Lange-Carter et al. 1993; Bagrodia et al. 1995; Manser et al. 1995; Martin et al. 1995; Teo et al. 1995). The PAK family (p21-activated kinases) consists of three conserved members which were first isolated by their ability to be autophosphorylated and activated upon binding to GTP-bound Rac1 and Cdc42Hs (reviewed in Lim et al. 1996). A direct implication of PAK1 in JNK/SAPK activation was observed using a constitutively activated PAK1 mutated protein in COS-7 cells (Brown et al. 1996). These results led to the hypothesis that the JNK/SAPK pathway in mammals followed a cascade, in which Ras activates Rac1/Cdc42Hs, which in turn activates PAK, which phosphorylates MEKK1, eventually leading to the activation of JNK/SAPK. Accordingly, such a scheme suggested that the transforming potential of Rac1 and Cdc42Hs might be mediated by the JNK/SAPK pathway. However, recent results do not support this.

First, whereas PAK1 interacts with components essential for Ras transformation, inhibition of PAK1 can be uncoupled from JNK but not ERK signalling. Two PAK1 mutants, R299-PAK which is catalytically inactive and L83L86R299-PAK, which is inactive and also unable to bind Rac1 or Cdc42Hs were constructed to assess the effect of PAK1 on transformation (Tang et al. 1997). Expression of R299-PAK inhibited Ras-mediated transformation in Rat-1 cells but not in NIH3T3 cells. It had no effect on Raf transforming activity but inhibited Ras- and Rac-dependent JNK activation. Expression of L83L86R299-PAK also inhibited Ras-dependent transformation but not JNK activation. Surprisingly, both mutants inhibited Ras-mediated ERK activation, suggesting that PAK might mediate signals from Ras to Raf.

Second, cooperation of Rac with Raf-CAAX in transformation, Rac/PAK interaction and JNK activation are all independent events (Joneson et al. 1996; Westwick et al. 1997). Rac proteins mutated in their effector regions were assessed for their abilities to modulate JNK activation and transformation (Khosravi-Far et al. 1996). One mutant (V12H40-Rac1) was found to be defective in binding to PAK-3 and in JNK activation, but could still mediate transformation and bind POR1 [a Rac-binding protein involved in membrane ruffling (Van Aelst et al. 1996)]. Conversely, a second mutant (V12L37-Rac1), which bound PAK-3 but not POR1, could activate JNK but was defective in transformation. A third Rac effector mutant (L61D43-Rac1) was unable to cooperate with Raf-CAAX or interact with PAK but could still activate the JNK pathway.

In conclusion, although the direct implication of PAK in JNK activation remains controversial, it appears that Rac- and Cdc42Hs-dependent activation of JNK does not require PAK. Instead, recent reports indicate that other kinases such as MLK (Mixed Lineage Kinase) and MEKK might be implicated (Teramoto et al. 1996; Tibbles et al. 1996; Fanger et al. 1997; Gerwins et al. 1997). Nevertheless, PAK activity is required for Ras- but not Raf-dependent transformation. Demonstration of cooperation in cell transformation between activated PAK or JNK with Ras would unambiguously settle this point. However, it appears now that JNK and PAK activations are distinct from the Rac-controlled pathway acting in synergy with Raf in cell transformation (Fig. 3). Whether JNK and PAK pathways are used by other Rho GTPases to cooperate with Raf remains to be investigated.

4.2 The Cytoskeleton and Rho Proteins

As mentioned earlier, oncogenic transformation is associated with changes in morphological characteristics of the cell, i.e. decreased anchorage dependent cell growth, invasiveness and reduction in cell contact inhibition, which appear controlled by Rho members. Various Rac1 and Cdc42Hs proteins mutated in their effector domains such as C40-Rac1 and C40-Cdc42Hs no longer interact with PAK and cannot activate the JNK pathway, but still induce cytoskeletal changes and G_1 cell cycle progression (Lamarche et al. 1996). Conversely, A37-Rac1 can bind PAK and activate the JNK pathway, but no longer interacts with POR1, and does not induce lamellipodia formation, G_1 cell cycle progression or cell transformation in association with Raf-CAAX (Joneson et al. 1996; Lamarche et al. 1996). In all instances, the formation of filopodia and lamellipodia induced by Cdc42Hs and Rac1 was associated with their ability to cooperate with Raf in cell transformation. However, whether both events are functionally linked remains to be determined.

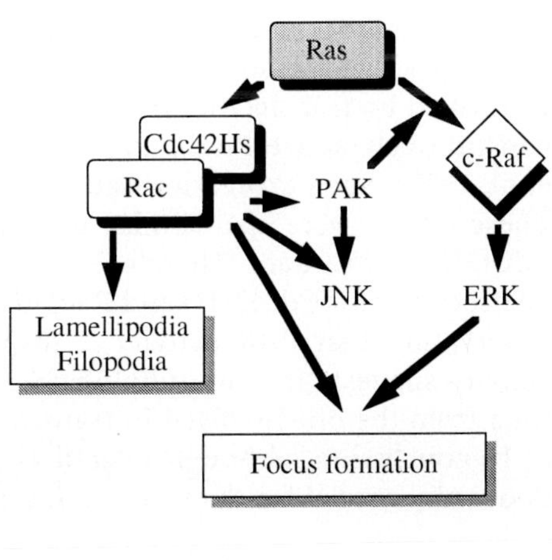

Fig. 3. Implication of pathways downstream of Rho proteins in focus formation. Selective activation of effectors was performed using specific Rac and Cdc42Hs variant proteins. Activation of PAK is required for Ras but not Raf transformation. JNK pathway is activated independently by PAK, Rac and Cdc42Hs, and is not involved in cooperation with Raf. Similarly, the formation of lamellipodia and filopodia elicited by Rac and Cdc42Hs respectively is independent from the cooperation process

Recent published work suggests that the relationships between morphological characteristics and transformation are more complex than initially suspected. Tiam1, dbl-like protein, was isolated as a result of its role in invasion and metastasis in T-lymphoma cells (Habets et al. 1994). Tiam1 expression was shown to induce Rac1-dependent extensive membrane ruffling in NIH3T3 cells, and to promote invasion in BW5147 T-lymphoma cells, like activated V12-Rac1 (van Leeuwen et al. 1995). However, cells expressing Tiam1 variant C-682, which lacks all N-terminal sequences upstream of the dbl-like domain, are still highly tumourigenic in nude mice, even though they do not show morphological transformation and membrane ruffling. Expression of Tiam1 variant N-385, which contains the first 385 amino acids but lacks the GEF domain, does not exhibit a tumourigenic effect in nude mice or invasion in NIH3T3 cells, but still induces invasion when expressed in T-lymphoma cells. Thus, although Tiam1 activates Rac1 both in vitro and in vivo, its ability to elicit cell invasion does not require Rac1-dependent ruffling, and in specific cells, is even independent of Rac1 activation.

A second example of uncoupling between cytoskeletal rearrangements and cell transformation was established from experiments with Cdc42Hs (Qiu et al. 1997). Low levels of N17-Cdc42Hs expression were shown to be sufficient to inhibit Ras and Raf-CAAX transformation, but did not reduce the formation of filopodia. Similar observations have been reported for Rac1, where L61V31 and V12L37 variants partially or completely inhibited lamellipodium formation, respectively, but only mildly suppressed Raf-CAAX-induced cell transformation (Westwick et al. 1997). Thus, the pathways leading to lamellipodia and filopodia formation appear distinct from the one leading to serum- and anchorage-independent cell growth involved in Ras- and Raf-mediated transformation.

A third, indirect, set of experiments dealing with PAK activity suggests that morphological changes might be dispensable for cooperation of Rho proteins in transformation. Indeed, as previously mentioned, PAK activation is thought to be required for Ras but not Raf-CAAX transformation (Tang et al. 1997). A GST-PAK fusion, constitutively active in vitro (Knaus et al. 1995), rapidly elicited filopodia and lamellipodia formation (Sells et al. 1997), similar to the effects observed upon introduction of V12-Cdc42Hs and V12-Rac1. This was confirmed by transfecting HeLa cells and fibroblasts with plasmids encoding constitutively active PAK, mutated at its autophosphorylation sites (Manser et al. 1997). Such expression caused loss of stress fibres and focal adhesions. These effects were also similar to those observed upon expression of V12-Cdc42Hs or V12-Rac1. Therefore both reports indicate that PAK activity might be necessary for Cdc42Hs- and Rac1-dependent actin reorganization. If so, the observation that PAK activity is dispensable for Raf-CAAX focus forming activity suggests that morphological modifications delineate a pathway distinct from the one involved in transformation (Fig. 3).

In conclusion, although overall changes in cellular morphology are undoubtedly crucial for the process of cell transformation, there is still a lack of evidence that Rho proteins use pathways leading to cytoskeletal reorganization to cooperate in cell transformation.

5 Rho Proteins and Apoptosis

Programmed cell death (also termed apoptosis) is a conserved active cellular mechanism involved in the control of many normal physiological processes, such as development and differentiation, or in pathological aspects such as tumourigenesis (Thompson 1995; Evan 1997). Apoptosis is characterized by cell membrane blebbing, phosphatidylserine externalization, cytoskeletal disruption, and nuclear chromatin condensation (reviewed in Martin et al. 1994). At later stages, the cytoplasm and nucleus become compartmentalised and form apoptotic bodies, which are submitted to phagocytosis by neighbouring cells. The resulting action of two types of signals is required for a cell to commit to apoptosis: (1) a number of external signals have been characterized, such as binding of specific ligands to receptors of the nerve growth factor family, ionising radiation, ceramides, cell-to-cell and cell-to-extra-cellular matrix contacts or serum starvation (Hale et al. 1996); (2) internal signals are also critical, including cell lineage, cell cycle stage, proto-oncogene expression, and metabolic state. Both types of signals also participate in other processes, such as cell proliferation and transformation.

Ras proteins were recently shown to inhibit apoptosis by selective activation of survival pathways. In fibroblasts, Ras expression suppresses c-myc-induced apoptosis through the activation of PI3K and the serine/threonine kinase PKB/Akt (Kauffmann-Zeh et al. 1997). Expression of Ras in epithelial cells, which normally undergo apoptosis upon detachment from the extracellular matrix, inhibits programmed cell death through activation of PI3K (Khwaja et al. 1997). Interestingly, opposing effects have been described in other cell systems. For example, Ras inhibition blocks Fas- or ceramide-induced apoptosis in Jurkat T-cells (Gulbins et al. 1995), and p120GAP gene inactivation in mice induces a dramatic apoptosis of neurons in the developing brain (Henkemeyer et al. 1995). Thus, opposed signals such as cell proliferation, differentiation and apoptosis can be conveyed through the activation of the same molecular switch, i.e. the Ras protein. This is illustrated by the activation of Ras in both IL-2 dependent proliferation and apoptosis of T cells (Gomez et al. 1997). Taking into account the regulatory cross-talk between Ras and Rho pathways, it was of importance to address the role of Rho GTPases in the process of programmed cell death.

It was first observed that overexpression of Rho proteins enhances apoptosis in serum deprived murine NIH3T3 fibroblasts (Jimenez et al. 1995). Such apoptotic activation is related to the production of ceramides (Esteve et al. 1995), which are sphingolipid breakdown products that function as inhibitors of cell growth (Fishbein et al. 1993; Jayadev et al. 1995) and inducers of apoptosis (Obeid et al. 1993; Obeid and Hannun 1995). Serum-starved but not serum-fed RhoA-overexpressing NIH3T3 cells show an increase in the intracellular level of ceramide, while no change could be detected in control cells or Ras expressing cells. In all situations, activation of ceramide production is associated with apoptosis. Addition of permeable ceramide has no apoptotic effect on control NIH3T3 cells, whereas it is sufficient to induce apoptosis in RhoA-expressing NIH3T3 cells, even in the presence of 10% serum. This led to the conclusion that Rho proteins control two complementary signals involved in apoptosis (Lacal 1997): one pathway leads to the production of ceramide, and is involved in the progression towards apoptosis and a second pathway, possibly the JNK/SAPK cascade, which makes the cell competent to interprete the ceramide-dependent apoptotic signal (Verheij et al. 1996). Recent reports strengthen this hypothesis. First, in T lymphoma cells, apoptosis requires the activation of Ras, Rac and RhoA proteins (Gulbins et al. 1996; Moorman et al. 1996). This points to the paradoxical role of the Ras/Rho pathway, which may control cell proliferation or programmed cell death, depending on internal or external co-signals. Then, activation of Jurkat cells via the Fas receptor or synthetic ceramides leads to a Ras- and Rac-dependent stimulation of JNK and RK/p38 (Brenner et al. 1997). The functional link between this pathway and apoptosis was established using transdominant inhibitory Ras, Rac and JNK, and a specific pharmacological blocker of RK/p38 kinase. Activation of JNK and RK/p38 was followed by phosphorylation of the transcriptional factor GADD153. Although the effect of such phosphorylation upon Fas receptor triggering is unknown, GADD153 (also termed CHOP or growth arrest and DNA damage-inducible gene 153) might inhibit C/EBP binding to DNA or redirect GADD153-C/EBP heterodimers to other DNA target sequences, thereby modifying gene expression (Wang et al. 1996; Bruhat et al. 1997). The implication of the Rho family in apoptosis has been recently extended to other members such as RhoA, whose activity is required for thrombin-induced apoptosis of cultured neurons and astrocytes (Donovan et al. 1997), and Rac2, whose constitutive activation enhances apoptosis in the thymus of transgenic mice (Lores et al. 1997).

In conclusion, Rho overexpression appears sufficient to render cells prone to initiating the apoptotic process. In contrast, Ras overexpression is not sufficient, although its activity is required for the apoptotic signal (Fig. 4). According to this scheme, one could speculate that the apparent low focus forming activity of Rho proteins might be explained by a high apoptotic index of Rho expressing cells. A cooperation with Ras or Raf-CAAX and Rho proteins in cell transformation might therefore be explained by a change in internal signals redirecting apoptotic signals towards proliferation.

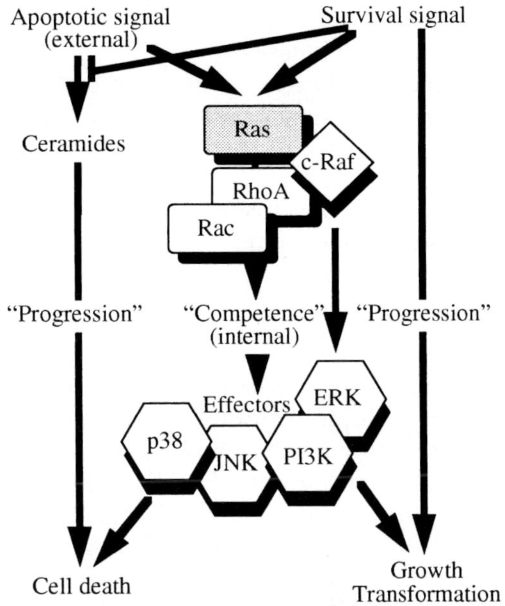

Fig. 4. Role of Ras and Rho proteins in the control of apoptosis. Cell fate is determined through a combination of external and internal signals. Apoptotic or survival signals use Ras/Rho pathways, which participate in the competence. Additional pathways such as ceramide production are required for the progression towards apoptosis. Conversely, survival signal inhibits ceramide production, and activates distinct pathways required for cell proliferation and transformation

6
Concluding Remarks

The Rho family of GTPases participate in the control of many basic physiological aspects of the cell, e.g. cell growth, differentiation, motility or apoptosis. The characterization of biological activities of the Rho proteins was mainly investigated by expressing mutated proteins. However, the use of such mutants suffers from several limitations:

1. Inasmuch as some regulatory proteins are able to control the activity of distinct GTPases, expression of N17-mutated proteins might inhibit more than one specific GTPase.
2. Taking into account the similarity in the effector domain of several GTPases, one could expect illegitimate activation of additional pathways upon overexpression of a given protein. Among the activated RK/p38, p46-JNK, p44-ERK and p70-S6K kinases, only expression of RK/p38 blocks NIH3T3 cells in G_1/S (Molnar 1997). In the same cell line, Cdc42Hs elicits RK/p38 activation and G_1/S arrest in NIH3T3 cells, whereas Rac expression does not cause cell cycle arrest, although activating RK/p38 in situ. This

suggests that the pattern of kinase activation in vivo might differ from that observed in kinase assays.

3. Specific mutations in the effector domain were reported to differentially decrease in vitro or in levuro (in yeast two-hybrid system), the affinity for particular target proteins. However, the effective loss of a specific binding within the cell (i.e. the absence of any activation) remains to be settled.
4. Expression of active and inhibitory mutated proteins sometimes produces identical phenotypic effects. Wild-type, N17- or V12-Cdc42Hs were reported to inhibit G_1/S progression to the same extent (Molnar et al. 1997), and expression of N19-RhoA was shown to promote stress fibre formation (Qiu et al. 1995b).

Despite these limitations, it emerges from this survey that Rho proteins control multiple pathways that cooperate in cell transformation. Two pathways are now firmly established: one is controlled by RhoG and Rac, and modulates cell contact inhibition, while the other is governed by Cdc42Hs and RhoA, and affects anchorage-independent cell growth. The precise molecular mechanisms underlying these biological features is still debated. The implication of other pathways is also suspected, although they remained to be characterized. Additionally, only a limited number of Rho family members have been studies in detail so far, which leaves open a wide range of unknown pathways that might also influence physiological parameters critical for cell transformation. Finally, recent data support a crucial function of Rho proteins in the signalling of apoptosis whose perturbation constitutes another way for a cell to escape growth control.

In all instances, activities of Rho proteins appear tightly linked each to one another, according to a complex network of regulatory cross-talk. Deciphering of such a network, and of relationships bridging Ras and Rho proteins will prove valuable for a better understanding of tumourigenesis.

Acknowledgements. I wish to thank C. Gauthier, P. Roux, and N. Taylor for helpful discussions and comments on the manuscript. This work was supported by institutional grants from the INSERM, the CNRS and the Université de Montpellier II, and grants from the Association pour la Recherche contre le Cancer, the Ligue Nationale Contre le Cancer and the CNRS (programme "Biologie Cellulaire").

References

Abo A, Pick E, Hall A, Totty N, Teahan CG, Segal AW (1991) Activation of the NADPH oxidase involves the small GTP-binding protein p21racl. Nature 353:668–670

Abo A, Webb MR, Grogan A, Segal AW (1994) Activation of NADPH oxidase involves the dissociation of p21rac from its inhibitory GDP/GTP exchange protein (rhoGDI) followed by its translocation to the plasma membrane. Biochem J 3:585–591

Adams AE, Johnson DI, Longnecker RM, Sloat BF, Pringle JR (1990) CDC42 and CDC43, two additional genes involved in budding and the establishment of cell polarity in the yeast *Saccharomyces cerevisiae.* J Cell Biol 111:131–142

Adra CN, Ko J, Leonard D, Wirth LJ, Cerione RA, Lim B (1993) Identification of a novel protein with GDP dissociation inhibitor activity for the ras-like proteins CDC42Hs and rac I. Genes Chromosom Cancer 8:253–261

Adra CN, Manor D, Ko JL, Zhu S, Horiuchi T, Van Aelst L, Cerione RA, Lim B (1997) RhoGDIgamma: a GDP-dissociation inhibitor for Rho proteins with preferential expression in brain and pancreas. Proc Natl Acad Sci USA 94:4279–4284

Aepfelbacher M, Vauti F, Weber PC, Glomset JA (1994) Spreading of differentiating human monocytes is associated with a major increase in membrane-bound CDC42. Proc Natl Acad Sci USA 91:4263–4267

Allen WE, Jones GE, Pollard JW, Ridley AJ (1997) Rho, Rac and Cdc42 regulate actin organization and cell adhesion in macrophages. J Cell Sci 110:707–720

Avraham H, Weinberg RA (1989) Characterization and expression of the human rhoH12 gene product. Mol Cell Biol 9:2058–2066

Bagrodia S, Derijard B, Davis RJ, Cerione RA (1995) Cdc42 and PAK-mediated signaling leads to Jun kinase and p38 mitogen-activated protein kinase activation. J Biol Chem 270:27995–27998

Bar-Sagi D, Feramisco JR (1986) Induction of membrane ruffling and fluid-phase pinocytosis in quiescent fibroblasts by ras proteins. Science 233:1061–1068

Barfod ET, Zheng Y, Kuang WJ, Hart MJ, Evans T, Cerione RA, Ashkenazi A (1993) Cloning and expression of a human CDC42 GTPase-activating protein reveals a functional SH3-binding domain. J Biol Chem 268:26059–26062

Brenner B, Koppenhoefer U, Weinstock C, Linderkamp O, Lang F, Gulbins E (1997) Fas- or ceramide-induced apoptosis is medicated by a Rac1-regulated activation of Jun N-terminal kinase/p38 kinases and GADD153. J Biol Chem 272:22173–22181

Brill S, Li S, Lyman CW, Church DM, Wasmuth JJ, Weissbach L, Bernards A, Snijders AJ (1996) The Ras GTPase-activating-protein-realted human protein IQGAP2 harbors a potential actin binding domain and interacts with calmodulin and Rho family GTPases. Mol Cell Biol 16:4869–4878

Brown JL, Stowers L, Baer M, Trejo J, Coughlin S, Chant J (1996) Human Ste20 homologue hPAK1 links GTPases to the JNK MAP kinase pathway. Curr Biol 6:598–605

Bruhat A, Jousse C, Wang XZ, Ron D, Ferrara M, Fafournoux P (1997) Amino acid limitation induces expression of CHOP, a CCAAT/enhancer binding protein-related gene, at both transcriptional and post-transcriptional levels. J Biol Chem 272:17588–17593

Burbelo PD, Miyamoto S, Utani A, Brill S, Yamada KM, Hall A, Yamada Y (1995) p190-B, a new member of the Rho GAP family, and Rho are induced to cluster after integrin cross-linking. J Biol Chem 270:30919-30926

Cano E, Mahadevan LC (1995) Parallel signal processing among mammalian MAPKs. Trends Biochem Sci 20:117–122

Cerione RA, Zheng Y (1996) The Dbl family of oncogenes. Curr Opin Cell Biol 8:216–222

Chan AM, McGovern ES, Catalano G, Fleming TP, Miki T (1994) Expression cDNA cloning of novel oncogene with sequence similiarity to regulators of small GTP-binding proteins. Oncogene 9:1057–1063

Chang EC, Barr M, Wang Y, Jung V, Xu HP, Wigler MH (1994) Cooperative interaction of *S. pombe* proteins required for mating and morphogenesis. Cell 79:131–141

Chardin P, Madaule P, Tavitian A (1988) Coding sequences of human rho cDNAs clone 6 and clone 9. Nucleic Acids Res 16:2717

Cicchetti P, Ridley AJ, Zheng Y, Cerione RA, Baltimore D (1995) 3BP-1, an SH3 domain binding protein, has GAP activity for Rac and inhibits growth factor-induced membrane ruffling in fibroblasts. EMBO J 14:3127–3135

Cooper JA, Kashishian A (1993) In vivo binding properties of SH2 domains from GTPase-activating protein and phosphatidylinositol 3-kinase. Mol Cell Biol 13:1737–1745

Coso OA, Chiariello M, Yu JC, Teramoto H, Crespo P, Xu NG, Miki T, Gutkind JS (1995) The small GTP-Binding proteins rac1 and cdc42 regulate the activity of the JNK/SAPK signaling pathway. Cell 81:1137–1146

Courjal F, Chuchana P, Theillet C, Fort P (1997) Structure and chromosomal assignment to 22q12 and 17qter of the ras-related Rac2 and Rac3 human genes. Genomics 44:242–246

Dallery E, Galiegue-Zouitina S, Collyn-d'Hooghe M, Quief S, Denis C, Hildebrand MP, Lantoine D, Deweindt C, Tilly H, Bastard C (1995) TTF, a gene encoding a novel small G protein, fuses to the lymphoma- associated LAZ3 gene by t(3;4) chromosomal translocation. Oncogene 10:2171–2178

Derijard B, Hibi M, Wu IH, Barrett T, Su B, Deng T, Karin M, Davis RJ (1994) JNK1: a protein kinase stimulated by UV light and Ha-Ras that binds and phosphorylates the c-Jun activation domain. Cell 76:1025–1037

Didsbury J, Weber RF, Bokoch GM, Evans T, Synderman R (1989) rac, a novel ras-related family of proteins that are botulinum toxin substrates. J Biol Chem 264:16378–16382

Donovan FM, Pike CJ, Cotman CW, Cunningham DD (1997) Thrombin induces apoptosis in cultured neurons and astrocytes via a pathway requiring tyrosine kinase and RhoA activities. J Neurosci 17:5316–5326

Drivas GT, Shih A, Coutavas E, Rush MG, D'Eustachio P (1990) Characterization of four novel ras-like genes expressed in a human teratocarcinoma cell line. Mol Cell Biol 10:1793–1798

Dutartre H, Davoust J, Gorvel J-P, Chavrier P (1996) Cytokinesis arrest and redistribution of actin-cytoskeleton regulatory components in cells expressing the Rho GTPase Cdc42Hs. J Cell Sci 109:367–377

Esteve P, del Peso L, Lacal JC (1995) Induction of apoptosis by rho in NIH 3T3 cells requires two complementary signals. Ceramides function as a progression factor for apoptosis. Oncogene 11:2657–2665

Evan G (1997) Cancer: a matter of life and and cell death. Int J Cancer 71:709–711

Fanger GR, Lassignal-Johnson N, Johnson GL (1997) MEK kinases are regulated by EGF and selectively interact with Rac/Cdc42. EMBO J 16:4961–4972

Fishbein JD, Dobrowsky RT, Bielawska A, Garrett S, Hannun YA (1993) Ceramide-mediated growth inhibition and CAPP are conserved in *Saccharomyces cerevisiae.* J Biol Chem 268:9255–9261

Flescher EG, Madden K, Snyder M (1993) Components required for cytokinesis are important for bud site selection in yeast. J Cell Biol 122:373–386

Foster R, Hu KQ, Lu Y, Nolan KM, Thissen J, Settleman J (1996) Identification of a novel human Rho protein with unusual properties: GTPase deficiency and in vivo farnesylation. Mol Cell Biol 16:2689–2699

Fukumoto Y, Kaibuchi K, Hori Y, Fujioka H, Araki S, Ueda T, Kikuchi A, Takai Y (1990) Molecular cloning and characterization of a novel type of regulatory protein (GDI) for the rho proteins, ras p21-like small GTP-binding proteins. Oncogene 5:1321–1328

Galley Y, Hagens G, Glaser I, Davis W, Eichhorn M, Dobbelaere D (1997) Jun NH2-terminal kinase is constitutively activated in T cells transformed by the intracellular parasite *Theileria parva.* Proc Natl Acad Sci USA 94:5119–5124

Garrett MD, Major GN, Totty N, Hall A (1991) Purification and N-terminal sequence of the p21rho GTPase-activating protein, rho GAP. Biochem J 276:833–836

Gerwins P, Blank JL, Johnson GL (1997) Cloning of a novel mitogen-activated protein kinase kinase kinase, MEKK4, that selectively regulates the c-Jun amino terminal kinase pathway. J Biol Chem 272:8288–8295

Gomez J, Martinez C, Fernandez B, Garcia A, Reboilo A (1997) Ras activation leads to cell proliferation or apoptotic cell death upon interleukin-2 stimulation or lymphokine deprivation, respectively [In Process Citation]. Eur J Immunol 27:1610–1618

Gulbins E, Bissonnette R, Mahboubi A, Martin S, Nishioka W, Brunner T, Baier G, Baier-Bitterlich G, Byrd C, Lang F et al. (1995) FAS-induced apoptosis is mediated via a ceramide-initiated RAS signaling pathway. Immunity 2:341–351

Gulbins E, Coggeshall KM, Brenner B, Schlottmann K, Linderkamp O, Lang F (1996) Fas-induced apoptosis is mediated by activation of a Ras and Rac protein-regulated signaling pathway. J Biol Chem 271:26389–26394

Haataja L, Groffen J, Heisterkamp N (1997) Characterization of RAC3, a novel member of the rho family. J Biol Chem 272:20384–20388

Habets GG, Scholtes EH, Zuydgeest D, van der Kammen RA, Stam JC, Berns A, Collard JG (1994) Identification of an invasion-inducing gene, Tiam-1, that encodes a protein with homology to GDP-GTP exchangers for Rho-like proteins. Cell 77:537–549

Hale AJ, Smith CA, Sutherland LC, Stoneman VE, Longthorne VL, Culhane AC, Williams GT (1996) Apoptosis: molecular regulation of cell death. Eur J Biochem 236:1–26

Hall C, Monfries C, Smith P, Lim HH, Kozma R, Ahmed S, Vanniasingham V, Leung T, Lim L (1990) Novel human brain cDNA encoding a 34 000 Mr protein n-chimaerin, related to both the regulatory domain of protein kinase C and BCR, the product of the breakpoint cluster region gene. J Mol Biol 211:11–16

Han J, Lee JD, Bibbs L, Ulevitch RJ (1994) A MAP kinase targeted by endotoxin and hyperosmolarity in mammalian cells. Science 265:808–811

Hart MJ, Eva A, Evans T, Aaronson SA, Cerione RA (1991a) Catalysis of guanine nucleotide exchange on the CDC42Hs protein by the dbl oncogene product. Nature 354:311–314

Hart MJ, Shinjo K, Hall A, Evans T, Cerione RA (1991b) Identification of the human platelet GTPase activating protein for the CDC42Hs protein. J Biol Chem 266:20840–20848

Hart MJ, Eva A, Zangrilli D, Aaronson SA, Evans T, Cerione RA, Zheng Y (1994) Cellular transformation and guanine nucleotide exchange activity are catalyzed by a common domain on the dbl oncogene product. J Biol Chem 269:62–65

Heisterkamp N, Stephenson JR, Groffen J, Hansen PF, de Klein A, Bartram CR, Grosveld G (1983) Localization of the c-abl oncogene adjacent to a translocation break point in chronic myelocytic leukaemia. Nature 306:239–242

Heisterkamp N, Kaartinen V, van Soest S, Bokoch GM, Groffen J (1993) Human ABR encodes a protein with GAPrac activity and homology to the DBL nucleotide exchange factor domain. J Biol Chem 268:16903–16906

Henkemeyer M, Rossi DJ, Holmyard DP, Puri MC, Mbamalu G, Harpal K, Shih TS, Jacks T, Pawson T (1995) Vascular system defects and neuronal apoptosis in mice lacking ras GTPase-activating protein. Nature 377:695–701

Herskowitz I (1995) MAP kinase pathway in yeast: fore mating and more. Cell 80:187–197

Hildebrand JD, Taylor JM, Parsons JT (1996) An SH3 domain-containing GTPase-activating protein for Rho and Cdc42 associates with focal adhesion kinase. Mol Cell Biol 16:3169–3178

Hill CS, Wynne J, Treisman R (1995) The rho family GTPases RhoA, rac1, and CDC42Hs regulate transcriptional activation by SRF. Cell 81:1159–1170

Hirata K, Kikuchi A, Sasaki T, Kuroda S, Kaibuchi K, Matsuura Y, Seki H, Saida K, Takai Y (1992) Involvement of rho p21 in the GTP-enhanced calcium ion sensitivity of smooth muscle contraction. J Biol Chem 267:8719–8722

Horii Y, Beeler JF, Sakaguchi K, Tachibana M, Miki T (1994) A novel oncogene, ost, encodes a guanine nucleotide exchange factor that potentially links Rho and Rac signaling pathways. EMBO J 13:4776–4786

Jayadev S, Liu B, Bielawska AE, Lee JY, Nazaire F, Pushkareva M, Obeid LM, Hannun YA (1995) Role for ceramide in cell cycle arrest. J Biol Chem 270:2047–2052

Jimenez B, Arends M, Esteve P, Perona R, Sanchez R, Ramon YSC, Wyllie A, Lacal JC (1995) Induction of apoptosis in NIH3T3 cells after serum deprivation by overexpression of rho-p21, a GTPase protein of the ras superfamily. Oncogene 10:811–816

Johnson DI, Pringle JR (1990) Molecular characterization of CDC42, a *Saccharomyces cerevisiae* gene involved in the development of cell polarity. J Cell Biol 111:143–152

Joneson T, McDonough M, Bar-Sagi D, Van Aelst L (1996) RAC regulation of actin polymerization and proliferation by a pathway distinct from Jun kinase. Science 274:1374–1376

Jullien-Flores V, Dorseuil O, Romero F, Letourneur F, Saragosti S, Berger R, Tavitian A, Gacon G, Camonis JH (1995) Bridging Ral GTPase to Rho pathways. RLIP76, a Ral effector with CDC42/Rac GTPase-activating protein activity. J Biol Chem 270:22473–22477

Katzav S, Martin-Zanca D, Barbacid M (1989) vav, a novel human oncogene derived from a locus ubiquitously expressed in hematopoietic cells. EMBO J 8:2283–2290

Kauffmann-Zeh A, Rodriguez-Viciana P, Ulrich E, Gilbert C, Coffer P, Downward J, Evan G (1997) Suppression of c-Myc-induced apoptosis by Ras signalling through PI (3)K and PKB. Nature 385:544–548

Khosravi-Far R, Solski PA, Clark GJ, Kinch MS, Der CJ (1995) Activation of Racl, RhoA, and mitogen-activated protein kinases is required for Ras transformation. Mol Cell Biol 15:6443–6453

Khosravi-Far R, White MA, Westwick JK, Solski PA, Chrzanowska-Wodnicka M, Van Aelst L, Wigler MH, Der CJ (1996) Oncogenic Ras activation of Raf/mitogen-activated protein kinase-independent pathways is sufficient to cause tumorigenic transformation. Mol Cell Biol 16:3923–3933

Khwaja A, Rodriguez-Viciana P, Wennstrom S, Warne PH, Downward J (1997) Matrix adhesion and Ras transformation both activate a phosphoinositide 3-OH kinase and protein kinase B/Akt cellular survival pathway. EMBO J 16:2783–2793

Knaus UG, Heyworth PG, Evans T, Curnutte JT, Bokoch GM (1991) Regulation of phagocyte oxygen radical production by the GTP-binding protein Rac 2. Science 254:1512–1515

Knaus UG, Heyworth PG, Kinsella BT, Curnutte JT, Bokoch GM (1992) Purification and characterization of Rac 2. A cytosolic GTP-binding protein that regulates human neutrophil NADPH oxidase. J Biol Chem 267:23575–23582

Knaus UG, Morris S, Dong HJ, Chernoff J, Bokoch GM (1995) Regulation of human leukocyte p21-activated kinases through G protein-coupled receptors. Science 269:221–223

Kyriakis JM, Avruch J (1996) Protein kinase cascades activated by stress and inflammatory cytokines. Bioessays 18:567–577.

Kyriakis JM, Banerjee P, Nikolakaki E, Dai T, Rubie EA, Ahmad MF, Avruch J, Woodgett JR (1994) The stress-activated protein kinase subfamily of c-Jun kinases. Nature 369:156–160

Lacal JC (1997) Regulation of proliferation and apoptosis by Ras and Rho GTPases through specific phospholipid-dependent signaling. FEBS Lett 410:73–77

Lamarche N, Tapon N, Stowers L, Burbelo PD, Aspenstrom P, Bridges T, Chant J, Hall A (1996) Rac and Cdc42 induce actin polymerization and G1 cell cycle progression independently of p65PAK and the JNK/SAPK MAP kinase cascade. Cell 87:519–529

Lange-Carter CA, Pleiman CM, Gardner AM, Blumer KJ, Johnson GL (1993) A divergence in the MAP kinase regulatory network defined by MEK kinase and Raf. Science 260:315–319

Leevers SJ, Paterson HF, Marshall CJ (1994) Requirement for Ras in Raf activation is overcome by targeting Raf to the plasma membrane. Nature 369:411–414

Lim L, Manser E, Leung T, Hall C (1996) Regulation of phosphorylation pathways by p21 GTPases. The p21 Ras-related Rho subfamily and its role in phosphorylation signalling pathways. Eur J Biochem 242:171–185

Lores P, Morin L, Luna R, Cacon G (1997) Enhanced apoptosis in the thymus of transgenic mice expressing constitutively activated forms of human Rac2GTPase. Oncogene 15:601–605

Luo L, Liao YJ, Jan LY, Jan YN (1994) Distinct morphogenetic functions of similar small GTPases: *Drosophila* Dracl is involved in axonal outgrowth and myoblast fusion. Genes Dev 8:1787–1802

Machesky L, Hall A (1996) Rho: a connection between membrane receptor signalling and the cytoskeleton. Trends Cell Biol 6:304–310

Madaule P, Axel R (1985) A novel ras-related gene family. Cell 41:31–40

Manser E, Leung T, Monfries C, Teo M, Hall C, Lim L (1992) Diversity and versatility of GTPase activating proteins for the p21rho subfamily of ras G proteins detected by a novel overlay assay. J Biol Chem 267:16025–16028

Manser E, Chong C, Zhao ZS, Leung T, Michael G, Hall C, Lim L (1995) Molecular cloning of a new member of the p21-Cdc42/Rac-activated kinase (PAK) family. J Biol Chem 270:25070–25078

Manser E, Huang HY, Loo TH, Chen XQ, Dong JM, Leung T, Lim L (1997) Expression of constitutively active alpha-PAK reveals effects of the kinase on action and focal complexes. Mol Cell Biol 17:1129–1143

Marais R, Marshall CJ (1996) Control of the ERK MAP kinase cascade by Ras and Raf. Cancer Surv 27:101–125

Marshall CJ (1996) Ras effectors. Curr Opin Cell Biol 8:197–204
Martin GA, Bollag G, McCormick F, Abo A (1995) A novel serine kinase activated by rac1/CDC42Hs-dependent autophosphorylation is related to PAK65 and STE20. EMBO J 14:1970–1978
Martin SJ, Green DR, Cotter TG (1994) Dicing with death: dissecting the components of the apoptosis machinery. Trends Biochem Sci 19:26–30
Michiels F, Habets GG, Stam JC, van der Kammen RA, Collard JG (1995) A role for Rac in Tiam1-induced membrane ruffling and invasion. Nature 375:338–340
Miki T, Smith CL, Long JE, Eva A, Fleming TP (1993) Oncogene ect2 is related to regulators of small GTP-binding proteins. Nature 362:462–465
Minden A, Lin A, McMahon M, Lange CC, Derijard B, Davis RJ, Johnson GL, Karin M (1994) Differential activation of ERK and JNK mitogen-activated protein kinases by Raf-1 and MEKK. Science 266:1719–1723
Molnar A, Theodoras AM, Zon LI, Kyriakis JM (1997) Cdc42Hs, but not Rac1, inhibits serum-stimulated cell cycle progression at G1/S through a mechanism requiring p38/RK. J Biol Chem 272:13229–13235
Moorman JP, Bobak DA, Hahn CS (1996) Inactivation of the small GTP binding protein Rho induces multinucleate cell formation and apoptosis in murine T lymphoma EL4. J Immunol 156:4146–4153
Murphy AM, Montell DJ (1996) Cell type-specific roles for Cdc42, Rac, and RhoL in *Drosophila* oogenesis. J Cell Biol 133:617–630
Murphy C, Saffrich R, Grummt M, Gournier H, Rybin V, Rubino M, Auvinen P, Lutcke A, Parton RG, Zerial M (1996) Endosome dynamics regulated by a Rho protein. Nature 384:427–432
Nobes CD, Hall A (1995) Rho, rac, and cdc42 GTPases regulate the assembly of multimolecular focal complexes associated with actin stress fibers, lamellipodia, and filopodia. Cell 81:53–62
Obeid LM, Hannun YA (1995) Ceramide: a stress signal and mediator of growth suppression and apoptosis. J Cell Biochem 58:191–198
Obeid LM, Linardic CM, Karolak LA, Hannun YA (1993) Programmed cell death induced by ceramide. Science 259:1769–1771
Olson MF, Ashworth A, Hall A (1995) An essential role for rho, rac, and cdc42 GTPases in cell cycle progression through G1. Science 269:1270–1272
Park SH, Weinberg RA (1995) A putative effector of Ral has homology to Rho/Rac GTPase activating proteins. Oncogene 11:2349–2355
Pasteris NG, Cadle A, Logie LJ, Porteous ME, Schwartz CE, Stevenson RE, Glover TW, Wilroy RS, Gorski JL (1994) Isolation and characterization of the faciogenital dysplasia (Aarskog-Scott syndrome) gene: a putative Rho/Rac guanine nucleotide exchange factor. Cell 79:669–678
Perona R, Esteve P, Jimenez B, Ballestero RP, Ramon Y, Cajal S, Lacal JC (1993) Tumorigenic activity of rho genes from *Aplysia californica*. Oncogene 8:1285–1292
Prendergast GC, Khosravi-Far R, Solski PA, Kurzawa H, Lebowitz PF, Der CJ (1995) Critical role of Rho in cell transformation by oncogenic Ras. Oncogene 10:2289–2296
Pronk GJ, de Vries-Smits AM, Ellis C, Bos JL (1993) Complex formation between the p21ras GTPase-activating protein and phosphoproteins p62 and p190 is independent of p21ras signalling. Oncogene 8:2773–2780
Qiu R-G, Chen J, Kirn D, McCormick F, Symons M (1995a) An essential role for Rac in Ras transformation. Nature 374:457–459
Qiu RG, Chen J, McCormick F, Symons M (1995b) A role for Rho in Ras transformation. Proc Natl Acad Sci USA 92:11781–11785
Qiu RG, Abo A, McCormick F, Symons M (1997) Cdc42 regulates anchorage-independent growth and is necessary for Ras transformation. Mol Cell Biol 17:3449–3458
Raitano AB, Halpern JR, Hambuch TM, Sawyers CL (1995) The Bcr-Abl leukemia oncogene activates Jun kinase and requires Jun for transformation. Proc Natl Acad Sci USA 92:11746–11750
Reinhard J, Scheel AA, Diekmann D, Hall A, Ruppert C, Bahler M (1995) A novel type of myosin implicated in signalling by rho family GTPases. EMBO J 14:697–704

Ridley AJ (1994) Membrane ruffling and signal transduction. Bioessays 16:321–327
Ridley AJ, Hall A (1992) The small GTP-binding protein rho regulates the assembly of focal adhesions and actin stress fibers in response to growth factors. Cell 70:389–399
Ridley AJ, Paterson HF, Johnston CL, Diekmann D, Hall A (1992) The small GTP-binding protein rac regulates growth factor-induced membrane ruffling. Cell 70:401–410
Ridley AJ, Self AJ, Kasmi F, Paterson HF, Hall A, Marshall CJ, Ellis C (1993) rho family GTPase activating proteins p190, bcr and rhoGAP show distinct specificities in vitro and in vivo. EMBO J 12:5151–5160
Rodrigues GA, Park M, Schlessinger J (1997) Activation of the JNK pathway is essential for transformation by the Met oncogene. EMBO J 16:2634–2645
Ron D, Zannini M, Lewis M, Wickner RB, Hunt LT, Graziani G, Tronick SR, Aaronson SA, Eva A (1991) A region of proto-db1 essential for its transforming activity shows sequence similarity to a yeast cell cycle gene, CDC24, and the human breakpoint cluster gene, bcr. New Biol 3:372–379
Roux P, Gauthier-Rouvière C, Doucet-Brutin S, Fort P (1997) The small GTPases Cdc42Hs, Rac1 and RhoG delineate Raf-independent pathways that cooperate to transform NIH3T3 cells. Curr Biol 7:629–637
Scherle P, Behrens T, Staudt LM (1993) Ly-GDI, a GDP-dissociation inhibitor of the RhoA GTP-binding protein, is expressed preferentially in lymphocytes. Proc Natl Acad Sci USA 90:7568–7572
Sells MA, Knaus UG, Bagrodia S, Ambrose DM, Bokoch GM, Chernoff J (1997) Human p21-activated kinase (PAK1) regulates actin organization in mammalian cells. Curr Biol 7:202–210
Settleman J, Narasimhan V, Foster LC, Weinberg RA (1992) Molecular cloning of cDNAs encoding the GAP-associated protein p190: implications for a signaling pathway from ras to the nucleus. Cell 69:539–549
Shinjo K, Koland JG, Hart MJ, Narasimhan V, Johnson DI, Evans T, Cerione RA (1990) Molecular cloning of the gene for the human placental GTP-binding protein Gp (G25K): identification of this GTP-binding protein as the human homolog of the yeast cell-division-cycle protein CDC42. Proc Natl Acad Sci USA 87:9853–9857.
Stowers L, Yelon D, Berg LJ, Chant J (1995) Regulation of the polarization of T cells toward antigen-presenting cells by Ras-related GTPase CDC42. Proc Natl Acad Sci USA 92:5027–5031
Tang Y, Chen Z, Ambrose D, Liu J, Gibbs JB, Chernoff J, Field J (1997) Kinase-deficient Pak1 mutants inhibit Ras transformation of Rat-1 fibroblasts. Mol Cell Biol 17:4454–4464
Tapon N, Hall A (1997) Rho, Rac and Cdc42 GTPases regulate the organization of the actin cytoskeleton. Curr Opin Cell Biol 9:86–92
Teo M, Manser E, Lim L (1995) Identification and molecular cloning of a p21cdc42/rac1 -activated serine/threonine kinase that is rapidly activated by thrombin in platelets. J Biol Chem 270:26690–26697
Teramoto H, Coso OA, Miyata H, Igishi T, Miki T, Gutkind JS (1996) Signaling from the small GTP-binding proteins Rac1 and Cdc42 to the c-Jun N-terminal kinase/stress-activated protein kinase pathway. A role for mixed lineage kinase 3/protein-tyrosine kinase 1, a novel member of the mixed lineage kinase family. J Biol Chem 271:27225–27228
Thompson CB (1995) Apoptosis in the pathogenesis and treatment of disease. Science 267:1456–1462
Tibbles LA, Ing YL, Kiefer F, Chan J, Iscove N, Woodgett JR, Lassam NJ (1996) MLK-3 activates the SAPK/JNK and p38/RK pathways via SEK1 and MKK3/6. EMBO J 15:7026–7035
Toksoz D, Williams DA (1994) Novel human oncogene lbc detected by transfection with distinct homology regions to signal transduction products. Oncogene 9:621–628
Tribioli C, Droetto S, Bione S, Cesareni G, Torrisi MR, Lotti LV, Lanfrancone L, Toniolo D, Pelicci P (1996) An X chromosome-linked gene encoding a protein with characteristics of a rhoGAP predominantly expressed in hematopoietic cells. Proc Natl Acad Sci USA 93:695–699
Van Aelst L, Joneson T, Bar-Sagi D (1996) Identification of a novel Rac1 interacting protein involved in membrane ruffling. EMBO J 15:3778–3786

van Leeuwen FN, van der Kammen RA, Habets GG, Collard JG (1995) Oncogenic activity of Tiam1 and Rac1 in NIH3T3 cells. Oncogene 11:2215–2221

Verheij M, Bose R, Lin XH, Yao B, Jarvis WD, Grant S, Birrer MJ, Szabo E, Zon LI, Kyriakis JM et al. (1996) Requirement for ceramide-initiated SAPK/JNK signalling in stress-induced apoptosis. Nature 380:75–79

Vincent S, Jeanteur P, Fort P (1992) Growth-regulated expression of rhoG, a new member of the ras homolog gene family. Mol Cell Biol 12:3138–3148

Wang DZ, Nur-E-Kamal MS, Tikoo A, Montague W, Maruta H (1997) The GTPase and Rho GAP domains of p190, a tumor suppressor protein that binds the Mr 120 000 Ras GAP, independently function as anti-Ras tumor suppressors. Cancer Res 57:2478–2484

Wang XZ, Lawson B, Brewer JW, Zinszner H, Sanjay A, Mi LJ, Boorstein R, Kreibich G, Hendershot LM, Ron D (1996) Signals from the stressed endoplasmic reticulum induce C/EBP-homologous protein (CHOP/GADD153). Mol Cell Biol 16:4273–4280

Westwick JK, Lambert QT, Clark GJ, Symons M, Van Aelst L, Pestell RG, Der CJ (1997) Rac regulation of transformation, gene expression, and actin organization by multiple, PAK-independent pathways. Mol Cell Biol 17:1324–1335

Whitehead I, Kirk H, Kay R (1995a) Retroviral transduction and oncogenic selection of a cDNA encoding Dbs, a homolog of the Dbl guanine nucleotide exchange factor. Oncogene 10:713–721

Whitehead I, Kirk H, Tognon C, Trigo-Gonzalez G, Kay R (1995b) Expression cloning of lfc, a novel oncogene with structural similarities to guanine nucleotide exchange factors and to the regulatory region of protein kinase C. J Biol Chem 270:18388–18395

Whitehead IP, Khosravi-Far R, Kirk H, Trigo-Gonzalez G, Der CJ, Kay R (1996) Expression cloning of lsc, a novel oncogene with structural similarities to the Dbl family of guanine nucleotide exchange factors. J Biol Chem 271:18643–18650

Wirth JA, Jensen KA, Post PL, Bement WM, Mooseker MS (1996) Human myosin-IXb, an unconventional myosin with a chimerin-like rho/rac GTPase-activating protein domain in its tail. J Cell Sci 109:653–661

Xu X, Heidenreich O, Kitajima I, McGuire K, Li Q, Su B, Nerenberg M (1996) Constitutively activated JNK is associated with HTLV-1 mediated tumorigenesis. Oncogene 13:135–142

Zalcman G, Closson V, Camonis J, Honore N, Rousseau-Merck MF, Tavitian A, Olofsson B (1996) RhoGDI-3 is a new GDP dissociation inhibitor (GDI). Identification of a non-cytosolic GDI protein interacting with the small GTP-binding proteins RhoB and RhoG. J Biol Chem 271:30366–30374

Zheng Y, Cerione R, Bender A (1994) Control of the yeast bud-site assembly GTPase Cdc42. Catalysis of guanine nucleotide exchange by Cdc24 and stimulation of GTPase activity by Bem3. J Biol Chem 269:2369–2372

Rho GTP-Binding Proteins as Targets for Microbial Pathogens

Patrice Boquet[1], Philippe J. Sansonetti[2], and Guy Tran Van Nhieu[2]

1 Introduction

Conceptual approaches in microbial pathogenesis studies have evolved quickly in the last 10 years, due to the development of cell biology. The merging of cell biology and microbial pathogenesis has been called cellular microbiology (Cossart et al. 1996), definitively sealing this seminal association.

To establish an infection, pathogens must circumvent, or destroy host defense mechanisms. Some pathogens may also hide from the immune system. To perform these tasks, pathogens have developed astonishingly clever strategies to divert host-cell processes for their own purposes. Among virulence determinants elaborated by microbes, production of toxins acting at a distance like hormones, or microbial invasion of non-phagocytic cells are particularly redundant features. In both examples, the cell actin cytoskeleton emerges as a preferential target for toxins and invasive virulence factors.

In recent years, an increasing number of studies on the Rho subfamily of p21 Ras GTP binding proteins has established these GTP-binding proteins as pivotal in the regulation of the actin cytoskeleton (Hall 1998). Interestingly, Rho proteins are also intracellular targets of choice for toxins and virulence factors. In this review we describe and analyze recent advances concerning the mode of action of bacterial toxins and invasion virulence factors on the Rho proteins.

2 Microbial Pathogens and the Epithelial Cell Actin Cytoskeleton

The outermost infection processes occur when pathogens are introduced and proliferate in host cavities (respiratory, intestinal or uro-genital). These

[1] INSERM U452, Faculté de Médecine, 248 Av de Valombrose 06107 Nice Cedex 2, France
[2] Institut Pasteur, 28 rue du Dr Roux, 75724 Paris Cedex 15, France

Progress in Molecular and Subcellular Biology, Vol. 22
Ph. Jeanteur (Ed.)

cavities are lined by epithelial cells. Epithelial cells adhere tightly to each other, forming a barrier that not only constitutes a separation between two compartments, but also allows polarized secretion. In epithelial cells, the actin cytoskeleton is essential for various processes which maintain epithelium integrity. Actin filaments control cell morphology, adherence to the extracellular matrix, opening or closing of the intercellular space (paracellular pathway), regulate microvilli absorbtion processes, and are probably involved in many other tasks yet to be characterized. Pathogens interact with epithelial cells with the ultimate goal of getting a better supply of food in order to multiply. In doing so, they can interact with epithelia via toxins to modify absorbtion of nutrients, or they may invade cells to gain access to the rich cytosolic medium. In many cases, the actin cytoskeleton of epithelial cells is manipulated by bacteria to perform these tasks.

2.1 Microbial Pathogens and Actin

Pathogens may induce their own phagocytosis by epithelial cells in a process involving actin polymerization. This is the case for enteropathogens such as *Listeria monocytogenes*, *Shigella flexneri* or *Salmonella typhimurium* (reviewed by Finlay and Falkow 1997). A non-specific increase in phagocytic properties of epithelial cells can be triggered by an *Escherichia coli* toxin-like molecule (CNF1; Falzano et al. 1993). Some pathogenic strains of *E. coli* (EPEC) induce the formation of an actin pedestal directly underneath the cell membrane, at the point of contact between the bacterium and the apical pole of the epithelial cell (Rosenshine et al. 1996). A similar actin pedestal structure is also observed on the surface of gastric epithelial cells infected with *Helicobacter pylori* (Segal et al. 1996). Some pathogens (of the genera *Listeria*, *Shigella*, *Rickettsia* or even the *Vaccinia* virus) can harness actin polymerization after invading the cell cytosol in an actin comet tail structure which drives their locomotion within infected cells (reviewed by Higley and Way 1997). *Clostridium difficile*, an anaerobic microbial pathogen causing outbreaks of diarrhea and inflammatory pseudomembranous colitis after antibiotherapy, produces toxin A (TCdA) and toxin B (TCdB), which are among the largest known toxins (reviewed by von Eichel-Streiber et al. 1996). TCdA and TCdB disrupt the perijunctional actin ring of epithelial intestinal cells, leading to opening of tight junctions and rupture of the epithelial barrier (reviewed by Thelestam 1997). Some strains of *Clostridium botulinum* (serotype C and D) produce a low molecular weight enzymatic toxin-like fragment (exoenzyme C3; reviewed by Aktories 1997) which can disrupt intestinal tight junctions by depolymerizing actin filaments (Nusrat et al. 1995).

2.2 Rho GTP-Binding Proteins

The Rho subfamily includes three members, Rho, Rac and Cdc42 (Hall 1998), which regulate a wide variety of cellular processes involving actin filaments. Among others, Rho regulates cell shape (Rubin et al. 1988), formation of actin stress fibers and focal adhesions (Chardin et al. 1989; Paterson et al. 1990; Ridley et al. 1992), cell motility (Stasia et al. 1991), smooth muscle contraction (Amano et al. 1996; Gong et al. 1996), neurite retraction (Jalink et al. 1994), and formation of tight junctions (Nusrat et al. 1995; Takaishi et al. 1997a). Rac regulates formation of lamellipodia (Ridley et al. 1992), and NADPH oxidase dependent superoxide production (Abo et al. 1991). Cdc 42 controls the formation of filopodia (Kozma et al. 1996; Nobes and Hall 1995). Many putative downstream effectors of Rho proteins have been proposed, but so far, only a few have been implicated in F-actin reorganization. For example Rho is probably involved in smooth muscle contraction by controlling myosin light chain phosphorylation (Kimura et al. 1996) via activation of the Rho kinase (Rock; Ishizaki et al. 1996; Matsui et al. 1996). By bundling actin filaments through myosin activation, Rho induces the formation of actin cables with a potential contractile property (Chrzanowska-Wodnicka and Burridge 1996). Rac, on the other hand, would control actin polymerization via uncapping of barbed filament ends (Hartwig et al. 1995). Uncapping of actin filaments may be achieved by the removal of capping proteins such as CapZ from the barbed ends, in a process involving PI 4-5 kinase activation by Rac and increased levels of PIP2 (Hartwig et al. 1995).

Rho GTP-binding proteins also regulate tight and adherens junction formation in epithelial cells. The first clue to this action was from treating polarized intestinal T84 epithelial cells with the chimeric toxin DC3B (Nusrat et al. 1995). These experiments indicated that Rho was selectively involved in the control of tight junctions, since cadherin-dependent junctions were not affected by C3 treatment (Nusrat et al. 1995). It was then shown that in keratinocytes (Braga et al. 1997) and in MDCK cells (Takaishi et al. 1997), Rho controlled both tight and adherens junctions. In both studies, however, Rac played the predominant role in the control of adherens junctions (Braga et al. 1997; Takaishi et al. 1997b). How might Rho and Rac regulate tight and adherens junctions? Treatment of T84 or MDCK cells with exoenzyme C3 leads to delocalization of ZO-1, a tight junction protein associated with occludin (Nusrat et al. 1995; Takaishi et al. 1997). ZO1 might link tight junctions to the perijunctional actin ring, a contractile element containing myosin. Contraction or relaxation of the perijunctional actin ring is supposed to open or close tight junctions (Madara 1988). Rho, by its activity on smooth muscle contraction, may control this process. Rac could determine the formation of adherens junctions (Braga et al. 1997; Takaishi et al. 1997), by regulating membrane ruffling at the level of cadherins.

Finally, Rho participates in the formation of focal adhesion points by regulating the activity of proteins of the ERM family (ezrin, moesin or radixin; Mackay et al. 1997). However, the precise mechanism by which Rho activates ERM proteins is not yet clear. Rho could activate ERM proteins by a mechanism involving PIP2, similar to the one described for vinculin (Gilmore and Burridge 1996), or by phosphorylation through activation of a kinase different from Rock. An interesting feedback mechanism derived from ERM proteins activation would be the trapping of Rho GDI by ERM complexed with their membrane targets (e.g. ezrin bound to CD44; Hirao et al. 1996). This mechanism would be a simple and efficient process for controlling the activation of Rho, Rac or Cdc42.

3 Interaction of Bacterial Protein Toxins with the Rho Subfamily of GTP-Binding Proteins

Two types of bacterial exotoxins (referred to here as toxins) interact with Rho GTP-binding proteins. These toxins modify precise domains of the GTP-binding proteins called "switches" which undergo structural changes when the GTPases are bound either to GTP or GDP (Wittinghofer and Nassar 1996). The first group inhibits the activity of Rho by modifying amino acids of the switch 1 domain (effector domain, see Fig. 1). The second group provokes the permanent activation of the GTP-binding protein by modifying an amino acid of the switch 2 domain (involved in GTPase activity, see Fig. 1).

3.1 Toxins Inhibiting Rho

The first group of toxins that inhibit Rho is the family of C3-like ADP-ribosyltransferases which catalyze ADP-ribosylation of RhoA, B and C at Asn-41 (Fig. 1). The second group consists of the large clostridial cytotoxins which mono-glucosylate the Rho subfamily proteins Rho, Rac and Cdc42 at Thr-37/35 (Fig. 1).

The C3-like family encompasses several C. *botulinum* C3 isoforms, the *C. limosum* exotoxin and transferases produced by *Bacillus cereus* and *Staphylococcus aureus* (Aktories 1997). The C3 transferase is a single-chained 23-kDa protein with a very basic isoelectric point. C3 lacks a specific cell binding and translocation domain, this results in rather inefficient cell penetration which probably occurs via fluid phase pinocytosis. Exoenzyme C3 may bind non-specifically to the negatively charged surface of eukaryotic cells due to its basic isoelectric point. Once internalized by cells (requiring at least 20 μg/ml of C3) or microinjection, C3 is cytotoxic and causes disorganization of actin filaments due to ADP-ribosylation of RhoA, B and C proteins at asparagine 41

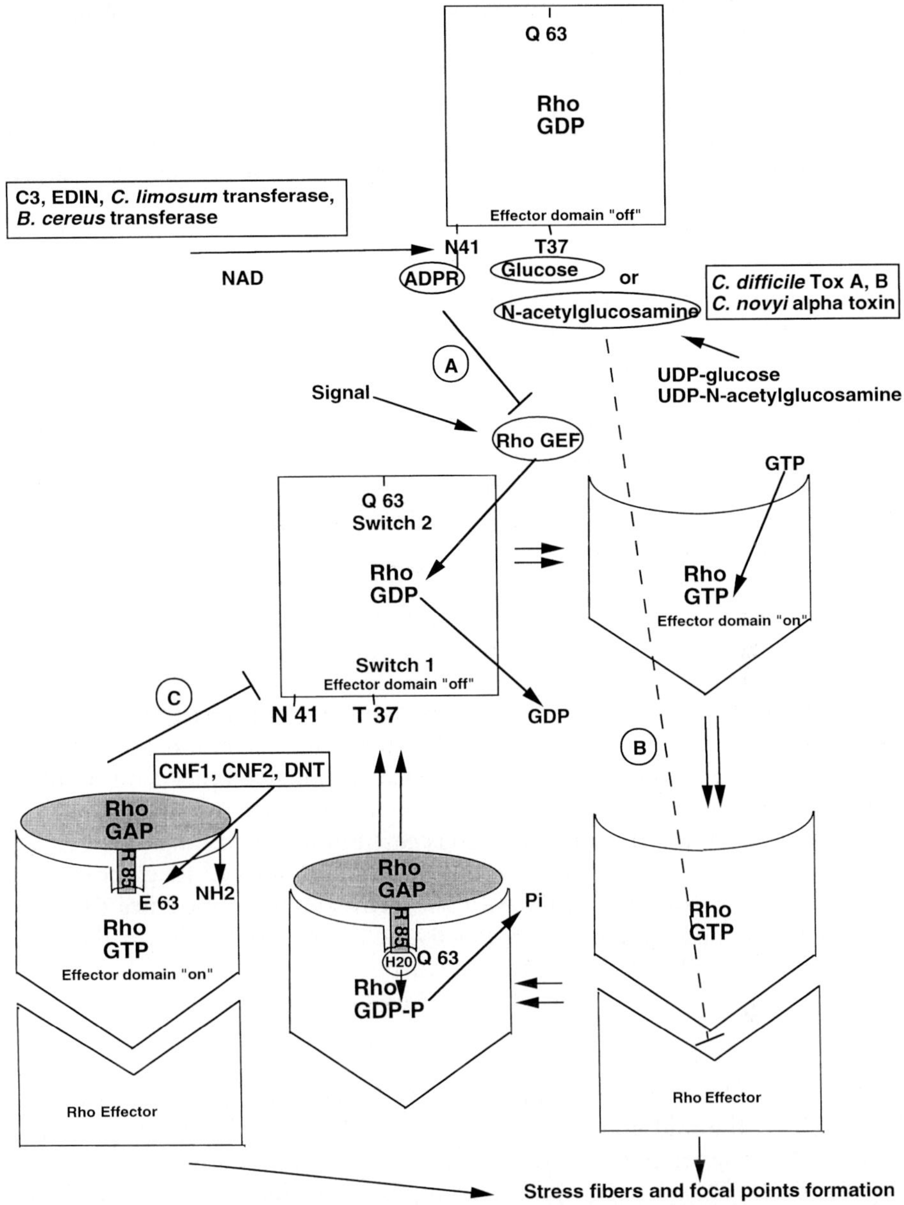

Fig. 1. Molecular mechanisms of toxins activating or inhibiting Rho GTP-binding proteins. The physiological activation/deactivation pathway of Rho is indicated by *double arrows*. **A** C3, EDIN, *C. limosum* and *B. cereus* transferases ADP-ribosylate Rho on asparagine 41, preventing Rho from being recruited to the cell membrane, and thus from reaching the Rho guanine exchange factor (RhoGEF). **B** *C. difficile* toxins A and B glucosylate Rho on threonine 37 whereas *C. novyi* alpha toxin transfers an N-acetylglucosamine on the same Rho residue. Glycosylation of Rho threonine 37 blocks the ability of Rho to bind and to activate its downstream effector. ADP-ribosylation and glycosylation of Rho modify the switch 1 domain of the GTP-binding protein. **C** CNF1, CNF2 and DNT deamidate Rho glutamine 63. Replacement of glutamine to glutamic acid on position 63 by CNFs or DNT results in inhibition of the intrinsic as well as the RhoGAP-mediated GTPase activities, and leads to permanent activation of Rho. Deamidation modifies the switch 2 domain of the GTP-binding protein

(Chardin et al. 1989; Paterson et al. 1990; Sekine et al. 1989). ADP-ribosylation of Rho at Asn-41 does not impair binding of GTP or RhoGAP on the GTPase, but rather alters the recruitment of Rho to the membrane (Fujihara et al. 1997). Rac and Cdc42 are poor substrates for C3 in vitro and are probably not targets in vivo.

Large clostridial cytotoxins are *C. difficile* toxins A and B (TCdA, TCdB), *Clostridium sordellii* lethal (LT) and hemorrhagic (HT) toxins and *Clostridium novyi* alpha toxin. TCdA and TCdB are single-chained exotoxins coproduced by pathogenic *C. difficile* strains. TCdA and TCdB have about 60% homology. They have been identified as causative agents of antibiotic-associated pseudomembranous colitis. TCdB is about a thousandfold more cytotoxic than TCdA to cultured cell lines (Sears and Kaper 1996). Both toxins induce disorganization of the microfilament system by their enzymic activity. TCdA and TCdB use UDP-glucose as a cofactor to transfer the glucose moiety to the Thr-37 of Rho, or to the Thr-35 of Rac and Cdc42, thus modifying the effector domain of the GTP-binding protein (Just et al. 1995a,b, switch 1; Fig. 1). Microinjection of glucosylated Rho induces the same effects on the cytoskeleton as the toxin itself, indicating that monoglycosylation of Rho, Rac and Cdc42 by toxin A and toxin B is probably the unique cause of their cytotoxicity (Just et al. 1995a,b).

C. sordellii LT and HT are identical to toxins A and B respectively (Popoff 1987; Genth et al. 1996), *C. sordellii* lethal toxin is a monoglucosyltransferase but has a different protein substrate specificity. Indeed, in addition to Rac, Ras proteins (Ras, Rap, Ral) are also modified (Just et al. 1996; Popoff et al. 1996) by the lethal toxin. This is an important finding since it was found that the protooncogenic function of Ras was blocked by glucosylation (Just et al. 1996; Popoff et al. 1996). *C. sordellii* HT toxin modifies Rho, Rac and Cdc42 by monoglucosylation exactly as TCdA and TCdB. On the other hand, *C. novyi* alpha-toxin catalyzes incorporation of N-acetyl-glucosamine into Rho proteins (threonine 37 of Rho, 35 of Rac and Cdc42; Selzer et al. 1996; Fig. 1).

Large clostridial cytotoxins, like most intracellular acting toxins, are composed of three functional domains: the N-terminal domain contains the catalytic activity, the middle part the putative membrane translocation domain, and the C-terminal part is involved in cell receptor binding (von Eichel-Streiber et al. 1996). A polypeptide encompassing amino acids 1–546 of TCdB, carries the catalytic monoglucosylating activity of the toxin (Hofman et al. 1997). The putative nucleotide binding site (aa 651–683) and the conserved cysteins are not needed for catalytic activity (Hofman et al. 1997). The receptor-binding domain is a repetitive structure which indicates a possible recruitment of several identical receptors per toxin molecule (von Eichel-Streiber et al. 1996). In the case of TCdA, the glycoprotein sucrase-isomaltase (260 kDa) has been identified as the cell receptor in rabbit ileal brush border (Pothoulakis et al. 1996). This cell receptor does not seem to be relevant because it is not expressed in many toxin-sensitive organs (e.g. human colon).

3.2 Toxins Activating Rho

Escherichia coli belongs to the normal microflora of the gastrointestinal tract, but some *E. coli* strains are associated with gastroenteritis, urogenital diseases, septicemia and pleural infections in humans. These pathogenic *E. coli* belong to a restricted number of pathovars defined by the presence of virulence factors that determine their host specificity and the type of disease they produce. Among these factors, production of toxins such as CNF1 is implicated in the pathogenesis of colibacillosis.

CNF1 was identified by Caprioli and colleagues as a cell-associated product of *E. coli* strains isolated from young children with diarrhea. It causes necrosis of rabbit skin and multinucleation of different types of tissue culture cells (Caprioli et al. 1983). A second type of CNF (CNF2) was then found in extracts of some *E. coli* strains isolated from calves with enteritis (De Rycke et al. 1987). CNF2 and CNF1 are immunologically related and have a similar apparent molecular weight (110–115kDa; De Rycke et al. 1990). CNF1, in the J96 uropathogenic *E. coli* strain, is chromosomally encoded in a 620-kb pathogenicity island (PAI II) containing the α-hemolysin and fimbriae genes (PRS; Blum et al. 1995) and many other *orf* with sequence similarities with known enterobacteria virulence factors (Swenson et al. 1996). The CNF2 gene is located on a large transmissible F-like plasmid (Vir).

A comparison of the amino acid sequence of CNFs with sequences in GenBank revealed that CNFs share regions of homologous amino acids with two other bacterial toxins: *Pasteurella multocida* toxin (PMT; Falbo et al. 1993; Oswald et al. 1994) and the dermonecrotic toxin of *Bordetella pertussis* (DNT; Walker and Weiss 1994). CNFs, PMT and DNT form the new family of dermonecrotic toxins (Walker and Weiss 1994; Lemichez et al. 1997).

CNF1 is able to provoke a remarkable reorganization of F-actin structures in cultured cells (Fiorentini et al. 1988). This toxin induces intense formation of stress fibers, focal contacts, membrane folding and retraction fibers. Reorganization of the F-actin cytoskeleton by CNF1 results in the cell's inability to undergo cytokinesis. Cell treatment for over 24h with CNF1, at a concentrations as low as 10^{-12}M gives rise to extremely flat and large multinucleated cells containing numerous stress fibers (Fiorentini et al. 1988).

Treatment of HEp-2 cells with CNF1, for increasing lengths of time (from 4 to 72h), induces an increased ability of the cell to ingest latex beads or non-invasive bacteria (Falzano et al. 1993). Macropinocytosis is totally blocked when CNF1-treated cells are incubated with the F-actin disrupting drug cytochalasin B, clearly demonstrating that the process is F-actin dependent (Falzano et al. 1993). Video time lapse studies of CNF1-induced macropinocytosis of latex beads in HEp-2 cells (Donelli et al. 1994) reveal a three step mechanism: (1) capture of beads by ruffles; (2) a short latency period; (3) entry of particles into the cytosol within membrane bound vacuoles that slowly migrate toward the perinuclear region.

CNF1 induces actin reorganization by irreversibly activating Rho. The first hint of such an activity was obtained when the cytosol from HEp-2 cells, previously incubated with CNF1, was ADP-ribosylated with exoenzyme C3. Under these conditions, it was observed that the molecular weight of Rho shifted to a slightly higher value. This result indicated a possible post-translational modification of the GTP-binding protein in CNF1 treated cells (Fiorentini et al. 1994; Oswald et al. 1994). Incubation of HEp-2 cells with CNF1 could also block cytotoxic activities of exoenzyme C3 and *Clostridium difficile* toxin B (Fiorentini et al. 1995), two bacterial toxins known to act on Rho proteins. CNF1 was then shown to direcly modify Rho in vitro without the need of cellular co-factors (Flatau et al. 1997). Microsequencing of CNF1-modified Rho showed that there was a single modification in the CNF1-treated GTP-binding protein, with glutamine 63 changed into a glutamic acid (Flatau et al. 1997). Therefore, CNF1 exerts a specific deamidase activity. An identical activity for CNF1 on Rho has been reported using mass spectrometry (Schmidt et al. 1997). Specific deamidation is a hitherto undescribed activity for a bacterial toxin. Indeed, toxin enzymatic activities described to date are ADP-ribosylation, depurination, metallo-protease, glucosyl transferase and adenylate cycling.

Rho glutamine 63 (glutamine 61 in p21 Ras) is known to be an important residue for the intrinsic and RhoGAP-mediated GTPase activity of Rho (Rittinger et al. 1997). Both CNF1-treated Rho and Rho mutated on glutamine 63 into glutamic acid (RhoQ63E) exhibit a mobility shift, upon electrophoresis, identical to CNF1-treated Rho. Nucleotide dissociation rates for CNF1-treated Rho and RhoQ63E were increased by two orders of magnitude, but the RhoGAP activity was totally impaired on both CNF1-treated Rho and RhoQ63E. Thus, deamidation of Rho on glutamine 63 by CNF1 allows Rho to be permanently bound with GTP, enhancing the activity of Rho on Rho effectors (Fig. 1). CNF1 deamidase activity is carried by thc 30 kDa carboxy-terminus end of the molecule (Lemichez et al. 1997), whereas the cell binding moiety of CNF1 is localized at the amino-terminus moiety of the molecule (Lemichez et al. 1997). CNF1 thus has a structure/activity relationship comparable to *Pseudomonas aeruginosa* exotoxin A (Allured et al. 1986). Sequence comparisons between the different dermonecrotic toxins CNF1, CNF2, DNT and PMT has established various levels of homology. CNF1 and CNF2 are highly homologous, sharing both comparable domains for enzymatic and cell binding activity. DNT however, has a high homology with CNFs in the catalytic domain but differs strongly from CNFs at the level of the cell binding domain. Recently, it has been shown that, like CNF1, DNT exhibits a deamidase activity towards Rho (Horiguchi et al. 1997). PMT on the other hand differs from CNFs or DNT at the level of its putative catalytic domain, but has a strong homology with the cell binding domain of CNFs (Lemichez et al. 1997).

4
Invasive Bacterial Factors and the Rho Subfamily of GTP-Binding Proteins

Some bacterial pathogens have the ability to enter normally non-phagocytic cells. This property is considered to be essential for the virulence of microorganisms that need to breach an epithelium to establish an infection, and has been studied in cultured cell systems for several bacterial pathogens. From these studies, it is becoming clear that invasive microorganisms have developed a wide variety of strategies for entering cells, involving various bacterial effectors and cellular receptors. In some instances, cell invasion is connected with a tight association of the bacterium with receptors on the cell surface. This is the case for *Yersinia pseudotuberculosis* binding to β1 integrin receptors via the invasin protein (Isberg 1991), or *Listeria monocytogenes* binding to cadherins via the internalin A protein (Mengaud et al. 1996). These receptors associate with the actin cytoskeleton through their cytoplasmic moiety, and have been implied in adhesion structures or adherens junctions. Although the formation of such adhesion structures is regulated in both cases by small G-proteins of the Rho subfamily, there is no evidence that activation of Rho is required for bacterial internalization. Other bacterial pathogens such as *Shigella*, the causative agent of bacillary dysentery in humans, or *Salmonella*, responsible for human infections with severity ranging from a mild gastroenteritis to systemic enteric fevers, however, induce their internalization by cultured cells in a very different way. For these enteropathogens, cell entry involves a drastic reorganization of the actin cytoskeleton and a strong adhesion to the cell surface does not appear to be a prerequisite for invasion. The entry process of these invasive bacteria presents homology with rufflings induced upon growth factor activation, although in this particular instance, the "ruffles" are localized to the vicinity of the bacterium interacting with the host cell membrane. It is believed that actin polymerization provides the driving force responsible for membrane deformation and the emergence of cell extensions rising above the bacterial body. These cell extensions eventually merge within minutes and lead to bacterial uptake in large vacuoles, in a process resembling macropinocytosis. Interestingly, as discussed below, the reorganization of the cytoskeleton induced by these invasive bacteria involves the activation of the Rho subfamily of small G proteins.

4.1
Bacterial Effectors of *Salmonella* or *Shigella*-Induced Internalization by Epithelial Cells

Genetic analysis of bacterial genes involved in the invasion process have underscored the similarity between both *Shigella* and *Salmonella* invasion apparatus. For both bacteria, invasion is dependent on a homologous type III, or cell contact dependent secretion apparatus called Mxi-spa for *Shigella* or Inv-spa for *Salmonella*. *Salmonella* also possesses another type III secretion apparatus, but it is likely that this latter may not perform a role that is linked to host cell invasion. Upon contact with host cells, secretion of bacterial effectors that are stored in the bacterial cytoplasm occurs within minutes. In the case of *Shigella*, the bacterial effectors responsible for cell entry have been identified as being the Ipa proteins, encoded by a single operon positioned in opposite orientation to the *mxi-spa* operon on the *Shigella* large virulence plasmid (Parsot and Sansonetti 1996). Three proteins, IpaB, C and D are essential for cell entry, whereas the IpaA protein optimizes the efficiency of the entry process. Interestingly, the Sip proteins encoded by a *Salmonella* chromosomally located operon, which are essential for *Salmonella* entry, present significant homologies with the *Shigella* Ipa proteins, with identity ranging from 57% to 27% (Hermant et al. 1995). Although the precise mechanism by which Sip or Ipa proteins promote bacterial entry remains unclear, both *Salmonella* and *Shigella* appear to have developed an original use for the type III secretion apparatus. Indeed, homologs of the *Shigella* or *Salmonella* type III secretion apparatus have also been identified in several other gram-negative bacterial pathogens, and such a secretion apparatus emerges as a way for the bacterium to directly inject bacterial effectors into host cells. For example, translocation of the Yop proteins via the *Yersinia* Ysc apparatus, a related type III secretion apparatus, is the best system characterized as yet. During cell contact, it is believed that YopB is secreted via the Ysc apparatus and forms a pore within the host cell membrane, through which other Yop proteins translocate in order to exert their antiphagocytic properties (Cornelis and Wolf-Watz 1997). Effectors of *Shigella* or *Salmonella* entry are also likely to translocate to the cell cyosol, in this latter case however, in order to induce the bacterial entry process. By translocating to the cell cytosol, these bacterial effectors can directly intercept the cellular machinary regulating the actin cytoskeleton and virtually bypass classical receptor-mediated signalling. Consistent with this, the *Salmonella* SipB and SipC proteins were shown to translocate from extracellularly located bacteria into the cell cytosol (Collazo and Galan 1997). Also, the *Shigella* IpaA protein translocates to the cell cytosol during the entry process and associates with the focal adhesion protein vinculin to enhance bacterial uptake (Tran Van Nhieu et al. 1997). Although *Salmonella* and *Shigella* present homologies in their mechanism of cell invasion, there may be differences in some aspects of the invasion process. For example, in contrast to the SipB and SipC proteins, the sublocalization of IpaB and IpaC during *Shigella* entry presents some ambiguity. It has been reported that the IpaB-C complex can associate with β1 integrins (Watarai et al. 1996) or with the CD44 hyaluronan receptor (Skoudy, unpubl.) on the host cell surface. Such interactions may be important for the entry process because binding of these *Shigella* proteins may transduce signals via these receptors, which could regulate the actin cytoskeleton. This, however, would be inconsistent with the moderate effects on cell entry observed after blocking of these receptors, or inhibition of downstream kinases activated upon receptor engagement. Alternatively, binding to these receptors may be a prerequisite for efficient translocation of the *Shigella* effectors in the cell cytosol. Perhaps IpaB and IpaC perform a dual role, by regulating the actin cytoskeleton from within the cell cytosol, as well as modulating the cell response by means of an extracellular assocation with cell receptors.

4.2 Cdc42 Is Involved in *Salmonella* Entry, Whereas Rho Proteins Are Required for *Shigella* Entry

To identify cellular pathways activated upon *Salmonella* entry, members of the Rho proteins were inhibited by microinjection into cultured HeLa cells of dominant negative forms of these proteins, or of the *Clostridium botulinum* exoenzyme C3 that specifically inactivates Rho, prior to challenge with *Salmonella*. From these studies, it appeared that inhibition of Rac of Rho had little effect on *Salmonella* induction of cytoskeletal reorganization (Jones et al. 1993). Transient expression of the dominant negative form of Cdc42, on the other hand, resulted in up to 80% inhibition of *Salmonella* induced-cytoskeletal reorganization and *Salmonella* uptake (Chen et al. 1996). In these transfection experiments, inhibition of Rac also resulted in about a 60% decrease in bacterial uptake. The discrepancy between the two sets of experiments may be explained by the difference in the technique used to inhibit Rho family members: transfection of dominant negative Rac may result in a more efficient inhibition of Rac than introduction of the inhibitor by microinjection. The preferential role of Cdc42 in *Salmonella* uptake argues for it being a primary target of bacterial effectors of entry. Consistent with a predominant role for Cdc42, stable expression of a constitutively active form of Cdc42 in epithelial cells complemented the invasion-defective *Salmonella invA* mutant that was unable to induce cell-contact secretion (Chen et al. 1996). Independent of cell invasion, *Salmonella* induces the production of IL-8, a pro-inflammatory cytokine with chemoattracting activity towards neutrophils. This process, which results from activation of the JNK/p38 MAP kinases, requires Cdc42 activation (Bagrodia et al. 1995). This suggests that *Salmonella* effectors induce the cytoskeletal reorganization that leads to bacterial uptake by activating Cdc42, and that activation of Cdc42 also results in JNK/p38 kinase activation, presumably following a pathway independent of actin polymerization (Lamarche et al. 1996; Fig. 2).

Interestingly, in contrast to *Salmonella* invasion, Shigella entry into epithelial cells requires activation of Rho. Cell pretreatment with the Rho inhibitor C3 (Watarai et al. 1997) or DC3B, a recombinant form of C3 that permeates live cells (Adam et al. 1996), resulted in inhibition of *Shigella*-induced ruffles and

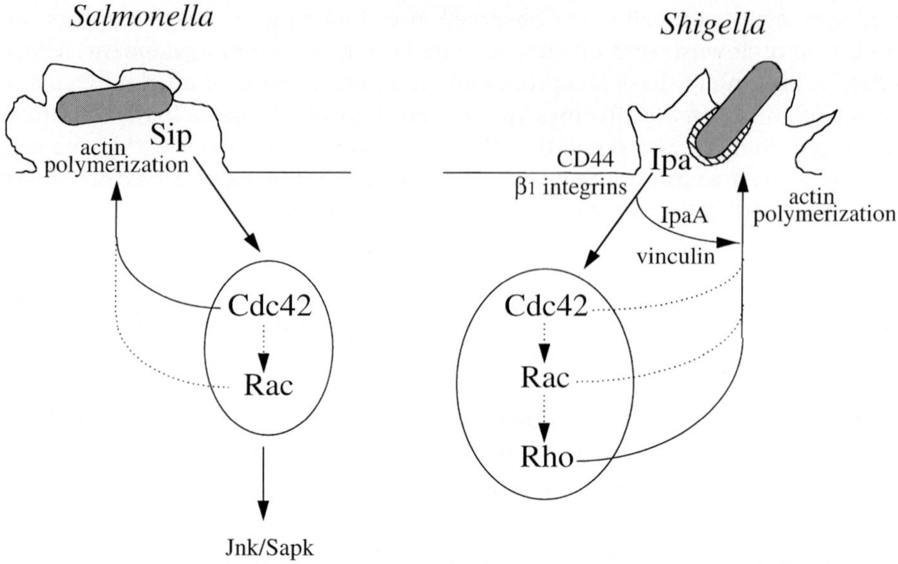

Fig. 2. *Salmonella* and *Shigella* require activation of small G proteins to initiate invasion. Upon cell contact, the *Salmonella* Sip or *Shigella* Ipa proteins are translocated via a type III secretion apparatus from the extracellular bacterium, and gain access to the cell cytosol. Although sharing significant homologies, Sip and Ipa proteins appear to activate different Rho proteins to induce actin cytoskeletal rearrangements that lead to bacterial uptake. *Salmonella* uptake requires primarily the activation of Cdc42 and of Rac to a lesser extent, whereas Rho does not seem to be required. In contrast, *Shigella* uptake is dependent on Rho activation, and probably also on Cdc42 and Rac. For *Shigella*, CD44 and b1 integrins may play a role during the early steps of the entry process, and some cytoskeletal rearrangements (*hatched area*) are dependent on an interaction between IpaA and vinculin

inhibition of Shigella entry by up to 90%. Rho inhibition, however, did not abolish *Shigella*-induced actin reorganization totally, as a coat of polymerized actin was still detectable in the close vicinity of the bacterium interacting with the host cell membrane. Although there is some evidence that Rac and Cdc42 may also be involved to different degrees in *Shigella* entry (Mounier et al. unpubl.), the fact that Rho itself is required for the entry process was unexpected because Rho is considered to participate mostly in the formation of stress fibers and focal adhesions, rather than ruffle-like structures (Ridley and Hall 1992). In agreement with a role for Rho during *Shigella* entry, various cytoskeletal proteins that have been shown to translocate to the membrane in response to Rho are recruited to the *Shigella*-induced entry structure. This is the case for ezrin, a member of the ERM family of proteins, which is presumed to allow anchorage of cell receptors such as CD44 to the actin cytoskeleton during the initial phase of cell adhesion (Tsukita et al. 1997). Ezrin, which actively participates in *Shigella* entry, is also recruited to the entry structure, in particular in the tip of the cell extensions that rise from the area delimitated by important actin polymerization induced by the bacterium (Skoudy, unpubl.). Focal adhesion components that cluster upon Rho activation are also recruited at the site of *Shigella* induced actin polymerization. For example, vinculin, a focal adhesion protein that acts as a molecular bridge between the cytoplasmic domain of β1 integrins and F-actin, is recruited at the area of actin polymerization induced by *Shigella*: in this case, however, vinculin shows a more intense recruitment in the close vicinity of the membrane adjacent to the bound bacterium, in a distribution that superimposes on a denser network of F-actin. This actin coat, however, appears to be dependent on expression of the *Shigella* IpaA invasin, presumably after direct binding of IpaA to vinculin (Tran Van Nhieu et al. 1997). Vinculin distribution in the *Shigella* entry structure is therefore the result of a combination of two responses: a primary response that results in vinculin recruitment at the site of *Shigella* entry, and an IpaA dependent response that allows further sublocalization of vinculin to the membrane juxtaposing the bacterium in the process of being internalized. This IpaA dependent response also allows for subsequent transformation of the site of actin polymerization from microspike-like to ruffle-like structures which are productive for *Shigella* internalization. Thus, *Shigella* entry appears to combine at least two types of cell responses which act in concert to form the entry structure (Fig. 2). Perhaps this may explain the seemingly paradoxical results of Rho involvement in *Shigella* entry: it is conceivable that primary responses to IpaB and IpaC lead to actin polymerization as well as Rho activation. This primary response is then modulated by IpaA to convert it into a ruffling-like response. In contrast to *Salmonella*, *Shigella* shows little cell adhesion. Perhaps Rho activation and formation of an adhesion-like structure provides a means for the latter bacterium to be stably anchored or even to be captured in the ruffles, induced at the site of entry, thus favouring its internalization.

What are the primary targets of *Salmonella* or *Shigella* effectors leading to Cdc42 or Rho activation, and responsible for the actin reorganization at the entry site? Treatment with various inhibitors such as genistein, a tyrosine kinase inhibitor, or H7, a serine/threonine kinase inhibitor, have little effect on the cytoskeletal rearrangements induced by both bacteria. This confirms the notion that these bacterial effectors do not use receptor-mediated pathways to activate Rho proteins, and that both Sip and Ipa protein targets are likely to be intracellular. Whether these targets consist of previously characterized upstream regulators of the Rho proteins, or of yet to be identified pathways of activation, remains to be defined. Obviously, important clues will be given by the ability to distinguish between the effects of bacterial effectors that act simultaneously to manipulate the cytoskeleton during the entry process. It is possible that modulation by different bacterial effectors leads to the different requirements for Rho proteins involved in *Salmonella* or *Shigella* entry, and that each of these pathogens may have evolved different integrated systems adapted to the pathology that they induce.

References

Abo A, Pick E, Hall A, Totty N, Teahan CG, Segal AW (1991) The small GTP-binding protein, p21 Rac, is involved in the activation of the phagocyte NADPH oxidase. Nature 353:668–670

Adam T, Giry M, Boquet P, Sansonetti PJ (1996) Rho-dependent membrane folding causes *Shigella* entry into epithelial cells. EMBO J 15:3315–3321

Aktories K (1997) Bacterial toxins that target Rho proteins. J Clin Invest 99:827–829

Allured VS, Collier RJ, Carrol SF, McKay DB (1986) Structure of exotoxin A of *Pseudomonas aeruginosa* at 3.0 Angstrom resolution. Proc Natl Acad Sci USA 83:1320–1324

Amano M, Ito M, Kimura K, Fukata Y, Chihara K, Nakano T, Matsuura Y, Kaibuchi K (1996) Phosphorylation and activation of myosin by Rho-associated kinase (Rho-kinase). J Biol Chem 271:20246–20249

Bagrodia S, Derijard B, Davis J, Cerione RA (1995) Cdc42 and PAK-mediated signalling leads to Jun kinase and p38 mitogen-activated protein kinase activation. J Biol Chem 270:27995–27998

Blum G, Falbo V, Alonso MP, Blanco JE, Gonzales EA, Garabal JI (1995) Gene clusters encoding the cytotoxic necrotizing factor type 1 PRS fimbriae and α hemolysin from the pathogenicity island II of the uropathogenic *Escherichia coli* strain J96. FEMS Microbiol Lett 126:189–196

Braga VMM, Machesky LM, Hall A, Hotchin N (1997) The small GTPases Rho and Rac are required for the establishment of cadherin-dependent cell-cell contacts. J Cell Biol 137:1421–1431

Caprioli A, Falbo V, Roda LG, Ruggeri FM, Zona C (1983) Partial purification and characterization of an *Escherichia coli* toxic factor that induces morphological cell alterations. Infect Immun 39:1300–1306

Chardin P, Boquet P, Madaule P, Popoff MR, Rubin EJ, Gill DM (1989) The mammalian G protein RhoC is ADPribosylated by *Clostridium botulinum* C3 and affects actin microfilaments in Vero cells. EMBO J 8:1087–1092

Chen LM, Hobbie S, Galan JE (1996) Requirement of Cdc42 for *Salmonella*-induced cytoskeletal and nuclear responses. Science 274:2115–2118

Chrzanowska-Wodnicka M, Burridge K (1996) Rho-stimulated contractility drives the formation of stress fibers and focal adhesions. J Cell Biol 6:1403–1415

Collazo CM, Galan JE (1997) The invasion-associated type III system of *Salmonella typhimurium* directs the translocation of Sip proteins into the host cell. Mol Microbiol 24:747–756

Cornelis G, Wolf-Watz H (1997) The *Yersinia* Yop virulon: a bacterial system for subverting eukaryotic cells. Mol Microbiol 23:861–867

Cossart P, Boquet P, Normark S, Rappuoli R (1996) Cellular microbiology emerging. Science 271:315–316

De Rycke J, Guillot JF, Boivin R (1987) Cytotoxins in non-enterotoxigenic strains of *Escherichia coli* isolated from faeces of diarrheic calves. Vet Microbiol 15:137–157

De Rycke J, Gonzalez EA, Blanco J, Oswald E, Blanco M, Boivin R (1990) Evidence for two types of cytotoxic necrotizing factor in human and animal clinical isolates. J Clin Microbiol 28:694–699

Donelli G, Fiorentini C, Falzano L, Pouchelet M, Oswald E, Boquet P (1994) Effects induced by the cytotoxic necrotizing factor 1 (CNF1) from pathogenic *E. coli* on cultured epithelial cells. Zentrabl Bakteriol Suppl 24:60–71

Falbo V, Pace T, Picci L, Pizzi E, Caprioli A (1993) Isolation and nucleotide sequence of the gene encoding cytotoxic necrotizing factor type 1. Infect Immun 61:4909–4914

Falzano L, Fiorentini C, Donelli G, Michel E, Kocks C, Cossart P, Cabanié L, Oswald E, Boquet P (1993) Induction of phagocytic behaviour in human epithelial cells by *Escherichia coli* cytotoxic necrotizing factor 1. Mol Microbiol 9:1247–1254

Finlay BB, Falkow S (1997) Common themes in microbial pathogenicity revisited. Mol Biol Rev 61:136–169

Fiorentini C, Arancia G, Caprioli A, Falbo V, Ruggeri FM, Donelli G (1988) Cytoskeletal changes induced in HEp-2 cells by the cytotoxic necrotizing factor of *Escherichia coli*. Toxicon 26:1047–1056

Fiorentini C, Giry M, Donelli G, Falzano L, Aullo P, Boquet P (1994) *E. coli* cytotoxic necrotizing factor 1 increases actin assembly via the p21 Rho GTPase. Zentrabl Bakteriol Suppl 24:404–405

Fiorentini C, Donelli G, Matarrese P, Fabbri A, Paradisi S, Boquet P (1995) *Escherichia coli* cytotoxic necrotizing factor 1: evidence for induction of actin assembly by constitutive activation of the p21 Rho GTPase. Infect Immun 63:3936–3944

Flatau G, Lemichez E, Gauthier M, Chardin P, Paris S, Fiorentini C, Boquet P (1997) Toxin-induced activation of the G-protein Rho by deamidation of glutamine. Nature 387:729–733

Fujihara H, Walker LA, Gong MC, Lemichez E, Boquet P, Somlyo AV, Somlyo AP (1997) Inhibition of RhoA translocation and calcium sensitivation by in vivo ADP-ribosylation of Rho with the chimeric DC3B. Mol Biol Cell 8:2437–2447

Genth H, Hofman F, Selzer J, Rex G, Aktories K, Just I (1996) Difference in protein substrates specificity between hemorrhagic toxin and lethal toxin from *Clostridium sordellii*. Biochem Biophys Res Comm 229:370–374

Gilmore AP, Burridge K (1996) Regulation of vinculin binding to talin and actin by phosphatidylinositol 4-5 biphosphate. Nature 381:531–535

Gong MC, Iizuka K, Nixon G, Browne JP, Hall A, Eccleston JF, Sugai M, Kobayashi S, Somlyo AV, Somlyo AP (1996) Role of guanine nucleotide binding proteins-Ras-family or trimeric proteins or both in Ca^{2+} sensitization of smooth muscles. Proc Natl Acad Sci USA 93:1340–1345

Hall A (1998) Rho GTPases and the actin cytoskeleton. Science 279:509–514

Hartwig JH, Bokoch GM, Carpenter CL, Jeanmey PA, Taylor LA, Toker A, Stossel TP (1995) Thrombin receptor ligation and activated Rac uncap actin filament barbed ends through phosphoinositide synthesis in permeabilized human platelets. Cell 82:643–653

Hermant D, Ménard R, Arricau N, Parsot C, Popoff MY (1995) Functional conservation of the *Salmonella* and *Shigella* effectors of entry into epithelial cells. Mol Microbiol 17:781–789

Higley S, Way M (1997) Actin and cell pathogenesis. Curr Opin Cell Biol 9:62–69

Hirao M, Sato N, Kondo T, Yonemura S, Monden M, Sasaki T, Takai Y, Tsukita S, Tsukita S (1996) Regulation mechanism of ERM (ezrin, radixin/moesin) protein/plasma association: possible involvement of phosphatidylinositol turnover and Rho-dependent signalling pathways. J Cell Biol 135:37–51

Hofman F, Busch C, Prepens U, Just I, Aktories K (1997) Localization of the glucosyltransferase activity of *Clostridium difficile* toxin B to the N-terminal part of holotoxin. J Biol Chem 272:11074–11078

Horiguchi Y, Inoue N, Masuda M, Kashimoto T, Katahira J, Sugimoto N, Matsuda M (1997) *Bordetella bronchiseptica* dermonecrotizing toxin induces reorganization of actin stress fibers through deamidation of Gln-63 of the GTP-binding Rho. Proc Natl Acad Sci USA 94:11623–11626

Isberg RR (1991) Discrimination between intracellular uptake and surface adhesion of bacterial pathogens. Science 252:934–938

Ishizaki T, Maekawa M, Fujisawa K, Okawa K, Iwamatsu A, Fujita A, Watanabe N, Saito Y, Kakizuka Y, Morii N, Narumiya S (1996) The small GTP-binding protein Rho binds to and activates a 160 kDa ser/thr protein kinase homologous to myotonic dystrophy kinase. EMBO J 15:1885–1893

Jalink K, van Gorven EJ, Hengeveld T, Morii N, Narumiya S, Moolenar WH (1994) Inhibition of lysophosphatidate- and thrombin-induced neurite retraction and neuronal cell rounding by ADP-ribosylation of the small GTP-binding protein Rho. J Cell Biol 126:801–810

Jones BD, Paterson HF, Hall A, Falkow S (1993) *Salmonella typhimurium* induces membrane ruffling by a growth-factor receptor independent mechanism. Proc Natl Acad Sci USA 90:10390–10394

Just I, Selzer J, Wilm M, von Eichel-Streiber C, Mann M, Aktories K (1995a) Glucosylation of Rho proteins by *Clostridium difficile* toxin B. Nature 375:500–503

Just I, Wilm M, Selzer J, Rex G, von Eichel-Streiber C, Mann M, Aktories K (1995b) The enterotoxin from *Clostridium difficile* (Tox A) monoglucosylates the Rho proteins. J Biol Chem 270:13932–13936

Just I, Selzer J, Hofman F, Green GA, Aktories K (1996) Inactivation of Ras *by Clostridium sordellii* lethal toxin-catalyzed glucosylation. J Biol Chem 271:10217–10224.

Kimura K, Ito M, Amano M, Chihara K, Fukuta Y, Nakafuku M, Yamamori B, Feng J, Nakano T, Okawa K, Iwamatsu A, Kaibuchi K (1996) Regulation of myosin phosphatase by Rho and Rho-associated kinases (Rho-kinase). Science 273:245–247

Kozma R, Ahmed S, Best A, Lim L (1996) The GTPase-activating protein n-chimerin cooperates with Rac1 and Cdc42Hs to induce the formation of lammelipodia and filopodia. Mol Cell Biol 16:5069–5080

Lamarche N, Tapon N, Stowers L, Burbelo PD, Aspenström P, Bridges T, Chan J, Hall A (1996) Rac and Cdc42 induce actin polymerization and G1 cell cycle progression independently of p65PAK and the JNK/SAPK MAP kinase cascade. Cell 87:519–529

Lemichez E, Flatau G, Bruzzone M, Boquet P, Gauthier M (1997) Molecular localization of the *Escherichia coli* cytotoxic necrotizing factor CNF1 cell-binding and catalytic domains. Mol Microbiol 24:1061–1070

MacKay DJG, Esch F, Furthmayr H, Hall A (1997) Rho-and Rac-dependent assembly of focal adhesion complexes and actin filaments in permeabilized fibroblasts: an essential role for ezrin/radixin/moesin proteins. J Cell Biol 138:927–938

Madara JL (1988) Tight junction dynamics: is paracellular transport regulated? Cell 53:497–498

Matsui T, Amano M, Yamamoto T, Chihara K, Nakafuku M, Ito M, Nakano T, Okawa K, Iwamatsu A, Kaibuchi K (1996) Rho-associated kinase, a novel serine/threonine kinase, as a putative target for the small GTP-binding protein Rho. EMBO J 15:2208–2216

Mengaud J, Ohayon H, Gounon P, Mège RM, Cossart P (1996) E-cadherin is the receptor for internalin, a surface protein required for entry of *L. monocytogenes* into epithelial cells. Cell 84:923–933

Nobes CD, Hall A (1995) Rho, Rac and Cdc42 GTPases regulate the assembly of multi-molecular focal complexes associated with actin stress fibres, lammelipodia and filopodia. Cell 81:53–62

Nusrat A, Giry M, Turner JR, Colgan SP, Parkos CA, Carnes D, Lemichez E, Boquet P, Madara JL (1995) Rho protein regulates tight junctions and perijunctional actin organization in polarized epithelia. Proc Natl Acad Sci USA 92:10629–10633

Oswald E, Sugai M, Labigne A, Wu HC, Fiorentini C, Boquet P, O'Brien A (1994) Cytotoxic necrotizing factor type 2 produced by virulent *Escherichia coli* modifies the small GTP-binding protein Rho involved in assembly of actin stress fibers. Proc Natl Acad Sci USA 91:3814–3818

Parsot C, Sansonetti PJ (1996) Invasion and the pathogenesis of *Shigella* infections. In: Miller VL (ed) Bacterial invasiveness. Springer, Berlin Heidelberg New York, pp 25–42

Paterson HF, Self AJ, Garrett MD, Just I, Aktories K, Hall A (1990) Microinjection of recombinant p21Rho induces rapid changes in cell morphology. J Cell Biol 111:1001–1007

Popoff MR (1987) Purification and characterization of *Clostridium sordellii* lethal toxin and cross reactivity with *Clostridium difficile* cytotoxin. Infect Immun 55:35–43

Popoff MR, Chaves-Olarte E, Lemichez E, von Eichel-Streiber C, Thelestam M, Chardin P, Cussac D, Antonny B, Chavrier P, Flatau G, Boquet P (1996) Ras, Rap and Rac small GTP-binding proteins are targets for *Clostridium sordellii* lethal toxin glucosylation. J Biol Chem 271: 10217–10224

Pothoulakis C, Gilbert RJ, Cladaras C, Castagliuolo I, Semenza G, Hitti Y, Moncrief JS, Linevsky J, Kelly CP, Nikulasson S, Lamont T (1996) Rabbit sucrase isomaltase contains a functional intestinal receptor for *Clostridium difficile* toxin A. J Clin Invest 98:641–649

Ridley AJ, Hall A (1992) The small GTP-binding protein Rho regulates the assembly of focal adhesions and actin stress fibers in response to growth factors. Cell 70:389–399

Ridley AJ, Paterson HF, Johnston CL, Diekmann D, Hall A (1992) The small GTP-binding protein Rac regulates growth factor-induced membrane ruffling. Cell 70:401–410

Rittinger K, Walker PA, Eccleston JF, Smerdon SJ, Gamblin SJ (1997) Structure at 1.65 Angstom of RhoA and its GTPase-activating protein in complex with a transition state analog. Nature 389:758–762
Rosenshine I, Ruschkowski S, Stein M, Reinscheid DJ, Mills SD, Finlay BB (1996) A pathogenic bacterium triggers epithelial cell signals to form a functional bacterial receptor that mediates actin pseudopod formation. EMBO J 15:2613–2624
Rubin EJ, Gill DM, Boquet P, Popoff MR (1988) Functional modification of a 21-kiloDalton G protein when ADP-ribosylated by exoenzyme C3 of *Clostridium botulinum*. Mol Cell Biol 8:418–426
Schmidt G, Sehr P, Wilm M, Selzer J, Mann M, Aktories K (1997) Gln 63 of Rho is deamidated by *Escherichia coli* cytotoxic necrotizing factor 1. Nature 387:725–728
Sears CL, Kaper JB (1996) Enteric bacterial toxins: mechanisms of action and linkage to intestinal secretion. Microbiol Rev 60:167–215
Segal DE, Falkow S, Tompkins LS (1996) *Helicobacter pylori* attachment to gastric cells induces cytoskeletal rearrangements and tyrosine phosphorylation of host cell proteins. Proc Natl Acad Sci USA 93:1259–1264
Sekine A, Fujiwara M, Narumiya S (1989) Asparagine residue in the Rho gene product is the modification site for botulinum ADP-ribosyltransferase. J Biol Chem 264:8602–8605
Selzer J, Hofman F, Rex G, Wilm M, Mann M, Just I, Aktories K (1996) *Clostridium novyi* alpha-toxin-catalyzed incorporation of GlcNAc into Rho subfamily proteins. J Biol Chem 271: 25173–25177
Stasia MJ, Jouan A, Bourmeyester N, Boquet P, Vignais P (1991) ADP-ribosylation of a small size GTP-binding protein in bovine neutrophils by the C3 exoenzyme of *Clostridium botulinum* and effect on cell motility. Biochem Biophys Res Comm 180:615–622
Swenson DL, Bukanov NO, Berg DE, Welch RA (1996) Two pathogenicity islands in uropathogenic *Escherichia coli* J96: cosmid cloning and sample sequencing. Infect Immun 64:3736–3743
Takaishi K, Sasaki T, Kotani H, Nishioka H, Takai Y (1997) Regulation of cell-cell adhesion by Rac and Rho small G proteins in MDCK cells. J Cell Biol 139:1047–1059
Thelestam M, Florin I, Chaves-Olartes E (1997) *Clostridium difficile* toxins. In: Aktories K (ed) Bacterial toxins. Tools in cell biology and pharmacology. Chapman and Hall, Weinheim, pp 131–141
Tran Van Nhieu G, Ben Ze'ev A, Sansonetti PJ (1997) Modulation of bacterial entry into epithelial cells by association between vinculin and the *Shigella* IpaA invasin. EMBO J 16:2717–2729
Tsukita S, Yonemura S, Tsukita S (1997) ERM proteins: head-to-tail regulation of actin-plasma membrane interaction. Trends Biochem Sci 22:53–58
Von Eichel-Streiber C, Boquet P, Sauerborn M, Thelestam M (1996) Large clostridial cytotoxins – a family of glycosyltransferases modifying small GTP-binding proteins. Trends Microbiol 375:375–382
Walker KE, Weiss AA (1994) Characterization of the dermonecrotic toxin in the members of the genus *Bordetella*. Infect Immun 62:3817–3828
Watarai M, Funato S, Sasakawa S (1996) Interaction of Ipa proteins of *Shigella flexneri* with $\alpha 5\beta 1$ integrin promotes entry of the bacteria into mammalian cells. J Exp Med 183:991–999
Watarai M, Kamata Y, Kozaki S, Sasakawa C (1997) Rho, a small GTP-binding protein, is essential for *Shigella* invasion of epithelial cells. J Exp Med 185:281–292
Wittinghofer A, Nassar N (1996) How ras-related proteins talk to their effectors. Trends Biochem Sci 21:488–491

Rho GTPases in Development

Jeffrey Settleman

1 Introduction

The rapidly expanding Rho family of evolutionarily conserved proteins comprise a subclass of the Ras superfamily of regulatory GTPases. Like their Ras relatives, the Rho GTPases cycle between conformationally distinct active and inactive states and thereby function as molecular switches that mediate signal transduction pathways in response to extracellular stimuli. Although the Rho family proteins have been implicated as important regulators of the actin cytoskeleton, their precise biological functions and the nature of their associated signaling pathways in vivo are only just beginning to be elucidated. In light of their ability to promote profound effects on cell morphology, it has been suggested that these proteins are likely to be critical regulators of the numerous morphogenetic events required of a developing multicellular organism. The development of most complex multicellular organisms procedes via a fundamentally similar mechanism. Following extensive cell proliferation, cell fates are specified, a body plan is established, and a variety of morphogenetic events take place that serve to elaborate the structure. This review will focus on published observations that support a role for the Rho family of proteins in these various morphogenetic processes required during development, with particular emphasis on recent studies conducted in genetically tractable organisms.

2 The Rho Family of Small GTPases

The Ras superfamily comprises a group of structurally related 21 kDa proteins that bind to and hydrolyze GTP, and appear to function as molecular switches for a variety of cellular processes [1]. These processes include the transduction of extracellular signals related to growth and differentiation, assembly and dissassembly of the actin cytoskeleton, protein trafficking, and gene expres-

Massachusetts General Hospital Cancer Center and Harvard Medical School, Building 149, 13th Street, Charlestown, Massachusetts 02129, USA

Progress in Molecular and Subcellular Biology, Vol. 22
Ph. Jeanteur (Ed.)

sion. This protein family has been further subdivided into several groups on the basis of primary structure similarity. These include the Ras, Rho, Rab, and Arf sub-families, each of which includes several identified genes. However, with the ongoing identification of many new related GTPases, the distinctions between these various sub-families are becoming increasingly blurred. Most of these genes have been highly conserved evolutionarily, and close relatives of several of the mammalian GTPases have been identified in such organisms as flies, worms, and yeast.

The Rho GTPases, which include the intensively studied Rho, Rac, and Cdc42 proteins, are about 30% identical at the amino acid level to the Ras GTPases. They appear, however, to perform quite distinct functions [2, 3]. The core consensus motifs that have been defined as being essential for GTP binding and hydrolysis by Ras are also maintained in the Rho GTPases, and these proteins exhibit similar biochemical properties [4, 5]. However, the putative effector domain of the Rho proteins significantly differs from that of Ras, suggesting that the cellular targets of the activated GTPase are different for Ras and Rho proteins. Indeed, as described below, this has been confirmed experimentally. In mammals, at least 12 distinct Rho family members have been identified (RhoA, RhoB, RhoC, RhoD, RhoE, Rac1, Rac2, Rac3, RhoG, Cdc42, TTF, and TC10). Related genes have been identified in yeast, worms, flies, slime molds, and plants, indicating a high degree of evolutionary conservation of the Rho subfamily of GTPases.

The encoded Rho proteins have recently received much attention as signaling intermediates for a variety of cellular functions. Most notably, numerous studies have revealed a role for these proteins in regulating cytoskeletal organization and cell morphology [2, 3, 6–8]. For example, introduction of activated Rho into cultured fibroblasts induces formation of actin stress fibers [6], Rac induces membrane ruffling [7], and Cdc42 induces filopodia formation [9, 10]. Moreover, specific inhibition of Rho or Rac in fibroblasts through the use of dominant negative mutants or Rho-specific toxins can block growth factor-stimulated cytoskeletal changes, suggesting that Rho family GTPases mediate cellular pathways that control the organization of the actin cytoskeleton in response to extracellular signals [6, 7].

Some additional functions have been ascribed to certain Rho family members. Accumulating evidence suggests a role for Rho proteins in cell motility, a function that is at least partly controlled by the actin cytoskeleton [11–13]. Other cellular functions that are affected by cytoskeletal reorganization, such as endocytosis [14], adhesion [11, 15–18], and axon guidance [19–24] also appear to be mediated by the various Rho family members. At least one highly specific function for a Rho family member, Rac2, has been identified. Rac2 is an essential component of the NADPH oxidase system in neutrophils and macrophages [25, 26]. In addition, Rho, Rac, and Cdc42 have each been implicated in other biological processes that may not be directly related to their roles in actin regulation, such as cell cycle control [27] and regulation of gene expression [28–30].

2.1 Rho-Mediated Signal Transduction

Regulation of the nucleotide state of the Rho family GTPases appears to be somewhat analogous to that which has been well documented for Ras. Several Rho-specific GTPase activating proteins (GAPs) and guanine nucleotide exchange factors (GNEFs) have been described, as well as so-called Rho-GDIs (guanine nucleotide dissociation inhibitors) that function to maintain the Rho GTPases in an inactive GDP-bound state by preventing dissociation of GDP [4]. Thus, it appears that, like Ras proteins, Rho GTPases cycle between two functional states, and thereby perform a "switching" function that constitutes a link in a cellular signal transduction cascade.

Several cellular protein targets of the active GTP-bound form of some of the Rho family GTPases have recently been identified. For example, a serine/threonine protein kinase, PAK (p21-activated kinase) has been reported to be a target of activated, GTP-bound Rac and Cdc42 proteins [31, 32]. This protein is homlogous to the *S. cerevisiae* STE20 protein, a MAPK pathway component of the yeast pheromone mating response [33]. At least three closely related PAK family members have now been identified. Activation of the PAKs is believed to lead in turn to activation of a group of MAP kinase-related proteins known as the JunK/SAPK/p38 family [34]. Activation of this kinase cascade by Rac and Cdc42 may be be unrelated to the effects of these proteins on cytoskeletal rearrangements [35], suggesting that the Rho family GTPases can utilize multiple effector pathways to elicit distinct cellular responses. Numerous other candidate targets of Rac and Cdc42 have also now been reported, although the biological role of these interactions has not yet been established [3].

Proteins that bind the activated Rho GTPase have also now been identified [2, 3]. Rho-GTP has been implicated in the activation of several phospholipid kinases, including PIP-4,5 kinase [36] and PI-3 kinase [37, 38], and phospholipase D [39]. Several protein kinases, including p120-PKN [40, 41], p140-PRK2 [42, 43], ROKα and ROKβ [44–46] have been reported to associate specifically with the activated GTP-bound form of Rho. These kinases, which share limited structural similarity with PKC and myotonic dystrophy kinase proteins, appear to be effector targets of the activated Rho GTPase. Thus, analogous to the Ras-Raf interaction, ROKα is translocated to the plasma membrane of cells in which an activated mutant form of RhoA is expressed [45]. This kinase was recently also reported to mediate Rho-induced actin stress fiber formation in cultured HeLa cells, suggesting that the Rho-kinase interactions are important for the biological functions of Rho proteins [44]. It is likely that the substrates of these kinases then serve as the next link in the signaling cascade that mediates the biological effects of the activated Rho protein. However, the organization of such effector pathways has yet to be elucidated.

2.2 Genetic Analysis of Signal Transduction in Simple Organisms

Much of what is now known about the organization of the Ras-mediated signaling pathway comes from the genetic analysis of Ras in relatively simple organisms, such as flies and worms. For example, the developing *Drosophila* eye has proved to be a powerful system for the genetic analysis of Ras regulation and function [47–49]. The *Drosophila* compound eye is a highly organized structure that consists of a discrete unit, the ommatidium, which is repeated about 800 times per eye. An attractive feature of the *Drosophila* eye as an experimental system is the fact that even very subtle disruptions in its normal structure and development are easily detectable as "rough eyes" by light microscopic observation of intact flies. This feature has served as the basis for mutational screens to identify genes that are essential for normal eye development. In one such analysis, a requirement for a Ras-mediated pathway to regulate differentiation of the R7 photoreceptor cell was revealed [47, 48]. Additional genetic and biochemical studies have led to the identification of several protein components of this pathway including Ras itself [47], an R8 cell-expressed ligand (Boss) [50], a transmembrane tyrosine kinase receptor (Sevenless) [51], a nucleotide exchange factor (Sos) [47], a GTPase activating protein (Gap1) [52], an adaptor protein (Drk) [53], several protein kinases (Raf, Dsor, Rolled, and Ksr) [54], and nuclear transcription factors (Yan, Pointed) [55]. One of the most important outcomes of these studies was the discovery that the components of this *Drosophila* Ras-mediated signaling pathway are each remarkably similar to proteins that appear to function in an analogous mammalian Ras pathway (Fig. 1). Thus, these results have validated the use of such genetic screens in simpler organisms to dissect the organization and function of signaling pathways.

Much of the work described in the following sections highlights the importance of genetic model systems, such as *Drosophila*, in elucidating the organization and function of signaling pathways, such as those mediated by the Rho family GTPases. There are two general strategies that are routinely used to explore gene function in *Drosophila*. The identification of mutations that specifically disrupt the normal function of a particular *Drosophila* gene allows the investigator to examine resulting phenotypes. Alternatively, the expression of dominant forms of a particular gene in transgenic flies can similarly be used to identify biological functions associated with specific proteins. An advantage of the transgenic approach is that expression can be targeted to specific tissues, such that the role of a protein within particular tissue types can be explored. Many studies in which this strategy has been used rely on the UAS-GAL4 expression system [56]. In this system, a gene of interest is placed under the control of the yeast UAS regulatory sequences, which are not normally active in *Drosophila*. However, transgenic flies that contain such constructs can be mated to flies that harbor any of a large number of characterized *Drosophila* lines that specifically express the yeast GAL4 transcriptional activa-

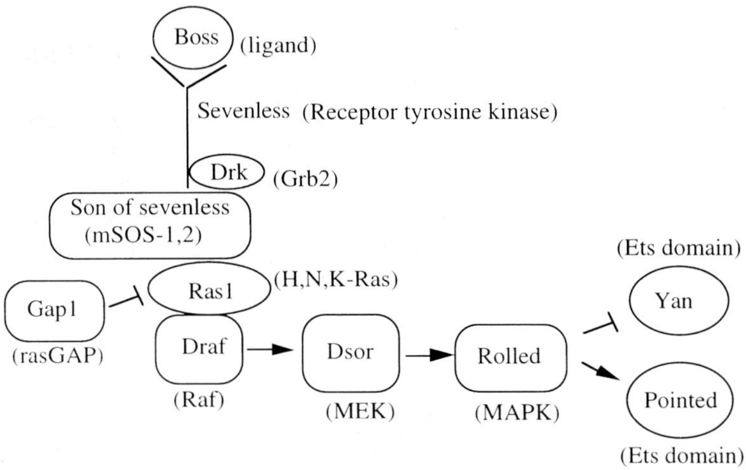

Fig. 1. Organization of the Ras-mediated signaling pathway that determines the fate of the R7 photoreceptor in *Drosophila*. Names in *parentheses* indicate mammalian homologs

tor (which recognizes the UAS sequence) in various tissues. As described in the following sections, many laboratories have now taken advantage of the UAS-GAL4 system to examine the role of Rho family GTPases in *Drosophila* development.

3
Genetic Analysis of Rho GTPases in *Drosophila* Development

3.1
Expression of Rho Family GTPases in *Drosophila* Embryogenesis

In *Drosophila*, five Rho family gene homologs, *rho1*, *rac1*, *rac2*, *cdc42*, and *rhoL* have now been identified [57–59]. The encoded GTPases are 70–90% identical to their mammalian counterparts, indicating substantial evolutionary conservation. Thus far, loss-of-function mutations for these genes have only been reported for *rho1* [60], and future efforts will certainly be directed toward the identification of mutants for each of these genes. The *Drosophila rho1* gene is expressed widely throughout embryogenesis, indicating a likely role in development [58]. Hypomorphic loss-of-function *rho1* mutant alleles have been described, and homozygous mutant *rho1* embryos exhibit an "anterior open" defect in dorsal closure (described in a later section) [60], suggesting a specific role for Rho1 in embryonic epidermal cells. The *Drosophila rac2* gene is also expressed widely throughout embryogenesis, whereas the expression of *rac1* is somewhat more restricted [57, 58]. Initially, *rac1* is also widely expressed, but is more specifically expressed in the mesoderm, the gut, and the nervous system later in development [57, 58]. *cdc42* expression is qualitatively similar to that of *rac1* [57]. The expression of *rhoL* is maintained through all developmental stages [59], although the tissue specificity has not yet been examined. Thus, the widespread expression of these various Rho family GTPases during *Drosophila* development, together with their well documented role as regulators of actin cytoskeleton suggests likely roles for these proteins in morphogenesis. Indeed, as described in the following sections, the requirement for the various *Drosophila* Rho family proteins in a variety of developmental processes has now been documented.

3.2 Oogenesis

Drosophila oogenesis provides a convenient model system in which to study actin-mediated developmental processes [61]. The *Drosophila* egg chambers are each comprised of 15 nurse cells and a single oocyte. These cells remain connected together throughout oogenesis by actin-rich bridges, referred to as ring canals. In addition to the ring canals, filamentous actin is also seen at high concentrations at the cortical regions of the nurse cells and the oocyte, and in adherens type junctions that form between the germ cells of the egg chamber and a layer of somatic epithelial cells known as follicle cells which surround the egg chamber.

During the process of oogenesis, several actin cytoskeleton-mediated morphogenetic events take place [61]. These include the "dumping" of nurse cell contents into the oocyte via the ring canals, which depends on the actin-myosin contractile system and is preceded by actin polymerization around the nuclei of the nurse cells. In addition, the follicle cells undergo characteristic migrations and cell shape changes during oogenesis that may be cytoskeleton-mediated. The identification of oogenesis-defective mutants associated with genes that encode actin-associated cytoskeletal components, such as Singed (fascin) [62], Quail (villin-like) [63], and Chickadee (profilin) [64], highlights the important role of the actin cytoskeleton in this developmental process. Thus, it was not surprising to find a role for several of the Rho family GTPases in oogenesis. These studies relied on the UAS-GAL4 system to drive expression of dominant negative GTPases in specific cell types. Specifically, dominant-negative forms of Rac1, Cdc42, and RhoL were expressed in subsets of follicle cells [59]. When expressed in border cells, a subset of follicle cells that normally migrate to the anterior tip of the egg chamber midway through oogenesis, dominant-negative Rac1, but not Cdc42 or RhoL, caused a nearly complete block in border cell migration. When dominant-negative Rac1 was expressed in other follicle cells, no effects were observed, suggesting a specific role for Rac1 in border cell migration. Expression of activated Rac1, on the other hand, does not cause detectable effects on migration.

The effects of mutant forms of Rac1, Cdc42, and RhoL on oogenesis were also examined by expressing them in the germline using a heat-shock promoter [59]. Expression of activated Cdc42 following heat shock causes a severe defect in nurse cell fusion that is associated with alterations in the normal distribution of filamentous actin at regions of cell-cell contact between adjacent nurse cells and in ring canals. A dominant-negative form of Cdc42 also disrupts nurse cell fusion as well as cytoplasmic transfer of nurse cell contents, whereas dominant-negative Rac1 blocks nurse cell transfer as well as border cell migration. Expression of activated or dominant-negative forms of the RhoL protein disrupts nurse cell-follicle cell contacts and nurse cell transfer. Taken together, these findings indicate that while all three of these GTPases are required for the normal transfer of nurse cell contents to the oocyte, these related proteins additionally perform distinct functions during oogenesis that are restricted to specific cell types.

3.3 Gastrulation

Early in *Drosophila* embryogenesis, immediately following the cellularization of the syncitial blastodermal embryo, the process of gastrulation is initiated [65]. This developmental event, which is evolutionarily conserved among higher eukaryotes, results in the internalization of mesodermal and endodermal tissue. In *Drosophila*, the major morphogenetic events associated with gastrulation include ventral furrow formation and invagination of the anterior and posterior midgut [65–67]. The ventral furrow forms in order to internalize the presumptive mesodermal cells in a process that is largely dependent on cytoskeleton-mediated cell shape changes within a group of mesodermal precursors located along the length of the mid-ventral surface of the embryo [68]. Subsequently, similar cell shape changes in the endodermal primordia give rise to invaginations of the anterior and posterior midgut [68]. The requisite cell shape changes in these processes have been well documented, and involve cytoskeleton-mediated constrictions of the apical membranes [69, 70], followed by a cell shortening event [66].

Recently, independently generated results from two groups established a role for a Rho-mediated signaling pathway in the cellular shape changes that drive these invaginations [71, 72]. Initially, genetic screens led to the identification of the gene, *dRhoGEF2*, which encodes a predicted RhoGEF, as being essential for the cell shape changes required during gastrulation. Moreover, it was observed that expression of a dominant-negative form of the Rho1 GTPase in early embryos similarly disrupts both ventral furrow formation and midgut invagination, suggesting that activation of the Rho1 GTPase by DRhoGEF2 is critical to the initiation of the required cell shape changes [71, 72].

Previous studies of *Drosophila* gastrulation revealed a role for two genes that also encode signaling molecules – *folded gastrulation* (*fog*) [73] and

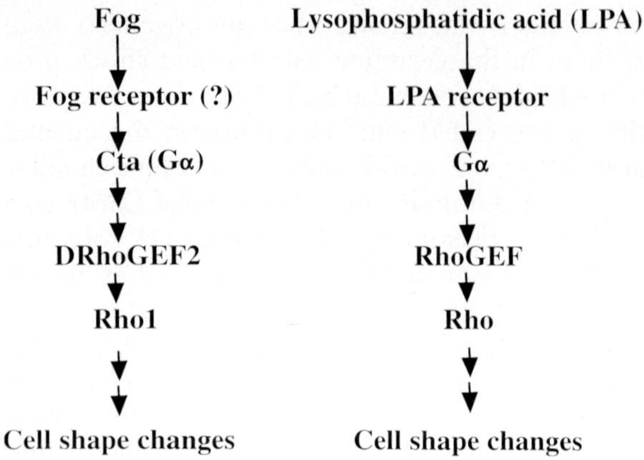

Fig. 2. Evolutionary conservation of a Rho-mediated signaling pathway for the cell shape changes required for *Drosophila* gastrulation (*left*) and for neuronal cell shape changes in mammals (*right*)

concertina (*cta*) [74]. Mutations in either of these genes give rise to gastrulation defects similar to those seen in the absence of DRhoGEF2. Fog is an extracellular ligand that binds to an as yet unidentified cell surface receptor and Cta is a probable Ga-subunit of a heterotrimeric protein that is believed to be coupled to the Fog receptor. Thus, it has been suggested that Fog initiates a signaling pathway mediated by a G-protein that includes Cta and which ultimately directs the appropriate cell shape changes [73]. Indeed, it has recently been demonstrated that cell shape changes induced by ectopic Fog expression are diminshed in the absence of Cta [75]. Moreover, the effects of ectopic Fog expression are eliminated in the absence of DRhoGEF2 [71]. Together, these observations suggest that the cell shape changes required during *Drosophila* gastrulation are regulated by a Rho-mediated signaling pathway in which a Fog-initiated signal leads to Rho activation via a G-protein mediated activation of RhoGEF activity. An analogous pathway has been described in mammalian cells in which lysophosphatidic acid (LPA) activates the G-protein-coupled LPA receptor, leading to subsequent activation of Rho via a RhoGEF [22]. Thus, it appears that evolution has conserved a Rho-mediated cell shape change pathway that is activated by G-protein-coupled cell surface receptors (Fig. 2). Future studies will be needed to determine the mechanism by which RhoGEF activity is promoted in response to activation of the cell surface receptor.

3.4 Dorsal Closure

Later in *Drosophila* embryogenesis, the process referred to as dorsal closure establishes the dorsal ectoderm [76, 77]. This major morphogenetic event results in the fusion of the two halves of lateral epidermis from each side of the embryo at the dorsal midline [77]. Following earlier major morphogenetic movements of the embryo, including gastrulation and germband extension and retraction, a hole is left in the dorsal side of the embryo that is covered over only by a thin transient epithelial structure, the amnioserosa, that ultimately degenerates. If dorsal closure fails to occur, an obvious hole remains in the ectodermal-derived dorsal cuticle of the embryo, giving rise to the so-called dorsal open phenotype, which has been observed in several *Drosophila* mutants [78].

The dorsal closure process requires cytoskeleton-mediated cellular shape changes in the presumptive ectoderm, and proceeds through a series of events that can be described as initiation, spreading, and fusion [78]. Significantly, like gastrulation, dorsal closure relies exclusively on cell shape changes as opposed to new cell divisions, highlighting the importance of cytoskeletal reorganization in the major morphogenetic events of *Drosophila* embryogenesis [77]. The isolation of *Drosophila* mutants that are associated with a dorsal-open phenotype has played an important role in identifying the protein components of the dorsal closure process [78]. Approximately 20 genes have already been identified that appear to be required for this process to occur normally. Some of these, such as *zipper* (non-muscle myosin heavy chain subunit) [79] and *coracle* (4.1 homolog) [80] encode cytoskeletal components that are likely to participate directly in the mechanical aspects of the required epidermal cell shape changes. Other genes required for dorsal closure include several components of a Jun kinase cascade that leads to new gene expression (described below) [78], indicating that signal transduction pathways regulate this process.

Early in the process of dorsal closure, a single row of ectodermal cells, the leading edge cells, which border the ectoderm and the amnioserosa, undergo an elongation along the dorsal-ventral axis. Subsequently, the more lateral ectodermal cells also begin to elongate, accompanied by a dorsal-ward movement of the lateral ectoderm on both sides of the embryo. Ultimately, the two rows of leading edge cells meet and fuse at the dorsal midline in a zippering-like process that results in the closing of the dorsal opening. This process, together with involution of the head, constitute the final major morphogenetic movements of *Drosophila* embryogenesis [76].

Strong evidence has recently been obtained that indicates that signal transduction pathways in both the leading edge cells and the more lateral ectodermal cells regulate the cell shape changes required for dorsal closure. Not surprisingly, these pathways appear to be mediated by memebers of the Rho GTPase family. Although loss-of-function mutants of the *Drosophila* Rac and Cdc42 GTPases have yet to be reported, expression of dominant-negative forms of Rac1 or Cdc42 either in the leading edge cells or throughout the ectoderm disrupt dorsal closure [81–83]. Moreover, in embryos expressing dominant-negative Rac1, a defect is observed in the accumulation of actin and non-muscle myosin in the leading edge cells [83], suggesting that the normal cell shape changes require a Rac-mediated signaling pathway that directs cytoskeletal reorganization.

Previously, it had been observed that mutants in several components of the MAP kinase related-Jun kinase (JNK) cascade, including Hemipterous (Jun kinase kinase), Basket (Jun kinase), Djun, and Dfos (substrates of JNK), are each associated with a similar dorsal open phenotype [78]. Together with the observation that mammalian Rac and Cdc42 GTPases can function as upstream activators of the mammalian JNK pathway, it seemed reasonable to postulate that the role of these GTPases in *Drosophila* is to activate this kinase cascade in dorsal closure. Indeed, expression of activated forms of Rac1 and Cdc42 during embryogenesis results in ectopic activation of the JNK pathway, as revealed by expression of the JNK pathway target genes, *dpp* and *puckered* [81]. In addition, the effects of ectopic JNK activation by Rac1 and Cdc42 are suppressed by a loss-of-function *hemipterous* mutant, suggesting that *Drosophila* Rac1 and Cdc42 function as upstream activators of the JNK pathway in the dorsal closure process [81]. In this postulated pathway, Rac and Cdc42 mediate a transcriptional activation signal that leads to the expression of particular genes in the leading edge cells. The product of one of these, Dpp, is secreted, and is able to activate yet another signaling pathway in neighboring lateral epidermal cells, via direct interaction with TGF-B-like receptors (Punt and Thick veins) [84], which are also required for normal dorsal closure [85]. Since expression of dominant-negative Rac or Cdc42 in the leading edge cells blocks their elongation, in addition to affecting gene expression, it is possible that Rac and Cdc42 also mediate a signaling pathway in the leading edge cells that affects actin rearrangements in a manner that does not require new gene expression.

A *Drosophila* homolog of the mammalian Rac/Cdc42 target, PAK (p21-activated kinase), named Dpak, has also been implicated in dorsal closure [83]. Although a *dpak* mutant has not yet been identified, Dpak protein has been reported to accumulate in the leading edge cells during dorsal closure [83]. Moreover, this accumulation is blocked in embryos expressing dominant-negative Rac, suggesting that Rac might signal to Dpak during dorsal closure, possibly affecting the subcellular localization of Dpak [83]. Whether or not Dpak, which may bind activated Rac directly, functions as an intermediate between Rac and Hemipterous to activate the JNK pathway for dorsal closure remains to be determined.

The *Drosophila* Rho1 GTPase has also been implicated in dorsal closure. *rho1* loss-of-function mutants have been identified, and embryos homozygously mutated for *rho1* (hypomorphic alleles) exhibit a dorsal-open phenotype that is restricted to the anterior region of the embryo [60]. This relatively milder dorsal closure defect may indicate a partial functional redundancy with other related Rho family GTPases, or alternatively, that Rho1 mediates a signaling pathway distinct from that mediated by Rac and Cdc42 that is also required for some aspect of dorsal closure. Although the precise role of Rho1 in this process is not yet understood, recent results from our laboratory have revealed that a *Drosophila* homolog of the mammalian Rho-specific target, PKN (Protein Kinase N), referred to as DPKN [86], is also required for dorsal closure, suggesting that DPKN mediates a signal from Rho1 during this process. Possibly, Rho1 mediates a cell shape change pathway that affects actin organization, and which is corrdinated with a Rac/Cdc42-mediated pathway that leads to new gene expression (Fig. 3).

Identifying the upstream signaling components that activate the Rho family proteins in the leading edge cells should provide some important clues regarding the regulation of dorsal closure. One or more GEFs for the Rho, Rac, and Cdc42 GTPases are likely to be involved, although the relevant GEF has yet to be identified. In addition, there may be cell surface receptors and/or adaptor proteins that are also involved. One such candidate is the protein, Myoblast City [87, 88], which is also required for dorsal closure, and, as described in a later section, appears likely to function as a specific Rac activator in the dorsal closure process.

3.5 Tissue Polarity

Epithelial cells of all organisms become polarized in order to localize regions of the cell that are responsible for specialized functions and to maintain their appropriately differentiated state [89]. The establishment of polarity appears to be dependent on cellular responses to positional cues from the environ-

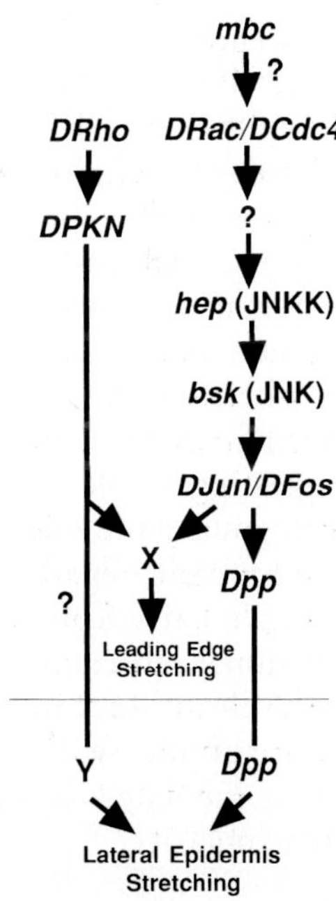

Fig. 3. Model for a putative signaling pathway in which distinct Rho- and Rac-mediated signals function coordinately to bring about cell shape changes in the leading edge cells of the *Drosophila* epidermis during the developmental process of dorsal closure. It has been suggested that the Myoblast city protein (mbc) serves as an upstream Rac activator in this pathway

ment. For example, in mammalian epithelial cells it has been demonstrated that extracellular matrix proteins provide a signal that is used to determine the apical-basal polarity axis [90] . In addition, epithelial cells organize themselves through the formation of intercellular junctions, such as cadherin-mediated adherens junctions, that are associated with dynamic actin structures [91]. Numerous studies with cultured epithelial cells, such as MDCK cells, have confirmed the role of specific actin rearrangements during the establishment of polarization [92]. More recently, the Rac and Rho GTPases have been directly implicated in this process with the demonstration that dominant-negative forms of these GTPases block cadherin-mediated cell-cell adhesion in cultured epithelial cells, and disrupt the reorganization of actin that normally occurs at adherens junctions [93].

In *Drosophila* development, epithelial polarity plays an important role in morphogenesis [94]. The epithelial tissues of *Drosophila* are derived from the invaginated embryonic ectoderm, and take the form of imaginal discs which differentiate during larval development, ultimately giving rise to the organ structures of the mature adult. In addition to the apical-basal polarity that can be observed in epithelial cells in an in vitro culture setting, cells in epithelial tissues also develop a planar polarity relative to the body axis [94]. For example, in the fly wing, the regular array of wing hairs is oriented distally on the wing blade due to the establishment of epithelial planar polarity. In studies in which mutant forms of the Rac1 and Cdc42 GTPases were expressed specifically in the wing imaginal disc in transgenic flies with the UAS-GAL4 system, distinct effects of these proteins on cellular processes that lead to the establishment of polarity were observed [95]. For example, Rac1 activity was found to be essential for the proper assembly of actin at adherens junctions as well as for the establishment of planar polarity. However, Cdc42 was found to be required for epithelial cell shape changes, but is not required for actin assembly at adherens junctions. Significantly, neither of these proteins was found to be required for the maintenace of polarity, once established.

Additional studies that further addressed the precise role of Rac and Cdc42 in planar cell polarity in wing hair formation revealed that Rac1 and Cdc42 regulate distinct aspects of hair formation by affecting different subsets of the cytoskeleton [96]. Specifically, expression of dominant-negative Cdc42 stunts wing hair growth, and such hairs fail to accumulate actin in their distal regions, suggesting a role for Cdc42 in polarized membrane outgrowth that may be analogous to the role of mammalian Cdc42 in filopodial formation. Expression of dominant-negative Rac1, on the other hand, results in the outgrowth of multiple hairs from a single cell. In addition, those hair cells exhibit gaps in junctional actin and a disorganization of their apical microtubule web, indicating a role for Rac1 in actin polymerization as well as microtubule organization. Future studies will certainly aim to elucidate the organization of Rac- and Cdc42-mediated signaling pathways that regulate these specific aspects of tissue polarity.

3.6
Eye Development

By targeting overexpression of wild type Rho1 to the cells of the developing *Drosophila* eye in transgenic flies, it was demonstrated that excessive Rho1 activity perturbs the normal development of the eye in a dosage sensitive manner [58]. With a single copy of a *rho1* transgene, the external surface of the eye is not detectably altered, however retinal sections revealed a mild disorganization of the ommatidial array. With two copies of the *rho1* transgene, the effects on eye development are more severe. Specifically, defects in the morphology of photoreceptors are observed, in addition to a defect in the post-mitotic elongation of photoreceptors that normally takes place near the completion of eye development. The lattice formed by secondary and tertiary pigment cells as well as the ommatidial architecture is completely disrupted. Furthermore, these defects are associated with a disruption of the normal appearance of polymerized actin in pupal eye discs, suggesting that the ability of Rho1 to affect actin organization plays an important role in the morphogenesis of the *Drosophila* eye [58]. Significantly, no alterations in cell fate determination or cell number were observed in the Rho1 transgenics, suggesting that Rho1 activity may be specifically required for cytoskeleton-mediated morphogenesis in the developing eye.

More recently, with the identification of loss-of-function alleles of the *rho1* gene, it has been possible to examine the precise requirement for this GTPase in eye development. Clonal analysis of *rho1* null alleles in imaginal discs revealed that Rho1 is essential for cell growth, however, hypomorphic alleles of *rho1* in somatic eye clones give rise to incorrectly rotated ommatidia, indicating a role for Rho1 in establishing tissue polarity [60]. Similarly, wing clones of hypomorphic *rho1* alleles were associated with abnormal wing hair polarity. Furthermore, examinantion of genetic interactions between Rho1 mutants and two genes previously associated with similar polarity defects, *frizzled* (*fz*) and *dishevelled* [97], revealed that Rho1 is likely to mediate a polarity signal transduced by the activated Fz receptor via the Dishevelled protein [60]. Since Fz appears to be a G-protein-coupled receptor [98], this pathway of Rho1 activation in *Drosophila* may be analogous to a proposed G-protein-mediated pathway for Rho activation, required for neuronal cell shape changes in mammalian neurons in response to LPA [22]. It also appears that the Jun kinase cascade mediates the Fz-Rho1-dependent tissue polarity signal [99], raising the possibility that Rac1 or Cdc42 (both of which can activate the Jun kinase cascade) may also function in establishing tissue polarity in the eye. Thus, it appears that the Rho, Rac, and Cdc42 proteins each participate in some aspect of tissue polarity.

3.7 Neural Development

A role for the Rho family GTPases in neural development has been suggested by a variety of cell culture studies in mammalian as well as insect systems [20, 24]. In both cases, it has been observed that the so-called growth cones that develop at the tips of growing axons are rich in highly organized actin that is quite dynamic during the growth and guidance of axons [100]. Indeed, the rearrangements of actin seen in migrating axonal growth cones are reminiscent of the growth factor-induced changes in actin organization in cultured fibroblasts [20, 100]. Moreover, growth cones are rich in filopodial and lamellapodial-like membrane protrusions similar to those seen in Cdc42- and Rac-injected fibroblasts, respectively. Recently, these analogies have been validated in microinjection experiments with cultured neurons, in microinjection experiments with cultured neurons, in which roles for Rho, Rac, and Cdc42 proteins were found in several growth cone functions, including retraction, collapse, outgrowth, and motility [19, 23, 24].

During later stages of *Drosophila* embryogenesis, Rac1 and Cdc42 are expressed at relatively high levels in the *Drosophila* nervous system [57], suggesting a possible role in neural development. Indeed, in studies in which dominant-negative Rac1 was expressed specifically in neuroblasts and in mature neurons of the central and peripheral nervous systems using the UAS-GAL4 system, embryo lethality was observed. The lethality is associated with severe axonal loss that can be partially rescued by expressing wild type Rac1 [57], suggesting a role for Rac1 in the initiation of axon outgrowth. Expression of activated Rac1 is associated with even greater axonal loss in neurons of both the central and peripheral nervous systems. By expressing activated Rac1 at later times in neural development using a different GAL4 driver, it was observed that excessive Rac1 activity affects axon elongation in addition to initiation [57]. However, other aspects of neural development, such as neuronal differentiation are unaffected by expression of mutant forms of Rac1. Interestingly, the effects of expressing Rac1 mutants in the nervous system are restricted to axons, since dendrites appear to be unaffected. Since neurons are highly polarized cells, this observation has been interpreted to indicate that Rac1 may play an early role in establishing neuronal asymmetry by specifically affecting the initiation of axonal outgrowth [57].

Significantly, expression of activated Rac1, but not dominant-negative Rac1, in neurons is associated with abnormal axonal accumulation of filamentous actin, suggesting that the role of Rac in axon outgrowth is to regulate actin organization [57]. In light of the known ability of Rac to affect actin organization in fibroblasts, and the role of dynamic actin containing structures in axonal growth cone function, it is likely that the observed effects of Rac mutants on axonal growth in the *Drosophila* nervous system reflect a critical role for Rac in regulating actin dynamics in the growth cone. Recently, in a separate study, a specific role for Rac1 in motor axon guidance was identified

[101]. In this report, it was observed that expression of a dominant-negative form of Rac1 in neurons causes axons associated with the intersegmental nerve b to bypass and extend beyond their normal synaptic targets. Moreover, this phenotype is enhanced in the presence of a mutation in the receptor-tyrosine phosphatase, DLAR, which has also been implicated in motor axon guidance [102]. Interestingly, it has been reported that the mammalian homolog of DLAR interacts directly with a protein called Trio, which contains two Rac/Rho GEF-like domains [16]. Thus, it is possible that a DLAR-mediated axon guidance signal at the cell surface activates a Trio-like GEF activity toward the Rac1 GTPase.

In addition to Rac1, the Cdc42 GTPase has also been implicated in nervous system development. Specific expression of activated Cdc42 in the nervous system using the UAS-GAL4 system results in 100% embryo lethality [57]. However, the effects of Cdc42 on neuronal morphology are qualitatively distinct from the effects of activated Rac1. Firstly, activated Cdc42, unlike Rac1, inhibits both axon and dendrite outgrowth. Moreover, the effects of activated Cdc42 appear to be more general, and include effects on neuronal position in addition to axon growth. Possibly, this reflects an additional specific role for Cdc42 in neuronal migration. Significantly, these results are consistent with distinct roles for the closely related Rac and Cdc42 GTPases in vivo. Because it is somewhat difficult to make firm conclusions regarding the normal functions of these proteins based on phenotypes observed upon expression of mutant forms under heterologous promoters, it will be very useful to determine the loss-of-function phenotypes associated with specific mutations in Rac1 and Cdc42, once they are generated.

Thus far, the signaling pathways that are mediated by the Rac and Cdc42 proteins during neural development have not been elucidated. However, a *Drosophila* RhoGEF family member, called Still life, has recently been identified in a mutational screen for motor activity defects [103]. Homozgous mutant *still life* flies exhibit reduced locomotion as well as male sterility. At late stages of embryogenesis, *still life* transcripts are only expressed in neurons, and antibody staining revealed that the protein is predominantly localized to synaptic regions, suggesting a potential role in synaptic differentiation. In transgenic flies that express a putative activating form of Still life specifically in neurons, defects were observed in motor axon elongation and on the formation of synaptic arbors on target muscles.

3.8 Muscle Development

As in vertebrates, a component of the myogenic program leading to normal muscle development in *Drosophila* is the fusion between myoblasts that gives rise to mature muscle fibers consisting of multinucleate syncitia. Previous studies have revealed a role for actin filament assembly in the myoblast fusion process, potentially implicating Rho family GTPases [104]. This hypothesis was indeed substantiated by the finding that expression of activated or dominant-negative forms of Rac1 in mesodermal tissue of transgenic flies gives rise to defects in myoblast fusion during embryogenesis [57]. Specifically, dominant-negative Rac1 expression causes a delay in myoblast fusion initially, followed by excessive and inappropriate fusion. On the other hand, expression of an activated Rac1 results in complete inhibition of myoblast fusion. Moreover, high levels of *rac1* mRNA are expressed in mesoderm at the relevant developmental stage, suggesting that the endogenous Rac1 protein normally mediates this process. Analagous experiments with mutant forms of Cdc42 did not reveal a role for Cdc42 in this process, indicating that, despite the strong similarity between the Rac and Cdc42 proteins, they mediate distinct biological functions [57]. Although these studies have indicated that a Rac-mediated signaling pathway is likely to be required for normal myoblast fusion, it has not yet been demonstrated that the role of Rac in this process is to affect the organization of the actin cytoskeleton. Thus, neither the precise role of Rac1 in this process nor its associated signaling pathway have yet been elucidated.

In addition to a requirement for Rac GTPase activity, several *Drosophila* loss-of-function mutants have been identified that are specifically associated with a defect in myoblast fusion. These include *rolling stone* [105], *blown fuse* [106], and *myoblast city* (*mbc*) [87]. Although the predicted coding sequences of blown fuse and rolling stone have not revealed clues regarding their biochemical or cellular function, *mbc*, as described earlier, appears to be the *Drosophila* homolog of the mammalian DOCK180 signaling protein [88, 107], and is also required for the Rac1-dependent developmental process of dorsal closure [88]. Thus, it appears that Rac1 and Mbc may function together in a signaling pathway utilized for both dorsal closure and myoblast fusion. Although the nature of their interaction is not yet clear, the association of the Mbc homolog, DOCK180, with the Crk adaptor protein, which also complexes with the mammalian RacGEF, Vav [108], raises the possibility that Mbc functions as an upstream activator of the Rac GTPase. Possibly, formation of an Mbc-containing signaling complex results in the juxtapositioning of the Rac1 GTPase and a RacGEF. Epistasis experiments with *rac1* and *mbc* have yet to establish their relationship in the signaling pathways that direct dorsal closure or myoblast fusion. The eventual identification of loss-of-function mutants of the Rac1 GTPase will certainly be useful for such analysis.

4
Rho Family GTPases in *C. elegans*

4.1
Conservation of Rho GTPases and their Regulators and Targets in Worms

The nematode *C. elegans* provides a powerful biological system that is amenable to both genetic and developmental studies. The rapid reproductive cycle and relatively simple body plan of the worm, combined with the availability of a complete cell lineage description and a nearly complete genome sequence provides a convenient experimental tool for the analysis of such fundamental biological processes as signal transduction, cell cycle regulation, and morphogenesis. Indeed, several components of a Ras-mediated signaling pathway that regulates vulval development in *C. elegans* have been identified through genetic analysis [109]. A significant feature of *C. elegans* development is that the embryo maintains a constant volume throughout embryogenesis [110]. Thus, the later stages of embryogenesis rely almost exclusively on morphogenetic rather than proliferative events. Moreover, much of the observed morphogenesis appears to be cytoskeleton-mediated [111, 112], suggesting a likely role for Rho family GTPases and their associated signaling pathways.

Thus far, five Rho family small GTPases have been identified in *C. elegans*. These include three Rac-like proteins, a Cdc42-related protein, and a Rho-related protein [113–117]. In addition, *C. elegans* homologs of the Rac/Cdc42 target kinase, PAK (CePAK) [118], and the Rho-kinase [119], have been identified and partially characterized. Database searching through the worm genome reveals numerous other Rho family regulators, including GAPs, GEFs, and GDIs that exhibit substantial sequence similarity to their mammalian counterparts. Thus, it seems that *C. elegans* will certainly be a very powerful model system for investigating the in vivo organization and biological functions of Rho-associated signaling pathways.

4.2
Embryo Elongation

One of the major morphogenetic events of *C. elegans* embryogenesis is embryo elongation. During this process, the hypodermal cells, which are a subset of epidermal cells that surround the embryo, "squeeze" the embryo in order to cause it to elongate [112]. These cells contain actin as well as microtubule bundles which undergo contraction in such a way as to maintain the shape of the embryo during elongation. In expression studies, it was demonstrated that CeRac1, CeCdc42, and CePAK all co-localize at hypodermal cell boundaries during embryo elongation [118]. Thus, it has been suggested that the worm Rac and Cdc42 GTPases may transduce a signal to the CePAK kinase that directs the requisite actin rearrangements.

Although mutational analysis of these GTPases and their associated signaling pathways in the worm is very much in its early days, the utility of this system in dissecting the role of such pathways in development has already been demonstrated. In studies aimed at identifying embryo elongation-defective *C. elegans* mutants, genes were identified that encode a homolog of the mammalian Rho-kinase, named *let-502*, and a homolog of the regulatory subunit of myosin light chain phosphatase, named *mel-11* [119]. Recently, studies of the corresponding mammalian proteins revealed that Rho-kinase phosphorylates and inactivates myosin light chain phosphatase as a regulatory mechanism for the actin-myosin based contractile system [120]. Significantly, *let-502* is highly expressed in the hypodermal cells at the time when embryo elongation is normally initiated and *let-502* mutant embryos fail to elongate due to defects in hypodermal cell shape changes [119]. Thus, these results raise the possibility that Rho transduces a signal to the Rho-kinase (let-502] that regulates hypodermal cell fusion by influencing the actin-myosin contractile system.

In a screen to identify extragenic suppressors of the *let-502* mutation, the *mel-11* gene (myosin light chain phosphatase regulatory subunit) was identified [119]. This finding suggests that, as in mammalian cells, the *C. elegans* Rho-kinase negatively regulates myosin light chain phosphatase. In the proposed scenario, a Rho-mediated developmental signal activates the let-502 kinase, leading to phosphorylation and inactivation of myosin light chain phosphatase (and possibly myosin regulatory light chain) and subsequent activation of the actin-myosin contractile system, thereby resulting in embryo elongation. Moreover, it appears that a delicate balance between the activities of myosin light chain kinase and phosphatase determines the state of contraction or relaxation of the hypodermal cells, and that Rho is a critical regulator of this balance.

4.3
Neural Development

The *C. elegans* nervous system is relatively simple, being comprised of 302 neurons, most of which send processes to the nerve ring, an information integrating center that encircles the pharynx [121]. Thus, the worm is ideally suited for genetic and developmental studies of the nervous system. Thus far, unlike for *Drosophila*, the necessary genetic tools have yet to be generated to determine the requirement for Rac and Cdc42 activity in *C. elegans* neural development. However, expression studies of the *C. elegans* homolog of the mammalian RhoA GTPase, CeRhoA, have implicated this protein in establishing the sensory circuitry during post-embryonic neural development [113]. Specifically, immunostaining studies revealed that the CeRhoA protein is enriched in cells of the anterior sensory organ at the tip of the head and in the nerve ring during the post-embryonic larval stage of development. Other neurons, such as the motor neurons and inter-neurons do not exhibit CeRhoA expression, suggesting a specific role for this GTPase in a subset of neurons.

In a mutational screen to identify *C. elegans* genes that are required for neuroblast migration, *mig-2* was isolated, and was found to encode a Rac-related GTPase [117]. Possibly, the mig-2 GTPase functions in cell motility in a manner analogous to that reported for the mammalian Rac GTPase. In the *C. elegans* screen, multiple alleles of *mig-2* were identified, some of which were found to be activated alleles that are predicted to encode proteins that are defective for GTPase activity or GAP sensitivity. Interestingly, both the activated and loss-of-function alleles of *mig-2* exhibit similar cell migration defects. However, in addition to the cell migration phenotype, activated *mig-2* alleles are also associated with defects in axon guidance. Specifically, the axonal pathfinding ability of the hermaphrodite-specific neuron (HSN) is disrupted, resulting in a "wandering axon" phenotype in some cases and an axonal truncation phenotype in others. Possibly, the mig-2 GTPase mediates both axonal outgrowth and guidance. Other axons in *mig-2* animals appear to be normal, suggesting that a requirement for mig-2 in axon guidance is potentially limited to a single neuron. Alternatively, redundant function of other Rho-related GTPases might mask potential phenotypes in additional neurons. Although the upstream regulators of mig-2 have not been identified, it is intriguing to consider that the *C. elegans* axon guidance defective mutant, unc-73, encodes a predicted RhoGEF activity [122], raising the possibility that unc-73 functions to activate the mig-2 GTPase. Indeed, this hypothesis was supported by the observation that double mutants of *unc-73* and *mig-2* exhibit enhanced cell migration defects [117].

5
Rho GTPase Function in Mammals

At the most recent count, there were 12 reported members of the Rho GTPase family in mammals. There are also at least 15 identified mammalian GEFs for the Rho GTPases, 12 RhoGAPs, and 3 RhoGDIs. Putative effector targets of the Rho family proteins number approximately 20 [3]. Although the expression patterns have not yet been reported for all of these genes, it is clear that several of the Rho GTPases and their associated regulators and targets are widely expressed during embryogenesis. Thus, it is likely that Rho-mediated signaling, as is the case in flies and worms, plays an important role in mammalian development. The obvious difficulty with studying the regulation and function of such signaling pathways in mammals is the complexity of the system. The large number of cell types combined with an even larger number of potentially relevant signaling molecules creates a formidable challenge to the elucidation of specific Rho-mediated signaling pathways required during mammalian development. In addition, the potential redundancy that results from the presence of multi-gene families that encode proteins of similar, or possibly identical, function provides yet an additional obstacle in many cases. Nevertheless, as described below, efforts to define these pathways in mammals are underway.

5.1
Knockout Mice

Presently, the best mammalian model system for the analysis of gene function during development is the knockout mouse [123]. The ability to disrupt a single gene through targeted homologous recombination and then assess the developmental consequences provides a powerful tool for dissecting the precise biological role of individual proteins in specific cell types. Although this system will ultimately yield much important information regarding the normal developmental function of Rho family GTPases and their associated signaling components, mouse knockouts of the vast majority of proteins that have been linked to Rho-mediated signaling have not yet been reported in the literature. The power of knockout technology in understanding the developmental requirement for Rho pathway components, however, has already been revealed. Genes corresponding to two of the Rho/Rac family GEFs, Vav and Bcr, have been knocked out in mice. Vav, which is expressed predominantly in hematopoietic cells, is essential very early in development, and homozygous Vav mutants die before the blastocyst stage, indicating a role in embryo implantation [124]. Interestingly, however, in experiments in which embryonal stem cells lacking Vav were used to reconstitute hematopoietic lineages in recipient animals, it was determined that Vav is not essential for either myeloid or erythroid development [125]. In other related knockout studies, a requirement for Vav in negative and positive thymocyte selection was revealed [126, 127].

Bcr, which contains both a GEF domain and functions as a Rac-specific GAP, is also expressed predominantly in hematopoietic lineages, and was found to be non-essential for normal development in mouse knockout studies [128]. However, neutrophils lacking Bcr exhibit an excessive respiratory burst that can lead to septic shock, suggesting a role for Bcr in regulating the activity of the Rac2 GTPase, which has previously been shown to be essential for the activation of NADPH oxidase in these cells. Thus, while Bcr is dispensible for development, it is required for a specific biological function in the mature animal. Mice lacking the lymphoid-specific RhoGDI, known as D4, do not display any detectable developmental defects in lymphoid development or thymocyte selection, suggesting that this molecule may perform a redundant function, or that its activity is only required for some physiological function in the mature animal that has not yet been tested [129].

5.2 Transgenic Mice

Generation of transgenic mice that express dominant forms of the Rho GTPases in restricted tissues may also be a useful way to examine the developmental requirement for particular GTPases in specific cell types. However, this approach, which has been eminently successful in *Drosophila* studies, as described earlier, has only been reported in two studies thus far in mice. In one study, it was found that transgenic mice that express an activated form of Rac1, specifically in the Purkinje cells of the cerebellum, develop ataxia, and the dendritic spines of their Purkinje cells are reduced in size but increased in number and are associated with supernumerary synapes [130]. Thus, this observation corroborates studies of the *Drosophila* Racl GTPase [57] and the *C. elegans* Rac-related mig-2 protein [117], which each appear to function in axonal outgrowth in the developing fly and worm nervous systems, respectively.

In another transgenic study, expression of an activated form of the Rac2 GTPase specifically in thymocytes resulted in thymic atrophy associated with excessive apoptosis, suggesting a role for normal rac activity in regulating cell death in the thymus [131]. In a related transgenic study, the Rho-inhibitory bacterial toxin, C3, was expressed specifically in thymocytes, and the resulting mice exhibited a substantial reduction in the number of thymocytes, indicating a role for Rho in the proliferation or cell survival of T cells [132]. Thus, transgenic mouse studies have revealed a role for both the Rho and Rac GTPases in normal thymus development.

5.3
Diseases of Human Development

In addition to identifying phenotypes associated with mouse knockouts of specific genes or in transgenic mice, occasionally there is an opportunity to learn about the normal function of a gene by linking it to a human disease. Such was the case for another Rho/Rac GEF family member, named FGD1. The FGD1 gene was identified, through a positional cloning effort, as the genetic locus responsible for an inherited X-linked human developmental syndrome known as Aarskog-Scott syndrome, or faciogenital dysplasia [133]. This syndrome is associated with a short stature and facial, skeletal, and urogenital abnormalities. Recently, the FGD1 protein has been found to function as a specific activator of the Cdc42 GTPase [134, 135], suggesting that appropriate Cdc42 regulation is essential for several processes during normal mammalian development. A variety of other developmental syndromes that are associated with defects in morphogenesis and that appear to be due to single gene defects have been described in humans. In light of accumulating evidence indicating a role for the Rho family GTPases in morphogenesis, it will certainly not be surprising to find that many of these syndromes are caused by mutations in genes that encode protein components of Rho-mediated signaling pathways.

6
Summary

It is becoming increasingly clear that the complex family of Rho-related GTPases and their associated regulators and targets are essential mediators of a variety of morphogenetic events required for normal development of multicellular organisms. It is worth noting that the results obtained thus far indicate that the Rho family proteins are largely associated with the regulation of morphogenesis, as opposed to other essential developmental processes such as cell proliferation and cell fate determination. Accumulating evidence also suggests that the role of these proteins and their associated signaling pathways in morphogenesis is in many, but not necessarily all, cases related to their ability to affect the organization of the actin cytoskeleton. Thus, these in vivo observations have served to corroborate similar findings in numerous cultured cell studies.

As described, the power of genetics, particularly in *Drosophila* and *C. elegans*, has been critical to the recent identification and functional characterization of several Rho family signaling components. Moreover, evidence suggests that the highly evolutionarily conserved structures of many of these proteins translate into conservation of function as well. Thus, it will be possible, in many cases, to extrapolate the findings in the simple systems described herein to higher eukaryotes, including humans. Expanding use of these genetic model systems to dissect Rho-mediated signaling pathways in vivo will undoubtedly lead to a flood of new insights into the organization and function of these pathways in the coming years, especially in development. As the *C. elegans* genome sequencing effort nears completion and with the *Drosophila* genome project well underway, the identification of novel relevant genes will proceed with even greater speed. In addition, the rapidly expanding use of mouse knockout strategies, combined with recent developments in the associated knockout technology, will also contribute greatly to the investigation of mammalain Rho signaling pathways and their roles in development.

References

1 Bourne HR, Sanders DA, McCormick F (1990) The GTPase superfamily: a conserved switch for diverse cell functions. Nature 348:125–132

2 Tapon N, Hall A (1997) Rho, Rac and Cdc42 GTPases regulate the organization of the actin cytoskeleton. Curr Opin Cell Biol 9:86–92

3 Van Aelst L, D'Souza-Schorey C (1997) Rho GTPases and signaling networks, Genes Dev 11:2295–2322

4 Nobes C, Hall A (1994) Regulation and function of the Rho subfamily of small GTPases. Curr Opin Gen Dev 4:77–81

5 LaMarche N, Hall A (1994) GAPS for rho-related GTPases. Trends Genet 10:436–440

6 Ridley AJ, Hall A (1992) The small GTP-binding protein rho regulates the assembly of focal adhesions and actin stress fibers in response to growth factors. Cell 70:389–399

7 Ridley AJ, Paterson HF, Johnston CL, Diekmann D, Hall A (1992) The small GTP-binding protein rac regulates growth factor-induced membrane ruffling. Cell 70:401–410

8 Ridley AJ (1995) Rho-related proteins: actin cytoskeleton and cell cycle. Curr Opin Gen Dev 5:24–30

9 Nobes CD, Hall A (1995) Rho, Rac, and Cdc42 GTPases regulate the assembly of multimolecular focal complexes associated with actin stress fibers, lamellipodia, and filopodia. Cell 81:53–62

10 Kozma R, Ahmed S, Best A, Lim L (1995) The Ras-related protein Cdc42Hs and bradykinin promote formation of peripheral actin microspikes and filopodia in Swiss 3T3 fibroblasts. Mol Cell Biol 15:1942–1952
11 Narumiya S (1996) The small GTPase Rho: cellular functions and signal transduction. J Biochem (Tokyo) 120:215–228
12 Takaishi K, Kikuchi A, Kuroda S, Kotani K, Sasaki T, Takai Y (1993) Involvement of rho p21 and its inhibitory GDP/GTP exchange protein (rhoGDI) in cell motility. Mol Cell Biol 13:72–79
13 Takaishi K, Sasaki T, Kato M, Yamochi W, Kuroda S, Nakamura T, Takeichi M, Takai Y (1994) Involvement of rho p21 small GTP-binding protein and its regulator in the HGF-induced motility. Oncogene 9:273–279
14 Lamaze C, Chuang T-H, Terlecky LJ, Bokoch GM, Schmid S (1996) Regulation of receptor-mediated endocytosis by Rho and Rac. Nature 382:177–179
15 Aepfelbacher M (1995) ADP-ribosylation of Rho enhances adhesion of U937 cells to fibronectin via the alpha 5 beta 1 integrin. FEBS Lett 3630:78–80
16 Debant A, Serra-Pages C, Seipel K, S OB, Tang M, Park SH (1996) The multidomain protein Trio binds the LAR transmembrane tyrosine phosphatase, contains a protein kinase domain, and has separate rac-specific and rho-specific guanine nucleotide exchange factor domains. Proc Natl Acad Sci USA 93:5466–5471
17 Narumiya S, Morii N, (1993) Rho gene products, botulinim C3 exoenzyme, and cell adhesion. Cell Signal 5:9–19
18 Takaishi K, Sasaki T, Kameyama T, Tsukita S, Takai Y (1995) Translocation of activated Rho from the cytoplasm to membrane ruffling area, cell-cell adhesion sites, and cleavage furrows. Oncogene 11:39–48
19 Jalink K, van Corven EJ, Hengeveld T, Morii N, Narumiya S, Moolenaar WH (1994) Inhibition of lysophosphatidate- and thrombin-induced neurite retraction and neuronal cell rounding by ADP ribosylation of the small GTP-binding protein Rho. J Cell Biol 126: 801–810
20 Mackay DJG, Nobes CD, Hall A (1995) The Rho's progress: a potential role during neuritogenesis for the Rho family of GTPases, Trends Neurosci 18:496–501
21 Tigyi G, Fisher DJ, Sebok A, Yang C, Dyer DL, Miledi R (1996) LPA-induced neurite retraction in PC12 cells: control by phosphoinositide-Ca^{2+} signaling and Rho. J Neurochem 66:537–548
22 Gebbink MF, Kranenburg O, Poland M, van Horck FP, Houssa B, Moolenaar WH (1997) Identification of a novel, putative Rho-specific GDP/GTP exchange factor and a RhoA-binding protein: control of neuronal morphology. J Cell Biol 137:1603–1613
23 Lamoureux P, Altun-Gultekin ZF, Lin C, Wagner JA, Heidemann SR (1997) Rac is required for growth cone function but not neurite assembly. J Cell Sci 110 (Pt 5): 635–641
24 Luo L, Jan L, Jan YN (1996) Small GTPases in axon outgrowth. Perspect Dev Neurobiol 4:199–204
25 Knaus UG, Heyworth PG, Kinsella BT, Curnutte JT, Bokoch GM (1992) Purification and characterization of rac2. A cytosolic GTP-binding protein that regulates human neutrophil NADPH oxidase. J Biol Chem 267:23575–23582
26 Diekmann D, Abo A, Johnston C, Segal AW, Hall A (1994) Interaction of rac with p67phox and regulation of phagocytic NADPH oxidase activity. Science 265:531–533
27 Olson MF, Ashworth A, Hall A (1995) An essential role for Rho, Rac, and Cdc42 GTPases in cell cycle progression through G1. Science 269:1270–1272
28 Minden A, Lin A, Claret F-X, Abo A, Karin M (1995) Selective activation of the JNK signaling cascade and c-Jun transcriptional activity by the small GTPases Rac and Cdc42Hs. Cell 81:1147–1157
29 Hill CS, Wynne J, Treisman R (1995) The Rho family GTPases RhoA, Rac1, and Cdc42Hs regulate transcriptional activation by SRF. Cell 81:1159–1170

30 Coso OA, Chiariello M, Yu J-C, Teramoto H, Crespo P, Xu N, Miki T, Gutkind JS (1995) The small GTP-binding proteins Racl and Cdc42 regulate the activity of the JNK/SAPK signaling pathway. Cell 81:1137–1146

31 Bagrodia S, Taylor SJ, Creasy CL, Chernoff J, Cerione RA (1995) Identification of a mouse p21Cdc42/Rac activated kinase. J Biol Chem 270:22731–22737

32 Martin GA, Bollag G, McCormick F, Abo A (1995) A novel serine kinase activated by Racl/ Cdc42Hs-dependent autophosphorylation is related to PAK65 and STE20. EMBO J 1970–1978

33 Ramer SW, Davis RW (1993) A dominant truncation allele identifies a gene, Ste20, that encodes a putative protein kinase necessary for mating in *Saccharomyces cerevisiae.* Proc Natl Acad Sci USA 1993:452–456

34 Lim L, Manser E, Leung T, Hall C (1996) Regulation of phosphorylation pathways by p21 GTPases. The p21 Ras- related Rho subfamily and its role in phosphorylation signalling pathways. Eur J Biochem 242:171–185

35 Lamarche N, Tapon N, Burbelo PD, Aspendröm P, Bridges T, Chant J, Hall A (1996) Rac and Cdc42 induce actin polymerization and G1 cell cycle progression independently of p65pak and the JNK/SAPK MAP kinase cascade. Cell 87:519–529

36 Chong LD, Traynor-Kaplan A, Bokoch GM, Schwartz MA (1994) The small GTP-binding protein rho regulates a phosphotidylinositol 4-phosphate 5-kinase in mammalian cells. Cell 79:507–513

37 Zhang J, King WG, Dillon S, Hall A, Feig L, Rittenhouse SE (1993) Activation of platelet phosphatidylinositide 3-kinase requires the small GTP-binding protein rho. J Biol Chem 268:22251–22254

38 Zheng Y, Bagrodia S, Cerione RA (1994) Activation of phosphoinositide 3-kinase activity by Cdc42Hs binding to p85. J Biol Chem 269:18727–18730

39 Bowman EP, Uhlinger DJ, Lambeth JD (1993) Neutrophil phospholipase D is activated by a membrane-associated Rho family small molecular weight GTP-binding protein. J Biol Chem 268:21509–21512

40 Watanabe G, Saito Y, Madaule P, Ischizaki T, Fujisawa K, Morii N, Mukai H, Ono Y, Kakizuka S, Narumiya S (1996) Protein kinase N (PKN) and PKN-related protein rhophillin as targets of small GTPase Rho. Science 271:645–648

41 Amano M, Mukai H, Ono Y, Chihara K, Matsui T, Hamajima Y, Okawa K, Iwamatsu A, Kaibuchi K (1996) Identification of a putative target for Rho as the serine-threonine kinase protein kinase N. Science 271:648–650

42 Quilliam LA, Lambert QT, Mickelson-Young LA, Westwick JK, Sparks AB, Kay BK, Jenkins NA, Gilbert DJ, Copeland NG, Der CJ (1996) Isolation of a NCK-associated kinase, PRK2, an SH3-binding protein and potential effector of Rho protein signaling. J Biol Chem 271:28772–28776

43 Vincent S. Settleman J (1997) The PRK2 kinase is a potential effector target of both Rho and Rac GTPases and regulates actin cytoskeletal organization. Mol Cell Biol 17:2247–2256

44 Leung T, Chen X-Q, Manser E, Lim L (1996) The p160 RhoA-binding kinase ROKα is a member of a kinase family and is involved in the reorganization of the cytoskeleton. Mol Cell Biol 16:5313–5327

45 Leung T, Manser E, Tan L, Lim L (1995) A novel serine/threonine kinase binding the Ras-related RhoA GTPase which translocates the kinase to peripheral membranes. J Biol Chem 270:29051–29054

46 Ishizaki T, Maekawa M, Fujisawa K, Okawa K, Fugita A, Watanabe G, Saito Y, Kakizuka A, Morii N, Narumiya S (1996) The small GTP-binding protein Rho binds to and activates a 160 kDa Ser/Thr kinase homologous to the myotonic dystrophy kinase. EMBO J 15:1885–1893

47 Simon MA, Bowtell DDL, Dodson GS, Laverty TR, Rubin GM (1991) Ras1 and a putative guanine nucleotide exchange factor perform crucial steps in signal transduction by the

sevenless protein tyrosine kinase receptor. Cell 67:701–716
48 Bowtell D, Fu P, Simon M, Senior P (1992) Identification of murine homologues of the *Drosophila* son of sevenless gene: potential activators of ras. Proc Natl Acad Sci USA 89:6511–6515
49 Greenwald I, Rubin GM (1992) Making a difference: the role of cell-cell interactions in establishing separate identities for equivalent cells. Cell 68:271–281.
50 Kramer H, Cagan RL, Zipursky SL (1991) Interaction of bride of sevenless membrane-bound ligand and the sevenless tyrosine-kinase receptor. Nature 352:207–212
51 Fortini ME, Somon MA, Rubin GM (1992) Signalling by the sevenless protein tyrosine kinase is mimicked by Ras1 activation. Nature 355:559–561
52 Gaul U, Mardon G, Rubin GM (1992) A putative Ras GTPase activating protein acts as a negative regulator of signaling by the sevenless receptor tyrosine kinase. Cell 68:1007–1019
53 Olivier JP, Raabe T, Henkemeyer M, Dickson B, Mbamalu G, Margolis B, Schlessinger J, Hafen E, Pawson T (1993) A *Drosophila* SH2-SH3 adaptor protein implicated in coupling the Sevenless tyrosine kinase to an activator of ras guanine nucleotide exchange. Cell 73:179–191
54 Therrien M, Michaud NR, Rubin GM, Morrison DK (1996) KSR modulates signal propagation within the MAPK cascade. Genes Dev 10:2684–2695
55 Wassarman DA, Therrien M, Rubin GM (1995) The Ras signaling pathway in *Drosophila.* Curr Opin Genet Dev 5:44–50
56 Brand AH, Perrimon N (1993) Targeted gene expression as a means of altering cell fates and generating dominant phenotypes. Development 118:401–415
57 Luo L, Liao YJ, Jan LY, Jan YN (1994) Distinct morphogenetic functions of similar small GTPases: *Drosophila* Drac1 is involved in axonal outgrowth and myoblast fusion. Genes Dev 8:1787–1802
58 Hariharan IK, Hu KQ, Asha H, Quintanilla A, Ezzell RM, Settleman J (1995) Characterization of rho GTPase homologues in *Drosophila melanogaster*: overexpressing *Rho1* causes a late developmental defect. EMBO J 14:292–302
59 Murphy AM, Montell DJ (1996) Cell type specific roles for Cdc42, Rac and RhoL in *Drosophila* oogenesis. J Cell Biol 133:617–630
60 Strutt DI, Weber U, Mlodzik M (1997) The role of RhoA in tissue polarity and Frizzled signalling. Nature 387:292–295
61 Cooley L. Theurkauf WE (1994) Cytoskeletal functions during *Drosophila* oogenesis. Science 266:590–596
62 Cant K, Knowles BA, Mooseker MS, Cooley L (1994) *Drosophila* singed, a fascin homolog, is required for actin bundle formation during oogenesis and bristle extension. J Cell Biol 125:369–380
63 Mahajan-Miklos S, Cooley L (1994) The villin-like protein encoded by the *Drosophila* quail gene is required for actin bundle assembly during oogenesis. Cell 78:291–301
64 Cooley L, Verheyen E, Ayers K (1992) chickadee encodes a profilin required for intercellular cytoplasm transport during *Drosophila* oogenesis. Cell 69:173–184
65 Leptin M (1995) *Drosophila* gastrulation: from pattern formation to morphogenesis. Annu Rev Cell Dev Biol 11:189–212
66 Leptin M, Grunewald B (1990) Cell shape changes during gastrulation in *Drosophila.* Development 110:73–84
67 Costa M, Sweeton D, Wieschaus E (1993) Gastrulation in *Drosophila*: cellular mechanisms of morphogenetic movements. The development of *Drosophila melanogaster* 425–465
68 Sweeton D, Parks S, Costa M, Wieschaus E (1991) Gastrulation in *Drosophila*: the formation of the ventral furrow and posterior midgut invaginations. Development 112:775–789
69 Young PE, Pesacreta TC, Kiehart DP (1991) Dynamic changes in the distribution of cytoplasmic myosin during *Drosophila* embryogenesis. Development 111:1–14
70 Thomas GH, Kiehart DP (1994) Beta heavy-spectrin has a restricted tissue and subcellular

distribution during *Drosophila* embryogenesis. Development 120:2039–2050

71 Barrett K, Leptin M, Settleman J (1997) The Rho GTPase and a putative RhoGEF mediate a signaling pathway for the cell shape changes in *Drosophila* gastrulation. Cell 91:905–915

72 Hacker U, Perrimon N (1998) DRhoGEF2 encodes a member of the Dbl family of oncogenes and controls cell shape changes during gastrulation in *Drosophila.* Genes Dev 12:274–284

73 Costa M, Wilson ET, Wieschaus E (1994) A putative cell signal encoded by the folded gastrulation gene coordinates cell shape changes during *Drosophila* gastrulation. Cell 76: 1075–1089

74 Parks S, Wieschaus E (1991) The *Drosophila* gastrulation gene concertina encodes a G alpha-like protein. Cell 64:447–458

75 Morize P, Christiansen A, Costa M, Parks S, Wieschaus E (1998) Hyperactivation of the folded gastrulation pathway induces specific cell shape changes. Dev (Suppl) 125:589–597

76 Campos-Ortega JA, Hartenstein V (1985) The embryonic development of *Drosophila melanogaster.* Springer, Berlin Heidelberg New York

77 Knust E (1997) Movements behind the edge. Curr Biol 7: R558–R561

78 Noselli S (1998) JNK signaling and morphogenesis in *Drosophila.* Trends Genet 14:33–38

79 Young PE, Richman AM, Ketchum AS, Kiehart DP (1993) Morphogenesis in *Drosophila* requires nonmuscle myosin heavy chain function. Genes Dev 7:29–41

80 Fehon R, Dawson I, Artavanis-Tsakonas S (1994) A *Drosophila* homologue of membrane-skeleton protein 4.1 is associated with septate junctions and is encoded by the coracle gene. Dev (Suppl) 120:545–557

81 Glise B, Noselli S (1997) Coupling of Jun amino-terminal kinase and Decapentaplegic signaling pathways in *Drosophila* morphogenesis. Genes Dev 11:1738–1747

82 Riesgo-Escovar JR, Hafen E (1997) Common and distinct roles of DFos and DJun during *Drosophila* development. Science 278:669–672

83 Harden N, Lee J, Loh HY, Ong YM, Tan I, Leung T, Manser E, Lim L (1996) A *Drosophila* homolog of the Rac- and Cdc42-activated serine/threonine kinase PAK is a potential focal adhesion and focal complex protein that colocalizes with dynamic actin structures. Mol Cell Biol 16:1896–1908

84 Ruberte E, Marty T, Nellen D, Affolter M, Basler K (1995) An absolute requirement for both the type II and type I receptors, punt and thick veins, for dpp signaling in vivo. Cell 80:889–897

85 Affolter M, Nellen D, Nussbaumer U, Basler K (1994) Multiple requirements for the receptor serine/threonine kinase thick veins reveal novel functions of TGF beta homologs during *Drosophila* embryogenesis. Development 120:3105–3117

86 Ueno N, Oishi I, Sugiyama S, Nishida Y, Minami Y, Yamamura H (1997) Identification of a novel *Drosophila* protein kinase highly homologous to protein kinase N (PKN). Biochem Biophys Res Commun 232:126–129

87 Rushton E, Drysdale R, Abmayr SM, Michelson AM, Bate M (1995) Mutations in a novel gene, myoblast city, provide evidence in support of the founder cell hypothesis for *Drosophila* muscle development. Development 121:1979–1988

88 Erickson MR, Galletta BJ, Abmayr SM (1997) *Drosophila* myoblast city encodes a conserved protein that is essential for myoblast fusion, dorsal closure, and cytoskeletal organization. J Cell Biol 138:589–603

89 Davies JA. Garrod DR (1997) Molecular aspects of the epithelial phenotype. Bioessays 19: 699–704

90 Berthiaume F, Moghe PV, Toner M, Yarmush ML (1996) Effect of extracellular matrix topology on cell structure, function, and physiological responsiveness: hepatocytes cultured in a sandwich configuration. Faseb J 10:1471–1484

91 Nusrat A, Giry M, Turner JR, Colgan SP, Parkos CA, Carnes D, Lemichez E, Boquet P, Madara JL (1995) Rho protein regulates tight junctions and perijunctional actin organization in polarized epithelia. Proc Natl Acad Sci USA 92:10629–10633

92 Ojakian GK, Schwimmer R (1988) The polarized distribution of an apical cell surface glycoprotein is maintained by interactions with the cytoskeleton of Madin-Darby canine kidney cells. J Cell Biol 107:2377–2387
93 Braga VM, Machesky LM, Hall A, Hotchin NA (1997) The small GTPases Rho and Rac are required for the establishment of cadherin-dependent cell-cell contacts. J Cell Biol 137:1421–1431
94 Eaton S (1997) Planar polarization of *Drosophila* and vertebrate epithelia. Curr Opin Cell Biol 9:860–866
95 Eaton S, Auvinen P, Luo L, Jan YN, Simons K (1995) Cdc42 and Rac1 control different actin-dependent processes in the *Drosophila* wing disc epithelium. J Cell Biol 131:151–164
96 Eaton S, Wepf R, Simons K (1996) Roles for Rac1 and Cdc42 in planar polarization and hair outgrowth in the wing of *Drosophila.* J Cell Biol 135:1277–1289
97 Krasnow RE, Wong LL, Adler PN (1995) Dishevelled is a component of the frizzled signaling pathway in *Drosophila.* Development 121:4095–4102
98 Park WJ, Liu J, Adler PN (1994) The frizzled gene of *Drosophila* encodes a membrane protein with an odd number of transmembrane domains. Mech Dev 45:127–137
99 Martin-Blanco E (1997) Regulation of cell differentiation by the *Drosophila* Jun kinase cascade. Curr Opin Genet Dev 7:666–671
100 Letourneau PC (1997) The cytoskeleton in nerve growth cone motility and axonal pathfinding. Perspect Dev Neurobiol 4:111–123
101 Kaufmann N, Wills Z, Vactor D (1998) *Drosophila* Rac1 controls motor axon guidance. Development 125:453–461
102 Krueger NX, Van Vactor D, Wan HI, Gelbart WM, Goodman CS, Saito H (1996) The transmembrane tyrosine phosphatase DLAR controls motor axon guidance in *Drosophila.* Cell 84:611–622
103 Sone M, Hoshino M, Suzuki E, Kuroda S, Kaibuchi K, Nakagoshi H, Saigo K, Nabeshima Y, Hama C (1997) Still life, a protein in synaptic terminals of *Drosophila* homologous to GDP-GTP exchangers [published erratum appears in Science 1997 Mar 7;275[5305]:1405]. Science 275:543–547
104 Holtzer H, Strahs K, Biehl J, Somlyo AP, Ishikawa H (1975) Thick and thin filaments in postmitotic, mononucleated myoblasts. Science 188:943–945
105 Paululat A, Burchard S, Renkawitz-Pohl R (1995) Fusion from myoblasts to myotubes is dependent on the rolling stone gene (rost) of *Drosophila.* Development 121:2611–2620
106 Doberstein SK, Fetter RD, Mehta AY, Goodman CS (1997) Genetic analysis of myoblast fusion: blown fuse is required for progression beyond the prefusion complex. J Cell Biol 136:1249–61
107 Hasegawa H, Kiyokawa E, Tanaka S, Nagashima K, Gotoh N, Shibuya M, Kurata T, Matsuda M (1996) DOCK180, a major CRK-binding protein, alters cell morphology upon translocation to the cell membrane. Mol Cell Biol 16:1770–1776
108 Smit L, van der Horst G, Borst J (1996) Sos, Vav, and C3G participate in B cell receptor-induced signaling pathways and differentially associate with Shc-Grb2, Crk, and Crk-L adaptors. J Biol Chem 271:8564–8569
109 Sundaram M, Han M (1996) Control and integration of cell signaling pathways during *C. elegans* vulval development. Bioessays 18:473–480
110 Emmons SW (1987) Mechanisms of *C. elegans* development. Cell 51:881–883
111 Strome S, Hill DP (1988) Early embryogenesis in *Caenorhabditis elegans*: the cytoskeleton and spatial organization of the zygote. Bioessays 8:145–149
112 Priess JR, Hirsh DI (1986) *Caenorhabditis elegans* morphogenesis: the role of the cytoskeleton in elongation of the embryo. Dev Biol 117:156–173
113 Chen W, Lim L (1994) The *Caenorhabditis elegans* small GTP-binding protein RhoA is enriched in the nerve ring and sensory neurons during larval development. J Biol Chem 51:32394–32404

114 Chen W, Lim HH, Lim L (1993) The Cdc42 homologue from *Caenorhabditis elegans.* Complementation of yeast mutation. J Biol Chem 268:13280–13285
115 Chen W, Yap SF, Lim L (1996) Isolation of the gene coding for *Caenorhabditis elegans* Rac2 homologue, a Ras-related small GTP-binding protein. Gene 180:217–219
116 Chen W, Lim HH, Lim L (1993) A new member of the ras superfamily, the rac1 homologue from *Caenorhabditis elegans.* Cloning and sequence analysis of cDNA, pattern of developmental expression, and biochemical characterization of the protein. J Biol Chem 268:320–324
117 Zipkin ID, Kindt RM, Kenyon CJ (1997) Role of a new Rho family member in cell migration and axon guidance in *C. elegans.* Cell 90:883–894
118 Chen W, Chen S, Yap SF, Lim L (1996) The *Caenorhabditis elegans* p21-activated kinase (CePAK) colocalizes with CeRac1 and Cdc42Ce at hypodermal cell boundaries during embryo elongation. J Biol Chem 271:26362–26368
119 Wissmann A, Ingles J, McGhee JD, Mains PE (1997) *Caenorhabditis elegans* LET-502 is related to Rho-binding kinases and human myotonic dystrophy kinase and interacts genetically with a homolog of the regulatory subunit of smooth muscle myosin phosphatase to affect cell shape. Genes Dev 11:409–422
120 Kimura K, Ito M, Amano M, Chihara K, Fukata Y, Nakafuku M, Yamamori B, Feng J, Nakano T, Okawa K, Iwamatsu A, Kaibuchi K (1996) Regulation of myosin phosphatase by Rho and Rho-associated kinase (Rho- kinase) [see comments]. Science 273:245–248
121 Duggan A, Chalfie M (1995) Control of neuronal development in *Caenorhabditis elegans,* Curr Opin Neurobiol 5:6–9
122 Run JQ, Steven R, Hung MS, van Weeghel R, Culotti JG, Way JC (1996) Suppressors of the unc-73 gene of *Caenorhabditis elegans.* Genetics 143:225–236
123 St-Jacques B, McMahon AP (1996) Early mouse development: lessons from gene targeting. Curr Opin Genet Dev 6:439–444
124 Zmuidzinas A, Fischer KD, Lira SA, Forrester L, Bryant S, Bernstein A, Barbacid M (1995) The vav proto-oncogene is required early in embryogenesis but not for hematopoietic development in vitro. EMBO J 14:1–11
125 Zhang R, Tsai FY, Orkin SH (1994) Hematopoietic development of vav-/- mouse embryonic stem cells. Proc Natl Acad Sci USA 91:12755–12759
126 Turner M, Mee PJ, Walters AE, Quinn ME, Mellor AL, Zamoyska R, Tybulewicz VL (1997) A requirement for the Rho-family GTP exchange factor Vav in positive and negative selection of thymocytes. Immunity 7:451–460
127 Fischer KD, Zmuldzinas A, Gardner S, Barbacid M, Bernstein A, Guidos C (1995) Defective T-cell receptor signalling and positive selection of Vav-deficient CD4+ CD8+ thymocytes. Nature 374:474–477
128 Voncken JW, van Schaick H, Kaartinen V, Deemer K, Coates T, Landing B, Pattengale P, Dorseuil O, Bokoch GM, Groffen J et al. (1995) Increased neutrophil respiratory burst in bcr-null mutants. Cell 80:719–728
129 Guillemot JC, Kruskal BA, Adra CN, Zhu S, Ko JL, Burch P, Nocka K, Seetoo K, Simons E, Lim B (1996) Targeted disruption of guanosine diphosphate-dissociation inhibitor for Rho-related proteins, GDID4: normal hematopoietic differentiation but subtle defect in superoxide production by macrophages derived from in vitro embryonal stem cell differentiation. Blood 88:2722–2731
130 Luo L, Hensch TK, Ackerman L, Barbel S, Jan LY, Jan YN (1996) Differential effects of the Rac GTPase on Purkinje cell axons and dendritic trunks and spines. Nature 379:837–840
131 Lores P, Morin L, Luna R, Gacon G (1997) Enhanced apoptosis in the thymus of transgenic mice expressing constitutively activated forms of human Rac2 GTPase. Oncogene 15:601–605
132 Henning SW, Galandrini R, Hall A, Cantrell DA (1997) The GTPase Rho has a critical regulatory role in thymus development. EMBO J 16:2397–2407

133 Pasteris NG, Cadle A, Logie LJ, Porteous ME, Schwartz CE, Stevenson RE, Glover TW, Wilroy RS, Gorski JL (1994) Isolation and characterization of the faciogenital dysplasia (Aarskog-Scott syndrome) gene: a putative Rho/Rac guanine nucleotide exchange factor. Cell 79:669–678
134 Olson MF, Pasteris NG, Gorski JL, Hall A (1996) Faciogenital dysplasia protein (FGD1] and Vav, two related proteins required for normal embryonic development, are upstream regulators of Rho GTPases. Curr Biol 6:1628–1633
135 Zheng Y, Fischer DJ, Santos MF, Tigyi G, Pasteris NG, Gorski JL, Xu Y (1996) The faciogenital dysplasia gene product FGD1 functions as a Cdc42Hs-specific guanine-nucleotide exchange factor. J Biol Chem 271:33169–33172

Subject Index

Printing: Saladruck, Berlin
Binding: Buchbinderei Lüderitz & Bauer, Berlin